# Theranostics and Image Guided Drug Delivery

**Drug Discovery Series**



*Titles in the Series:*

1: Metabolism, Pharmacokinetics and Toxicity of Functional Groups
2: Emerging Drugs and Targets for Alzheimer's Disease; Volume 1
3: Emerging Drugs and Targets for Alzheimer's Disease; Volume 2
4: Accounts in Drug Discovery
5: New Frontiers in Chemical Biology
6: Animal Models for Neurodegenerative Disease
7: Neurodegeneration
8: G Protein-Coupled Receptors
9: Pharmaceutical Process Development
10: Extracellular and Intracellular Signaling
11: New Synthetic Technologies in Medicinal Chemistry
12: New Horizons in Predictive Toxicology
13: Drug Design Strategies: Quantitative Approaches
14: Neglected Diseases and Drug Discovery
15: Biomedical Imaging
16: Pharmaceutical Salts and Cocrystals
17: Polyamine Drug Discovery
18: Proteinases as Drug Targets
19: Kinase Drug Discovery
20: Drug Design Strategies: Computational Techniques and Applications

21: Designing Multi-Target Drugs
22: Nanostructured Biomaterials for Overcoming Biological Barriers
23: Physico-Chemical and Computational Approaches to Drug Discovery
24: Biomarkers for Traumatic Brain Injury
25: Drug Discovery from Natural Products
26: Anti-Inflammatory Drug Discovery
27: New Therapeutic Strategies for Type 2 Diabetes: Small Molecules
28: Drug Discovery for Psychiatric Disorders
29: Organic Chemistry of Drug Degradation
30: Computational Approaches to Nuclear Receptors
31: Traditional Chinese Medicine
32: Successful Strategies for the Discovery of Antiviral Drugs
33: Comprehensive Biomarker Discovery and Validation for Clinical Application
34: Emerging Drugs and Targets for Parkinson's Disease
35: Pain Therapeutics; Current and Future Treatment Paradigms
36: Biotherapeutics: Recent Developments using Chemical and Molecular Biology
37: Inhibitors of Molecular Chaperones as Therapeutic Agents
38: Orphan Drugs and Rare Diseases
39: Ion Channel Drug Discovery
40: Macrocycles in Drug Discovery
41: Human-based Systems for Translational Research
42: Venoms to Drugs: Venom as a Source for the Development of Human Therapeutics
43: Carbohydrates in Drug Design and Discovery
44: Drug Discovery for Schizophrenia
45: Cardiovascular and Metabolic Disease: Scientific Discoveries and New Therapies
46: Green Chemistry Strategies for Drug Discovery
47: Fragment-Based Drug Discovery
48: Epigenetics for Drug Discovery
49: New Horizons in Predictive Drug Metabolism and Pharmacokinetics
50: Privileged Scaffolds in Medicinal Chemistry: Design, Synthesis, Evaluation

51: Nanomedicines: Design, Delivery and Detection
52: Synthetic Methods in Drug Discovery: Volume 1
53: Synthetic Methods in Drug Discovery: Volume 2
54: Drug Transporters: Role and Importance in ADME and Drug Development
55: Drug Transporters: Recent Advances and Emerging Technologies
56: Allosterism in Drug Discovery
57: Anti-aging Drugs: From Basic Research to Clinical Practice
58: Antibiotic Drug Discovery: New Targets and Molecular Entities
59: Peptide-based Drug Discovery: Challenges and New Therapeutics
60: Drug Discovery for Leishmaniasis
61: Biophysical Techniques in Drug Discovery
62: Acute Brain Impairment Through Stroke: Drug Discovery and Translational Research
63: Theranostics and Image Guided Drug Delivery

*How to obtain future titles on publication:*
A standing order plan is available for this series. A standing order will bring delivery of each new volume immediately on publication.

*For further information please contact:*
Book Sales Department, Royal Society of Chemistry, Thomas Graham House, Science Park, Milton Road, Cambridge, CB4 0WF, UK
Telephone: +44 (0)1223 420066, Fax: +44 (0)1223 420247
Email: booksales@rsc.org
Visit our website at books.rsc.org

# *Theranostics and Image Guided Drug Delivery*

Edited by

**Maya Thanou**
*King's College London, UK*
*Email: maya.thanou@kcl.ac.uk*

ROYAL SOCIETY
OF CHEMISTRY

THE QUEEN'S AWARDS
FOR ENTERPRISE:
INTERNATIONAL TRADE
2013

Drug Discovery Series No. 63

Print ISBN: 978-1-78262-466-0
PDF ISBN: 978-1-78801-059-7
EPUB ISBN: 978-1-78801-393-2
ISSN: 2041-3203

A catalogue record for this book is available from the British Library



The Royal Society of Chemistry is a charity, registered in England and Wales, Number 207890, and a company incorporated in England by Royal Charter (Registered No. RC000524), registered office: Burlington House, Piccadilly, London W1J 0BA, UK, Telephone: +44 (0) 207 4378 6556.

For further information see our web site at www.rsc.org

Printed in the United Kingdom by CPI Group (UK) Ltd, Croydon, CR0 4YY, UK

# Preface

Future and emerging technologies in drug therapy may include a certain element of imaging. The recent advances in imaging have opened the route for the development of a new type of non-invasive treatment named interventional radiology. Interventional Radiology includes a range of therapies which rely on the use of image guidance, such as fluoroscopy, ultrasound, or magnetic resonance imaging (MRI) to precisely target treatments. Most of these techniques are minimally invasive or non-invasive treatments. Usually termed as image guided surgery these techniques combine imaging with radiofrequency ablation, electromagnetic radiation, microwaves, light, and/or high intensity focused ultrasound for the focal modulation of the temperature of the target (hyperthermia and/or ablation) with the aim to remove lesions and unhealthy tissues. Image guided drug delivery is a much-related novel concept and another emerging field that aims to optimise drug targeting using information provided by real-time medical imaging of the drug and or its carrier. Image guided drug delivery can be used to monitor biodistribution, as well as drug carrier behaviour such as drug release. In the clinic, image guided drug delivery can be used to substantially improve pharmacokinetics, and therefore the safety and the therapeutic potential of the drug treatment. If the drugs can be imaged while they are in the body, then it is easy to control their pharmacokinetics and tissue distribution. Similar to image guided surgery, image guided drug delivery combines imaging and drug therapy. Image guided drug delivery employs either sound, light or electro-magnetic energy to change the state of the drug molecules and activate them in order to activate their pharmacological action only where it is required in the body. This provides an excellent method for targeted drug delivery and therapy. For diseases such as cancer this becomes important as it is now possible to direct potent cytotoxic drugs to act only in tumours. For this type of therapy researchers are seeking the development of methods that do not use ionising radiation for imaging. Furthermore, it is important

Drug Discovery Series No. 63
Theranostics and Image Guided Drug Delivery
Edited by Maya Thanou

Published by the Royal Society of Chemistry, www.rsc.org

to use radiation sources that can be focused in a small defined volume in the tissue. Theranostics are compounds that are used to diagnose and treat diseases. Theranostics is a term derived from the words diagnosis and therapy and was coined to define efforts in clinics to develop specific, individualized therapies, and to combine diagnostic and therapeutic capabilities into a single agent. Theranostics may be one molecule that carries the probe for imaging coupled onto the therapeutic agent (*e.g.* radiolabelled antibodies) or a nanoparticle that is composed of the carrier the imaging probe or material and the therapeutic. Emphasis has been given to nanosized theranostics due to their potential in cancer treatments. Theranostics play a significant role in the development of image guided drug delivery treatments.

The scope of the book is to introduce the topic of image guided drug delivery to the scientific community and highlight its interdisciplinary approach, its efficiency and its translational ability. As such the book covers all recent and novel research efforts that address issues of theranostics and image guided drug delivery. Emphasis has been given to therapeutic concepts that reached clinical trials. Some of these methods are likely to emerge as disruptive technologies in healthcare.

The book begins by presenting advances in MRI guided focused ultrasound methods and its application in drug delivery. Focused ultrasound mediated drug delivery has been applied in several studies with various drug molecules to improve their local distribution in diseased tissues. In this case an imaging modality such as MRI or ultrasound is used to assess the method and to provide evidence of drug release. The increased temperature induced by focused ultrasound (hyperthermia) can improve blood supply in tissues and therefore improve drug distribution.

Multifunctional nanoagents that combine diagnostic and therapeutic functions for antitumor treatments and respond to near infra-red light are presented. In particular, combining real-time imaging with spatially precise photothermal therapy mediated by nanoparticles responsive to near-infrared (NIR, $\lambda$ = 700–1100 nm) light through conversion of photo energy into heat. This therapeutic strategy is characterised by its simplicity, safety, and non-invasiveness, as well as targeting and remote-control release properties.

The book also describes the magnetic nanoparticles that have found application in the clinic, not only as imaging agents, but as a potent tool for treating tumours. These magnetic nanoparticles have been utilised as imaging probes for several years. When incorporated with therapeutic agents, they are specially designed to concentrate at the target site with the aid of magnetic force, resulting in theranostic function (therapy and diagnosis). MRI of magnetic nanoparticles is the most studied imaging application and has been used in the clinic with high resolution. Furthermore, their use in combination with magnetic field can provide magnetic hyperthermia that can treat tumours specifically, a concept currently being tested in clinical trials.

The emerging technique of photodynamic therapy is presented and described. Photodynamic therapy is a clinical technique for the treatment of cancers, microbial infections and other medical conditions by means of light-induced generation of reactive oxygen species using photosensitising drugs. The intrinsic fluorescence of many such drugs make them potential theranostic agents for simultaneous diagnosis and therapy. The design and development of these photosensitizers and their carriers will play an important role in designing new photodynamic treatments.

The book also introduces methods of making theranostic nanomedicines for positron emission tomography (PET) imaging. The high sensitivity and spatio-temporal resolution of PET makes this non-invasive imaging technique ideal for the *in vivo* tracking of liposomal nanomedicines in the clinical setting. This image guided therapeutic approach may eventually allow clinicians to select patients that would benefit from the nanomedicinal treatment and, by doing so, enhance the clinical value/efficacy of this promising treatment. This is a simple approach that can be used for personalised medicine and efficient treatments.

A special chapter is dedicated to the description and preparation of thermosensitive liposomes, a tool that is widely used in image guided drug delivery. The use of thermosensitive liposomes for anticancer treatment, which was first described in the seventies, has gained an increasing amount of attention over recent years. Various thermosensitive liposome formulations have been designed and tested in many different ways, all having various advantages and disadvantages. The composition and methods of preparation and characterisation are presented in this section.

Within the topic of formulation the book also presents recent advances on combining pharmaceutical carriers such as cyclodextrins and image guided focused ultrasound for the targeted release of chemotherapeutics. Ultrasound-mediated therapeutic drug delivery is one of various methods for local drug release and it can also influence cell permeability. Furthermore, combining it with MRI provides a controllable system for drug release and impact assessment. The chapter describes the design and development of carriers for ultrasound activated drug delivery.

For the first time the concept of image guided drug delivery in the gastrointestinal tract is presented. Special devices that can be ingested can provide energy in a controlled and focused manner to effect the tissue. These ingestible devices could promote drug therapies. Imaging and drug activation/delivery mechanism can be combined in a miniaturised piece of equipment that can be used in a way similar to capsule endoscopy. This capsule aims to apply drug therapy as well. This may be the only approach where image guidance and drug delivery are designed to act in the gut.

The last chapter provides the novel tool of microwaves and describes their potential as a cost-effective equipment for the development of imaging sensing and image guided drug delivery. It is interesting to see that early studies indicate that microwave energy interacts with nanoparticles that can affect the dielectric properties of the biological tissue. It is expected that the theranostic nanoparticles for this technology may be different compared to the ones mentioned above used by other techniques.

Overall, this book aims to present the most advanced concepts designed and currently tested in image guided drug delivery. There are concepts that are more studied than others and are now in clinical trials, such as thermosensitive liposomes and magnetic nanoparticles. There are however novel concepts such as photothermal and microwave therapy that are still at very early stage. For these the optimal chemical formulations have yet to be developed and delivered.

Although the concept of image guided drug delivery is a very recent one there are hundreds of studies presented and several strategies have moved rapidly to clinical trials to tackle unmet clinical needs. Image guided drug delivery methods hold promise to provide revolutionised methods of treating diseases.

Maya Thanou

# Contents



Drug Discovery Series No. 63
Theranostics and Image Guided Drug Delivery
Edited by Maya Thanou

Published by the Royal Society of Chemistry, www.rsc.org

CHAPTER 1

# *Image Guided Focused Ultrasound as a New Method of Targeted Drug Delivery*

M. WRIGHT[a], M. CENTELLES[a], W. GEDROYC[b] AND M. THANOU*[a]

[a]Institute of Pharmaceutical Science, King's College London, London, UK;
[b]Department of Experimental Medicine, Imperial College London, UK
*E-mail: maya.thanou@kcl.ac.uk

## 1.1 Introduction to Image Guided Focused Ultrasound Drug Delivery

Focused Ultrasound (FUS) is promoting the deposition of energy inside the human body in a non-invasive way.[1] (http://www.fusfoundation.org/for-researchers/mechanisms-of-action). Focused ultrasound energy can be deposited in tissues and lesions with a diameter as little as 1 mm providing a substantial advantage for induction of heat.[2,3] This in turn has a lot of advantages for drug targeting due to the locally increased temperature. When ultrasound is applied in biological systems it can induce local tissue heating, cavitation, and radiation force, which can be used to initiate local (focal) drug delivery, increased molecule permeation through membranes and enhanced diffusivity of drugs, only at the site of sonication therefore allowing control of local drug delivery.[4] Delivery of certain therapeutics largely

Drug Discovery Series No. 63
Theranostics and Image Guided Drug Delivery
Edited by Maya Thanou

Published by the Royal Society of Chemistry, www.rsc.org

benefits from local drug delivery. Cytotoxics, immune suppressive drugs and certain biologicals would have an improved therapeutic index when administered locally with limited exposure to the healthy tissues. Smart drug delivery forms that encapsulate the drug can be designed to deliver their cargo in tissues with increased temperature. FUS (or HIFU: high intensity focused ultrasound) induced local hyperthermia can be the trigger of drug release from thermosensitive carriers.[5]

The ability of FUS waves to induce thermal or mechanical effects at a defined location in living tissue was first described in 1942, when Lynn *et al.* tested FUS in the brain.[6] In the 1950s the Fry brothers developed a clinical device to treat patients with Parkinson disease. They used an ultrasound system in combination with X-rays (imaging) to determine the target location relative to skull and to focus the ultrasound beam through a craniotomy into the deep brain for functional neurosurgery.[7] Later on, in the 1980's the first FDA-approved FUS system, Sonocare CST-100, was developed to treat ocular disorders such as glaucoma and many patients were clinically treated with this system.[8] More recently technological developments have delivered new FUS equipment coupled most of the time with an imaging device such as diagnostic ultrasound and/or MRI. Current research and development aim to design and develop novel transducer technology and array designs to achieve rapid delivery of focal sonication, to improve transducer accessibility (smaller devices) or devices to conform/fit to certain parts of the body such as a helmet of arrays for brain focal treatment of diseases.[9,10]

Several FUS devices are currently in clinical practice either for approved treatments or for research purposes. These devices are combined with either ultrasound (US) imaging or MR imaging for guidance and thermometry.[11] Insightec manufactures the ExAblate2000® which uses MRI for extracorporeal treatment of uterine fibroids (FDA approved) with significant success, and extensive current research focuses on investigating its application in other parts of the body.[12,13] Recently the FDA approved the transcranial MR-guided focused ultrasound for the treatment of essential tremor.[14] This is a hemispheric phased-array transducer (ExAblate Neuro; InSightec Ltd, Tirat Carmel, Israel) with each element directed separately, providing individual correction of skull distortion as well as electronic guidance. The Ablatherm HIFU/US consists of a transrectal probe for prostate treatment and has CE mark approval. In this case imaging is performed with ultrasound.[15] The Sonablate 500, an ultrasound guided system, uses a transrectal FUS probe to carry out prostate cancer focal ablation surgery.[16] The Sonalleve HIFU/MR is an MR compatible device developed to examine applications such as fibroids and bone metastases.[17] The device has been used for the treatment of neuropathic pain and essential tremor and there is also promise of possible application for brain tumours.[18–20] Essential tremor non-interventional functional neurosurgery treatment has shown an immense potential of transcranial MRgFUS application to induce lesions focally and treat patients non-surgically.[21,22] MRgFUS devices are constantly developed. Currently there are 32 manufacturers of image guided FUS worldwide that develop equipment for ablative treatments. There is an on-going interest to use these devices in combination with novel formulations for image guided targeted drug delivery.

### 1.1.1 Fundamentals of Focused Ultrasound Treatment in Living Tissues

Ultrasound propagates as mechanical vibrations that make molecules within their medium oscillate around their positions and in the direction of the waves' propagation. As a result the molecules form compressions and rarefactions that propagate the wave. The ultrasound energy is attenuated exponentially through the tissue.[23] The rate of energy flow through a unit area, in the direction of the wave propagation, is called acoustic intensity. At 1 MHz the ultrasound wave is attenuated approximately 50% and it propagates through 7 cm of tissue. The attenuated energy is then transformed into temperature elevation; heat in the tissue.[24,25]

Ultrasound waves are transmitted from one soft tissue to another adjacent tissue. In soft tissues a small amount of waves are reflected back. However, in the soft tissue–bone interface almost one-third of the incident energy is reflected back. In addition, the amplitude attenuation coefficient of ultrasound waves is 10–20 times higher in bone than in the soft tissues. This causes the transmitted ultrasound beam to be absorbed rapidly within the bone, leading to high temperature increase.[26] When the ultrasound beams are focused at one point, a focal diameter of 1 mm can be achieved at 1.5 MHz. The length of the focus is 5–20 times larger than the diameter (cigarette or rice shaped focus point). If the ultrasound beam is transmitted from an applicator 2–3 cm in diameter, the ultrasound intensity at the millimetre-sized focal spot can be several hundred times higher than in the overlying tissues. The ultrasound exposure drops off rapidly across the area within the sonication path and focusing helps overcoming attenuation losses and to concentrate energy deep in the body avoiding the surrounding tissues.[27]

The fact that focusing of ultrasound energy is so defined is important for the design of smart materials that can respond or transform to temperature change. Hyperthermia applied in a focused way could affect the phase transition of these materials and induce release of the therapeutic payload only at the heated site. Focussing can help the release of the drug only at the desired site.[28]

Focusing of ultrasound energy is significantly improved with the use of transducer arrays that are driven by signals having the necessary phase difference to obtain a common focal point.[29] The advantage of phased transducer arrays is that the focal spot can be guided and controlled. In addition, the electronically focussed beam allows multiple focal points to be induced simultaneously or fast electronic scanning of the focal spot which increases the area of the focal region. This feature allows shorter treatment time.[30,31] Therefore when focused ultrasound is coupled with imaging and theranostics it could target treatments in areas inside the body independent of the size or location.

### 1.1.2 Image Guided Focused Ultrasound Mediated Drug Delivery

During the last 15 years emphasis has been given to nanosized carriers for cancer therapy. Nanomedicine is a topic that investigates the effects of nanotechnology in healthcare and has introduced a series of novel drug delivery systems, among which are multifunctional chemical entities called nanotheranostics.[32,33] These are designed to simultaneously detect, image and treat tumours due to the fact that nanoparticles preferably accumulate in tumours compared to other tissues. If these nanoparticles can provide contrast enhancement and deliver their therapeutic cargo then these nanoparticles act as theranostics.[34] These systems can be engineered using biocompatible and biodegradable materials and nanomaterials, or by "labelling" previously developed nanoparticles.[35] The recent advances of imaging modalities enable nanotheranostics to bind onto lesions or biomarkers on specific cells opening the doors to personalised cancer therapy. The concept of "scan and treat" at the same time can provide clinicians the opportunity to adapt treatments according to imaging data in real time. Nanotheranostics can be used to detect and treat metastases.[36,37] The ability of these drug carrying nanoparticles to reach small metastatic lesions and detect them can offer a substantial advantage to image guided treatments. Nanotheranostics were introduced in the field of nanomedicine research about a decade ago, however their clinical potential is yet to be seen.[38] Nanotheranostics can be specially designed to respond to focused ultrasound and an imaging modality.

## 1.2 Requirements of Image Guided FUS Triggered Drug Delivery Systems

The combination of high-intensity focused ultrasound together with high-resolution MR guidance has created a platform that can produce focussing of the ultrasound energy deep within solid organs without invasive steps. Being non-invasive, focused ultrasound gives several advantages to modify the temperature of deep tissues and in combination with smart temperature triggered nanotheranostics to create a powerful tool for killing cancer cells. Accurate targeting and detailed accurate thermal mapping are provided by MRI and provide accurate deposition of energy in tissues that can be altered in response to near real-time thermal imaging produced by MRI, so that the variation in tissue response that is otherwise observed can be avoided.[39,40]

In recent years two clinical imaging modalities have been combined with FUS to provide guidance within the targeted area in the tissues, and this is mainly used to ablate lesions.[40] Ultrasound-guided FUS was the first image-guided system, however MR-guided HIFU or MRgFUS has been suggested as it shows several inherent advantages, such as superior anatomic detail and real-time thermometry during the thermoablation process, and it has demonstrated positive results in the ablation of both benign and malignant tumours.[41] FUS has been employed in the cases of hepatocellular carcinoma, prostate cancer, uterine myomas, and breast tumours, and has shown success in palliative pain management in pancreatic cancer and bone tumours.[40] Ultrasound and MRI are widely used clinical imaging modalities that have been combined with FUS for Image guided FUS treatments. Ultrasound microbubbles for ultrasound imaging or contrast enhancing agents for MRI imaging provide the required contrast enhancement for the application of FUS, and although gadolinium based contrast enhancement agents have been found to interfere with MRI thermometry, this may be dependent on concentration and type of agent used.[42] On the other hand, paramagnetic liposomes have been suggested as indicators of temperature based on release of their content at 42 °C. These liposomes are loaded with $Gd^{3+}$ contrast enhancing agents and their release at this temperature indicates a method of confirming focal hyperthermia (Figure 1.1).[43]

MRI contrast enhancing agents have been suggested as a method to assess drug release from thermosensitive liposomes. In a recent study thermosensitive liposomes were doubly loaded with iron oxide nanoparticles and Gd-chelate. At low temperatures, the transverse relaxivity of the liposomes was high, allowing detection of TSLs in tissues. This is important due to the fact that these theranostics need to be imaged in tissues before intervention with FUS. After temperature increase and thermal liposomal membrane instability the longitudinal relaxivity steeply increased indicating release of the Gd-chelate contents. By choosing the appropriate MR sequences, availability and release could be assessed without interference of one contrast agent with the other.[44]

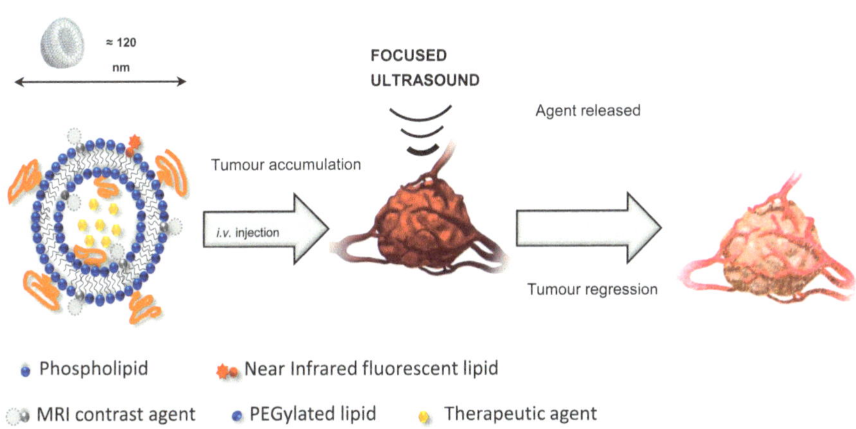


**Figure 1.1** Schematic of thermally triggered theranostic nanoparticle methods *in vivo*. iTSLs are injected *i.v.* and allowed to accumulate in the tumour due to the EPR effect, while monitoring by imaging. Once the concentration is thought optimal, the therapeutic agent is locally released with hyperthermia. Exposure of the cancer cells to a high local concentration of agent within a short period of time maximises cytotoxicity.

In a separate study Peller *et al.* introduced the combination of therapeutic (doxorubicin thermosensitive liposomes, DOX-TSL) and imaging (contrast enhancement thermosensitive, CA-TSL) to confirm drug release in the tumour.[45]

When using nanocarriers sensitive to mechanical forces (similar to the oscillating ultrasound pressure waves) and/or sensitive to temperature, ultrasound can induce either cavitation and/or hyperthermia and the cargo can be released locally. Although these are two different concepts and have different suggested mechanisms their interplay and combination has also been considered.[46]

The most widely tested concept of combining ultrasound with triggered release is the use of MRgFUS with thermosensitive liposomes (Figure 1.2). Thermosensitive liposomes have been suggested for local drug release in combination with local hyperthermia more than 25 years ago.[47]

The use of microbubbles may be designed specifically to enhance cavitation effects. Cavitation can lead to sonoporation of cell membranes and

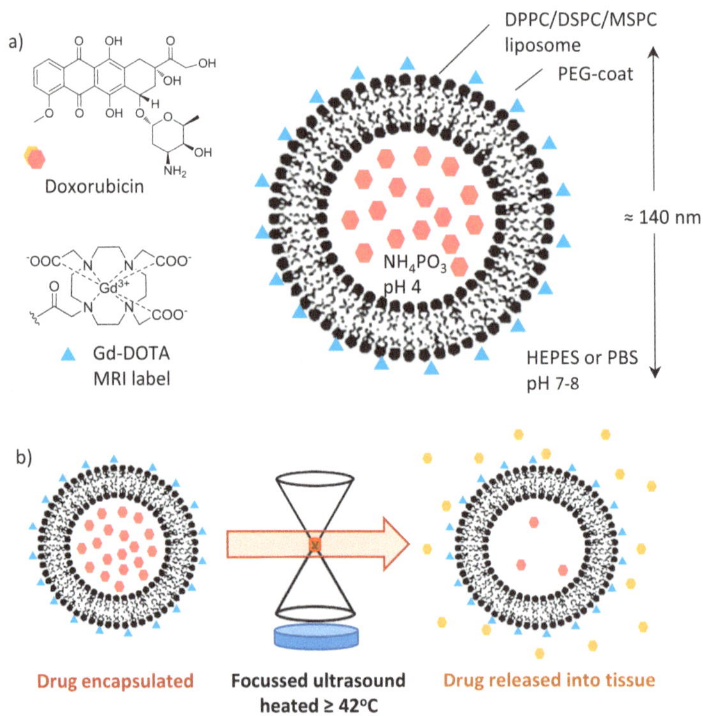


**Figure 1.2** The liposome envelope is formed of neutral phospholipids, covered in a PEG stabilising coat and decorated with a gadolinium MRI label; (a) Doxorubicin is loaded *via* pH gradient at ≈38 °C; (b) The resulting nanoparticle in stable under normal *in vivo* conditions but rapidly releases the encapsulated drug when surrounding tissue is heated to ≈42 °C.

transfer of therapeutics and or their carriers.[48,49] Real-time imaging methods, such as magnetic resonance, optical and ultrasound imaging have led to novel insights and methods for ultrasound triggered drug delivery. Image guidance of ultrasound can be used for: (a) targeted tissue identification; (b) spatio-temporal guidance of energy that translates to heat to release or activate the drugs and/or permeabilise membranes; (c) imaging can assist evaluation of bio-distribution, pharmacokinetics and pharmacodynamics; and (d) physiological read-outs to evaluate the therapeutic efficacy.

Image guided focused ultrasound targeted drug delivery requires the following: (a) a clinically approved imaging modality; (b) a contrast enhancing agent for this imaging modality; (c) a drug delivery system that responds or coordinates with the energy of the focused ultrasound deposited at the site of action; (d) feedback from the targeted tissue usually provided as a temperature map or a change in the contrast enhancement; (e) the right focused ultrasound equipment for the respective targeted part of the body. If the carrier is a nanosized liposomal system that responds to temperature the attributes and mechanism of drug release should be similar to the one shown in Figure 1.2.

Ideally the drug delivery system should be non-toxic and inert to other tissues and not act by delivering the drug anywhere other than the targeted site. The drug delivery system can also provide feedback of the localisation of the pathology and response to the treatment of the focused ultrasound.

## 1.3 FUS Induced Increase in Temperature for Tissue Specific Drug Release from Thermosensitive Carriers

Thermosensitive liposomal carriers were first developed almost four decades ago. The concept of a temperature modified liposomal membrane had triggered scientists to use liposomes as models to understand lipid membranes behaviour at different temperatures. Yatvin *et al.* first described the effect of increased temperature (hyperthermia) on liposomes in 1978.[50] However, development of temperature sensitive liposomes (TSL) for therapeutic reasons, such as cancer treatment, was first introduced in 1999 by Needham's group, who suggested that lipid phase transition enhanced lipid membrane permeability.[51] The team performed *in vivo* data using cancer mice models. The authors described a new advanced lipid formulation containing the drug doxorubicin with thermosensitive properties. The formulation was tuned to be stable at 37 °C, but became porous at mild hyperthermic temperatures (39 °C to 40 °C) achievable in the clinic, leading to very rapid and sharp doxorubicin release. This new TSL formulation, in combination with hyperthermia (42 °C), was found to be significantly more efficient in reducing tumour growth in a human squamous cell carcinoma xenograft than free drug or previous liposome formulations, indicating that thermally triggered release was advantageous.[52] These low temperature-sensitive liposomes (LTSL) were later tested in dogs having canine tumours and showed a tumour growth efficiency indicating that the concept of thermally triggered release is an attractive mechanism for targeted drug delivery of chemotherapeutics.[53,54] A formulation made by these thermosensitive liposomes took the brand name Thermodox® and was later developed by Celsion corporation, an organisation that currently leads the field of FUS thermally triggered doxorubicin drug release. Thermodox® or LTSL liposomes have been used by several groups in combination with induced hyperthermia. They can be triggered to release their pharmaceutical payload by any heat-based treatment such as radiofrequency thermal ablation (RFA), microwave hyperthermia (MWH), and HIFU. Results from a Phase I study using Thermodox® were recently published.[55] Researchers used escalating dose of Thermodox® and concluded that Thermodox® can be safely administered at 50 mg $m^{-2}$ in combination with RFA. Thermodox® in combination with RFA has been tested in a large Phase I study to treat hepatocellular carcinoma.[56] Analysis of the results from this study indicated the imaging features following RFA with LTLD were different from those after standard RFA, the increased size of the LTLD-treated ablation zone after RFA indicated that the doxorubicin drug-induced effects in the treated tissues were ongoing.[57]

There is a clear shift in Celsion's research pipeline to use HIFU as a more efficient and targeted approach of locally increasing the temperature.

The concept of using liposomes and FUS was introduced recently, in 2006 when Frenkel *et al.* used liposomal doxorubicin (Doxil®) in combination with pulsed HIFU exposures in a murine breast cancer tumour model. Doxil® is a stable liposomal preparation that has no response to increased temperature.[58] This was developed to minimise doxorubicin's cardiotoxicity, by encapsulating doxorubicin within stealth liposomes. Although Doxil® achieves a long circulation of doxorubicin with minimum cardiotoxicity it does not rapidly release the drug within the tumour, instead it facilitates the drug's accumulation in tumours due to EPR effect and allows the drug to be slowly diffused from the liposomes in the tumour. In another study, pulsed-HIFU exposures were not found to enhance the therapeutic delivery of doxorubicin and did not induce tumour regression. However, a fluorescent dextran showed blood vessels to be dilated as a result of the exposures.[59] This indicates that hyperthermia induces vasodilation and helps extravasation of nanoparticles and drugs. In that study experiments with polystyrene nanoparticles of similar size to the liposomes showed a greater abundance to be present in the treated tumours.[59] Although this study did not achieve or prove a therapeutic advantage of the use of FUS with chemotherapeutic liposomes, it showed clearly that pulsed HIFU has the potential of improving drug distribution in the tumour and induces a substantial increase of permeation of macromolecules (fluorescent dextran) and nanoparticles (fluorescent microspheres) through the tumour blood vessels.

In 2007 Dromi *et al.* presented the first study on thermosensitive liposomes LTSL and HIFU hyperthermia. The authors investigated pulsed-HIFU as a source of heat in combination with thermosensitive liposomes to enhance delivery and therapeutic efficacy of doxorubicin in murine adenocarcinoma tumours. *In vitro* treatments simulating the pulsed-HIFU thermal dose (42 °C for 2 min) triggered drug release of 50% of doxorubicin from the thermosensitive liposomes; however, no detectable release from the non-temperature sensitive liposomes (similar to Doxil®) was observed, as expected. Similarly, *in vivo* experiments showed that pulsed-HIFU exposures combined with the LTSL-Dox resulted in a more rapid release of doxorubicin, as well as significantly higher concentration within the tumour when compared with LTSL-Dox alone or the non-thermosensitive liposomes (doxil type), with or without exposures.[60] This study showed that FUS can induced focal hyperthermia that affects the phase transition of lipid bilayers of thermosensitive liposomes.

The same team developed MR imageable thermosensitive liposomes (iLTSL), with the objective to characterise drug release in phantoms and *in vivo*. An MRI contrast agent (ProHance® Gd-HP-DO3A) and doxorubicin were loaded in liposomes and drug release was quantified by spectroscopic and fluorescence techniques, respectively. The drug release after application of FUS under MR guidance was examined in tissue-mimicking phantoms containing iLTSL and in a VX2 rabbit tumour model usually used in interventional radiology, iLTSLs demonstrated consistent size and doxorubicin release kinetics. Release of doxorubicin and ProHance® from iLTSL was minimal at 37 °C, but faster when heated to 41.3 °C. The MRI relaxivity of iLTSL increased significantly from 1.95 +/− 0.05 to 4.01 +/− 0.1 $mMs^{-1}$ when liposomes were heated above the phase transition temperature, indicating the release of ProHance® from liposomes and its exposure to the aqueous surroundings (enhanced MRI contrast). Importantly, the signal increase corresponded spatially and temporally to MRgFUS-heated locations in the phantoms used in this experiment. *In vivo*, the investigators confirmed MRI signal after administration of iLTSL injection and after each 10 min heating, with the greatest increase in the heated tumour region. The authors concluded that MRgFUS combined with iLTSL may enable real-time monitoring and spatial control of drug release from liposomes.[61] This is important as it provides control of the administration and offers the opportunity for the development of personalised treatments.

In a follow up study the same authors investigated the effect of iLTSL in rabbits bearing VX2 tumours. In that study image-guided non-invasive hyperthermia was applied for a total of 30 min, completed within 1 hour after LTSL infusion and quantified doxorubicin release in tumours with HPLC and fluorescence microscopy. Sonication of VX2 tumours resulted in accurate and spatially homogenous temperature control in the target region. LTSL + MR-HIFU resulted in significantly higher tumour doxorubicin concentrations (3.4-fold greater compared with LTSL respectively, $p < 0.05$, Newman-Keuls). The authors observed that both free doxorubicin and LTSL treatments appeared to deliver more drug in the tumour periphery as compared to the tumour core. This indicated that FUS induced hyperthermia and LTSL increases the permeability of doxorubicin as doxorubicin was found in both the tumour periphery and core.[62] The group further developed a heating algorithm using the same rabbit tumour model proving that the use of the binary feedback algorithm results in accurate and homogenous heating within the targeted area.[63] A computational model that simulated the tissue heating with FUS treatments and the resulting hyperthermia in tissues (similar to the one that leads to drug release) were developed by Haemmerich *et al.*[64] In this model a spatio-temporal multicompartmental pharmacokinetic model simulated the drug release in the blood vessels and its transport into the interstitial space, as well as cell uptake. Two heating schedules were simulated each lasting 30 min, the first corresponding to hyperthermia (HT; 43 °C) and the second corresponding to HT followed by a high temperature (50 °C) for a 20s pulse, (HT+). Using the computational model (validated in rabbit VX2 tumours) the authors found that the cellular drug uptake is directly related to hyperthermia duration. However, HT+ enhanced drug delivery by 40% compared to HT.[64] The study indicated the importance of simulations in the application of drug delivery mechanisms to tumours and that understanding hyperthermia effects can support the design of MRgFUS treatments for patients.

In addition to the progress in the understanding of the physical mechanism of drug delivery from well validated thermosensitive liposomes carrying doxorubicin, researchers further investigated the chemical composition of such liposomes in response to MRgFUS induced hyperthermia.

De Smet *et al.* compared thermosensitive liposomes carrying doxorubicin and ProHance® (gadolinium based contrast enhancing agent). Two temperature-sensitive systems composed of the following lipids DPPC : MPPC : DPPE-PEG$^{2000}$ (LTSL) and DPPC : HSPC : cholesterol : DPPE-PEG$^{2000}$ (traditional temperature-sensitive liposomes, TTSL) were investigated for their stability and release profile at 37 °C and 42 °C in phantoms using MRI [lipids; 1,2-dipalmitoyl-*sn*-glycero-3-phosphocholine (DPPC), 1-palmitoyl-*sn*-glycero-3-phosphocholine (MPPC), 1,2-dipalmitoyl-*sn*-glycero-3-phosphoethanolamine-*N*-[methoxy(polyethyleneglycol)-2000](DPPE-PEG$^{2000}$), hydrogenated-L-α-phosphatidylcholine (HSPC)]. The LTSL system showed a higher leakage of doxorubicin at 37 °C, but a faster release of doxorubicin at 42 °C compared to the TTSL system, indicating that lipid composition plays an important role in the stability and release profile.[65] The authors further investigated the more stable traditional temperature sensitive liposomes carrying doxorubicin and ProHance® *in vivo* in rats bearing 9L gliosarcoma tumours. A clinical MRgFUS system was applied in a proof-of-concept study to induce local hyperthermia for 30 min. The local temperature-triggered release of ProHance® was monitored with interleaved $T_1$ mapping of the tumour. A good correlation between the $\Delta R_1$, (change in longitudinal relaxation rate $\Delta R_1 = \Delta(1/T_1)$), and the intratumour doxorubicin and gadolinium concentration was found, implying that the *in vivo* release of doxorubicin from the thermosensitive liposomes in the tumours can be probed (imaged) *in situ* with the longitudinal relaxation time of the co-released MRI contrast agent (dose painting).[65]

Temperature sensitive liposomes release their encapsulated drugs at the melting phase transition temperature ($T_m$) of the lipid bilayer. At this $T_m$ the lipid membrane changes its structure as it transfers from a gel to the liquid crystalline phase.[66] When the liposomal membranes are in the gel phase they show less permeability to molecules and water compared to the liquid crystalline phase.

The liposomes transition to the liquid crystalline phase can be achieved with the incorporation of a lyso-phospholipid such as MSPC ($R = -C_{17}H_{35}$). This lipid is also used in the Thermodox® formulation.[47] A potential disadvantage of MSPC containing liposomal formulations is the their rapid doxorubicin leakage at 37 °C.[64] Tagami *et al.* prepared temperature sensitive liposomes using nonionic surfactants Brij which are PEG-ylated lysolipids resembling the chemical structures of MSPC and DSPE-PEG$^{2000}$. Results indicated that the optimal acyl chain length of the surfactant was between C(16) and C(18) with a saturated carbon chain and a PEG repeating unit ranging between 10 and 100 with a molecule weight above 600 Da. In the panel of surfactants tested, Brij78 was optimal and could be incorporated into the liposomes by the thin film hydration or the post-insertion method with an optimal range of 1 to 8 mol%.[67] The authors continued with *in vivo* experiments in mice bearing mammary carcinoma cells EMT-6, investigating $Gd^{3+}$DTPA (diethylene triamine pentaacetic acid) release with NMR relaxometry. The authors observed a good correlation between relaxation enhancement in the heated tumour and the inhibition of tumour growth at day 21 post treatment.[68]

Kono *et al.* investigated the effect of poly[2-ethoxy(ethoxyethyl)vinyl ether] chains (having a lower critical solution temperature) and polyamidoamine G3 dendron-based lipids with $Gd^{3+}$ chelate residues into PEGylated liposomes. These theranostic liposomes exhibited excellent ability to shorten the longitudinal proton relaxation time (MRI). When administered intravenously into tumour-bearing mice, accumulated liposomes in tumours increased with time, reaching a constant level 8 hours after administration by following $T_1$-weighted MRI signal intensity. Liposome size affected the liposome accumulation efficiency in tumours; liposomes of about 100 nm diameter were accumulated more efficiently than those with about 50 nm diameter. Tumour growth was strongly suppressed when liposomes loaded with doxorubicin were administered intravenously into tumour-bearing mice and the tumour heated mildly at 44 °C for 10 min, 8 hours after administration.[69]

Ultrasound sonication using microbubbles has been suggested to enhance the delivery of nanoparticles into the treated tumours.[70] Tumour-bearing mice were injected with microbubbles intravenously and sonicated before being injected with the PEGylated liposomal doxorubicin. It was demonstrated that FUS sonication with microbubbles can significantly enhance drug accumulation in the sonicated tumour at 24 hours after treatment. A significant hindrance to tumour growth was additionally achieved indicating that cavitation can enhance extravasation and drug tumour distribution.[70]

HIFU has also been tested in combination with ultrasound imaging and microbubbles (ultrasound contrast agent/cavitation agent) and chemotherapy drugs. The effect of HIFU (1.0 MHz; 12 999 W $cm^{-2}$ in continuous waves), in the presence of hematoporphyrin and/or microbubbles, on the anticancer potency (in cells) of 5-fluorouracil, cisplatin, paclitaxel, mitomycin C or adriamycin, was investigated. Insonated adriamycin resulted in a lower death rate of human cancer cells HO-8910 (45.85 ± 2.65% *vs.* 34.84 ± 1.21%, $p < 0.05$), which was exacerbated when employing hematoporphyrin (34.84 ± 1.21% *vs.* 23.09 ± 7.82%, $p < 0.05$) or hematoporphyrin combined with microbubbles (34.84 ± 1.21% *vs.* 8.79 ± 3.69%, $p < 0.05$); the therapeutic activity was not affected when adding microbubbles alone. Overall the authors suggested to avoid the use of microbubbles as the enhanced ultrasonic cavitation produces a large amount of free radicals that affect the integrity of the anticancer agents.[71] This suggests that a thorough study of the mechanisms involved in image guided FUS is required before selecting the imaging modality as well as the theranostic system. However, the selection of the chemotherapy appears to be most important.

Micelles have been also suggested as drug delivery carriers for FUS induced hyperthermia. Polymeric micelles (pluronics; P-105) were used to encapsulate doxorubicin.[72] These polymers were suggested as sensitisers of multidrug resistant (MDR) cells to chemotherapeutic drugs. The authors suggested that upon the accumulation of drug-loaded micelles at the tumour site, ultrasound treatment released the drug from micelles. This enhanced the intracellular uptake of both the released and encapsulated drug, possibly as a result of the induced extravasation and drug release from the micelles.[72] This work suggested that formulations other than thermosensitive liposomes could be used in combination with the ultrasound induced hyperthermia. Nevertheless such formulations have been less explored during recent years.

Targeted liposomes have also been investigated in combination with FUS. This was attempted to achieve therapeutically effective drug concentrations in tumours while avoiding healthy tissue damage (targeted drug delivery). In a recent work, a novel tumour-targeting peptide iRGD (CCRGDKGPDC) was used to modify the drug-loaded LTSL (iRGD-LTSL-DOX) and to investigate the anti-tumour effects when HIFU was applied to iRGD-LTSL-DOX specifically targeting $\alpha_v\beta_3$-positive cells and release the encapsulated doxorubicin (DOX) after triggering. *In vivo* results showed that DOX from iRGD-LTSL-DOX was intravascularly released and rapidly penetrated into the tumour interstitial space after FUS-triggered treatment, overcoming the limited tumour penetration of anticancer drugs. Strong anti-tumour efficacy further supported the effective combination of iRGD-LTSL-DOX and FUS-induced hyperthermia.[73]

In our research project we aim to design theranostic liposomes. We investigated the potential of a labelled phospholipid/lysolipid containing liposomes to accumulate in tumours and release the drug under conditions of mild hyperthermia. We have prepared liposomal nanoparticles and we have investigated the potential of labelling for imaging. We developed optically labelled TSLs for image guidance drug delivery (iTSLs). These iTSLs were used to encapsulate the drug topotecan and iTSL-topotecan liposomes were administered in tumour bearing mice. These iTSLs allow the simultaneous, real-time diagnostic imaging of nanoparticle biodistribution using a near-infrared (NIR; 750–950 nm) fluorophore coupled to the lipidic component of the lipid bilayer. When combined with multispectral fluorescence analysis, this allowed for specific and high sensitivity tracking of the nanoparticles *in vivo*.

Application of hyperthermia indicated higher accumulation in tumours and concomitant drug release in the heated tumour.[74]

Near Infrared Fluorescence (NIRF) labelling of the liposomes proved to be an excellent tool of monitoring the liposome accumulation in tumours and indicated that the liposomes signal increases with time (Figure 1.3).

When these liposomes were administered intravenously in mice with tumours and mild hyperthermia was applied for a short period of time, drug release was observed as assessed by fluorescence microscopy (Figure 1.4). Tumour sections that were isolated from tumours treated with hyperthermia clearly showed improved drug distribution within the tumour, compared to non-heated tumours (Figure 1.4). This study indicated that when formulations can be imaged (theranostics) modulation by means of triggered release is possible.

MRI labelling of the liposome nanoparticles can be done with a lipid that consists of a DOTA [1,4,7,10-tetraazacyclododecane-1,4,7,10-tetraacetic acid] headgroup (Figure 1.2).[75,76] Introducing the imaging lipid in the lipid bilayer provides better and clearer monitoring of liposomal particle kinetics and a better knowledge of the time required for maximum nanoparticles accumulation in tumours (monitored by MRI in the MRgFUS).

Combination of macromolecules, such as polymer drug conjugates with hyperthermia could also be a method of increased drug accumulation in

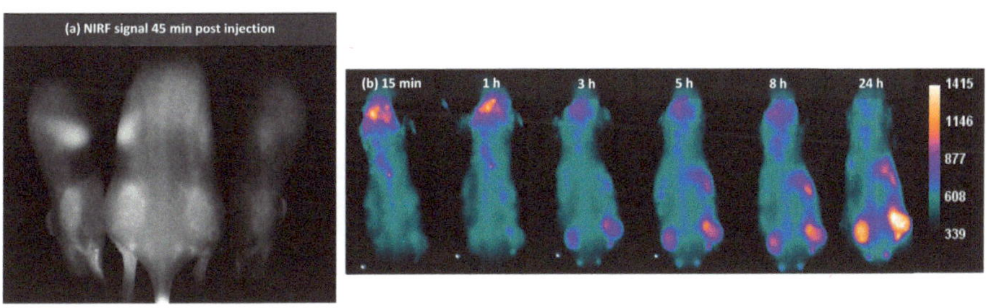


**Figure 1.3** *In vivo* kinetics of thermosensitive liposomes iTSLs in tumours (IGROV-1 ovarian cancer cells) after intravenous administration. (a) Illustration of the highlighted vasculature due to iTSLs 45 min after injection. (b) Illustration of the accumulation of iTSLs in tumours over time, in one representative animal. The images were acquired using the same NIR acquisition setting at each time point; images were unmixed using the Maestro software, quantified using pixel intensity, and false coloured using ImageJ. Adapted from ref. 74. Copyright (2015) American Chemical Society.

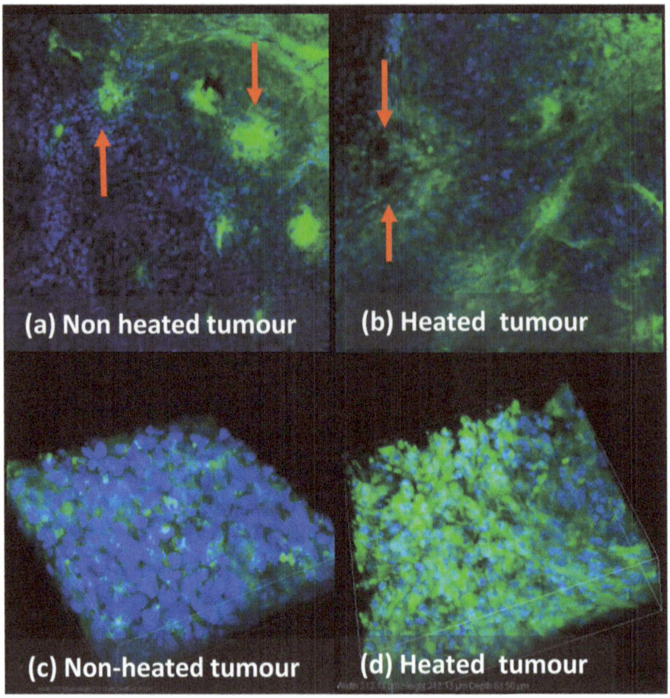


**Figure 1.4** Drug delivery of iTSL-topotecan and efficacy upon heating; (a) and (b) show the low magnification (20×), and (c) and (d) show the high magnification (100×) of tumour sections (z-stack) imaged by confocal microscopy. The topotecan gives the green fluorescence while the nucleus of the cells is counterstained by DAPI (blue). Reprinted from ref. 74. Copyright (2015) American Chemical Society.

the tumours that are treated with FUS. In a recent study researchers have introduced the use of heat shock (HS)-targeting towards tumour tissues as a method to enhance the drug accumulation and the retention, and improve therapeutic outcomes. HIFU was applied to generate hyperthermia in prostate tumour tissue in mice. When hyperthermia is applied an upregulation of the cell surface HS receptor glucose regulated protein 78 kDa (GRP78) was observed. This receptor was the further targeted by specific HS-targeting peptides attached to polymer drug conjugates. It was shown that HIFU-mediated HS-targeted copolymer–docetaxel conjugates improved treatment efficacy in a murine prostate tumour xenograft model. The study showed that the use of HIFU hyperthermia in combination with polymer therapeutics has potential to improve therapeutic outcomes in prostate cancer treatment.[77]

In a similar study MRgFUS was found to enhance the delivery of both Evans blue dye (EBD) and gadolinium-chelated *N*-(2-hydroxypropyl)methacrylamide (HPMA) copolymers (polymer therapeutic). The EBD accumulation in the heated tumours increased by nearly two-fold compared to unheated tumours, an observation consistent with similar FUS treatments.

The gadolinium-chelated HPMA copolymers also showed significant enhancement in accumulation over control as evaluated through MRI T1-mapping measurements.[78] These labelled drug conjugates indicate that macromolecules can also function as theranostics, of which delivery can be modulated.

Our recent results using NIRF labelling of an antibody indicates that the concept of theranostics can find application on established anticancer therapies. We have recently labelled trastuzumab with a near infrared label and monitored its distribution within tumours with or without repeated applications of FUS hyperthermia.[79] We have used a custom made FUS (Philips Research) that mimics the transducers used in the clinic (Figure 1.5).

Using this application of FUS as a non-invasive method of hyperthermia we increased the tumour temperature to 42 °C for a short period of time (3–5 min) and we observed an increased accumulation of labelled antibody. Repetition of focused ultrasound induced hyperthermic treatment further increased the accumulation of the antibodies in the tumour (Figure 1.6). This treatment also augmented the accumulation of other macromolecules non-specific to the tumour, such as IgG and albumin.[79] Image guided FUS could be used to enhance the therapeutic efficiency of antibodies and/or targeted nanoparticles.

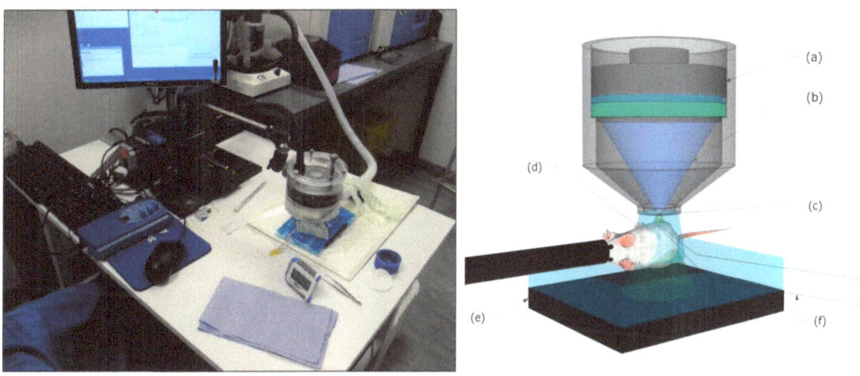


**Figure 1.5** Preclinical focused ultrasound; (left) overview of the equipment showing the water-filled transducer chamber, the thermocouple interface, and the control PC; (right) schematic of the *in vivo* configuration with the transducer: (a) raised such that the ultrasound biconic; (b) focuses just above the skin surface over the tumour; (c) the mouse is surrounded with warmed, degassed ultrasound gel; (d) and placed on an ultrasound absorbing mat; (e) to prevent reflections off the table. Temperature monitoring is *via* two or three fine-wire thermocouples (f) implanted around the tumour. Reprinted from *Pharmacological Research*, **114**, Miguel N. Centelles, Michael Wright, Wladyslaw Gedroyc, Maya Thanou, Focused ultrasound induced hyperthermia accelerates and increases the uptake of anti-HER-2 antibodies in a xenograft model, 144–151, Copyright 2016, with permission from Elsevier.

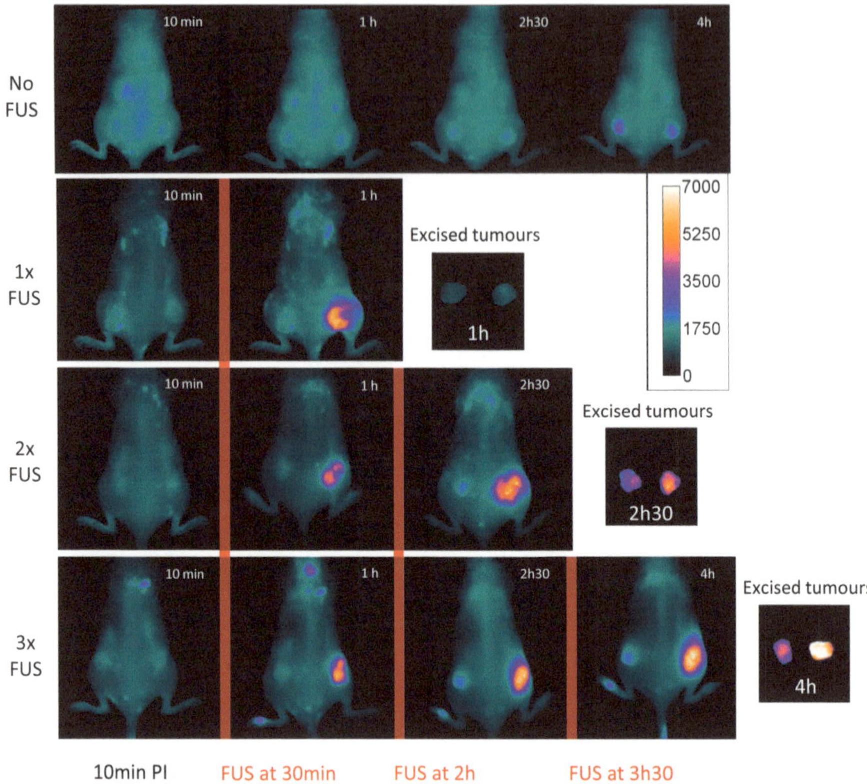


**Figure 1.6** Comparisons of *in vivo* and excised IGROV-1 tumours labelled XL750-herceptin NIRF from mice treated with no, 1, 2, or 3 rounds of FUS hyperthermia to the right hand tumour. The animals were sacrificed at 1 hour, 2 hours 30 min, and 4 hours post injection. Reprinted from *Pharmacological Research*, **114**, Miguel N. Centelles, Michael Wright, Wladyslaw Gedroyc, Maya Thanou, Focused ultrasound induced hyperthermia accelerates and increases the uptake of anti-HER-2 antibodies in a xenograft model, 144–151, Copyright 2016, with permission from Elsevier.

Assessment of the current research in this field shows that the number of applications of image guided drug delivery that employ the focused ultrasound as a method of drug modulation is expanding. New methods and formulations are suggested that can maximise the effect of drug delivery. Combination of nanobubbles with TSLs proved to be an efficient method of delivering doxorubicin in mouse tumour model avoiding drug delivery in other tissues.[80]

Triggered drug delivery using an external physical force provides the required control of drug deposition in certain tissues avoiding exposure of healthy tissues to toxic concentrations. The triggered induced delivery should be immediate and the effect induced on non-targeted tissues non-damaging. Hyperthermia induced by a means such as ultrasound can be exploited as an external trigger in drug delivery.[4,81]

Mild hyperthermia can be induced by pulsed FUS that can reduce extreme tissue heating by allowing the tissue to cool down between US exposures.[82] The increase in temperature can be 3–5 °C (hyperthermia) despite the high energy deposited in the tissue. Hyperthermia applied in tumours can increase blood flow and enhances vascular permeability. Studies with canine soft tissue sarcoma and human tumour clinical studies have also demonstrated that hyperthermia improves tumour oxygenation, and enhances response of such tumours to radiotherapy or chemoradiotherapy. The increased blood flow and vascular permeability caused by temperatures such as 42 °C may also improve the delivery of chemotherapy drugs, immunotherapeutic agents and genes to tumour cells.[83] FUS exposures in pulsed mode lowers the rates of energy deposition and generate primarily mechanical effects for enhancing tissue permeability to improve local drug delivery. These pulsed exposures can be modified for low-level hyperthermia as an enhancement of drug delivery that would lead to better drug deposition and better therapeutic effect.[84] Mild hyperthermia of 42 °C can improve the degree of nanocarrier extravasation as shown by Kong *et al.*[85] The reason that this leads to increased extravasation maybe due to down regulation of VE-cadherin that contributes to vascular integrity as it was shown in HUVEC endothelial cells.[86] It is clear that hyperthermia can provide a boost to extravasation and drug deposition in tumours. This should provide an adjuvant effect when nanocarriers are used and accumulate in tumours due to enhanced permeation and retention effect. It would be interesting to investigate the effect of hyperthermia on tumour/tissue drug clearance.

### 1.3.1 Ultrasound and Bubbles to Increase Drug Permeability in Tissues

FUS can also induce non-thermal effects on tissues. Acoustic cavitation can be induced using microbubbles exposed to US.[87] Acoustic cavitation can be defined as the growth, oscillation and collapse of gas containing bubbles under the influence of the varying pressure field of sound waves in a medium and can have an effect on the permeability of a biological tissue.[87–89] There are two types of acoustic cavitation; non-inertial and inertial cavitation.[90] The non-inertial (stable) cavitation occurs when bubbles persist for a number of acoustic cycles. In this case the bubble's size increases and decreases (the bubble expands and contracts) according to the applied ultrasound frequency. Inertial (transient cavitation) occurs when bubbles grow very fast and expand two or three-fold their resonant size, oscillate in an unstable manner and collapse in one compression half cycle.[88] It has been considered that the primary mechanism to affect the structure of intact cells is inertial cavitation that can induce irreversible damage as well as increase cell membrane permeability.[91,92] This has been tested as a hypothesis of drug permeation improvement.[93–96]

An important application of FUS and bubbles lies in the area of altering the permeability of the blood brain barrier.[97]

In 2002, Mesiwala *et al.* observed that HIFU could alter BBB permeability. HIFU induced reversible, non-destructive, BBB disruption in a targeted area and this opening reversed after 72 hours. The authors showed with microscopy that HIFU either preserved brain architecture while affecting (opening) the BBB, or generated tissue damage in a small volume within the affected region of BBB. Further electron microscopy suggested that HIFU disrupted the BBB by widening capillary endothelial cell tight junctions, a mechanism used to open BBB.[98]

The effect of FUS on the integrity of tight junctions was later confirmed in a study investigating the rat brain microvessels after this BBB disruption. The authors used immune-electron microscopy, to identify tight junctional proteins such as occludin, claudin-1, claudin-5, and sub-membranous ZO-1 junctional protein after sonication. They found substantial redistribution and loss of occludin, claudin-5 and ZO-1 indicating that the tight junction protein complex is affected. However, claudin-1 seemed less involved. Monitoring the leakage of horse radish peroxidase (large protein, permeability marker; MW 40 kDa) the authors observed that the BBB disruption appears to last up to 4 hours after sonication.[99] In a later study the role of caveolin in the mechanism of FUS BBB enhanced permeation was suggested. In a study investigating caveolae density it was found that caveolae and caveolin-1 were primarily localised in the brain microvascular endothelial cells of all the animals tested (rats) regardless of treatment, and that caveolin-1 expression was highest in the rats treated with both FUS and microbubbles. The authors in that study concluded that caveolin-1-mediated transcellular transport pathway may cooperate with other transport pathways (*e.g.* tight junctional disruption) to induce the opening of the BBB and the increased permeation of drug molecules.[100]

Hynynen and colleagues investigated the BBB FUS enhanced permeability in rabbits. Rabbit brains were exposed to pulsed focused ultrasound while micro-bubbles were intravenously administered. The BBB opening was measured by an MRI contrast agent evaluating the local enhancement of permeation in the brain using MRI. The authors found that low ultrasound powers and pressure amplitudes were found to cause focal enhancement of BBB permeability to this contrast enhancing agent. Trypan blue injected before animals were sacrificed indicated blue spots in the areas of the sonicated locations.[101] The authors concluded that HIFU disruption of BBB could be used to enhance drug delivery to the brain.[102]

McDannold *et al.* tested the safety of this method by searching for ischemia and apoptosis in areas with BBB disruption induced by pulsed ultrasound in the presence of gas bubbles and by looking for post treatment effects up to one month after sonication. Pulsed ultrasound exposures (sonications) were performed in the brains of rabbits under monitoring by MRI. BBB disruption was confirmed with contrast-enhanced MR images. Whole brain histologic examination was performed using staining for ischemic neurons and TUNEL staining for apoptosis. Tiny regions of extravasated red blood cells scattered around the sonicated locations, indicated capillaries. Despite these vasculature effects, only a few cells in some of the sonicated areas showed evidence of apoptosis or ischemia. The authors found that ultrasound-induced BBB disruption is possible without inducing substantial vascular damage that would result in ischemic or apoptotic death to neurons.[103]

The method could find application in the delivery of large therapeutic molecules that do not normally permeate the BBB. Herceptin (trastuzumab), a humanised anti-human epidermal growth factor receptor 2 (HER2/c-erbB2) monoclonal antibody, was delivered locally and noninvasively into the mouse central nervous system through the BBB under image guidance by using an MRI-guided FUS. The amount of Herceptin delivered to the target tissue was correlated with the extent of the MRI-monitored barrier opening, making it possible to estimate indirectly the amount of Herceptin delivered. The method could be used to treat breast cancer metastases to the brain.[104] It was further shown that dopamine D(4) receptor-targeting antibody could also be delivered using the same technique in the brain.[105,106]

Delivery of small molecules can also be enhanced with the use of FUS and disruption of the BBB through cavitation. Treat *et al.* demonstrated relatively high concentrations of doxorubicin in the brain with minimal healthy tissue damage effects and effect that would benefit the delivery of small drugs that do not cross the BBB. The authors observed that doxorubicin accumulation in the non-targeted contralateral brain tissue remained significantly lower showing the efficiency of the method. MRI signal enhancement in the sonicated region correlated strongly with tissue doxorubicin concentration, suggesting that contrast-enhanced MRI could perhaps indicate drug penetration during image-guided interventions.[107]

Konofagou and co-workers assessed the spatial permeability of the BBB-opened region using dynamic contrast-enhanced MRI (DCE-MRI) in mice. The authors generated permeability maps and $K_{\text{trans}}$ (the transfer rate constant from the intravascular system to the extracellular extravascular space) values were estimated for a predefined volume of interest in the sonicated and the control area for each mouse. The results demonstrated that $K_{\text{trans}}$ in the BBB-opened region was at least two orders of magnitude higher when compared to the contralateral (control) side confirming the hypothesis that FUS can transiently open the BBB.[108]

There are several parameters that affect the level of BBB enhanced permeability and the endothelial tight junctions disruption; the pulse sequence comprising short bursts, the spacing between bursts or the rate of infusion of the microbubbles, and the size of microbubbles were found to affect the effect on BBB disruption.[109,110]

The method could be applied for a number of therapeutic applications. The brain-derived neurotrophic factor (BDNF) was delivered to the left hippocampus in mice through the noninvasively disrupted BBB using FUS. The BDNF bioactivity was found to be preserved following delivery as assessed quantitatively by immunohistochemical detection of the pTrkB receptor and activated pAkt, pMAPK, and pCREB in the hippocampal neurons. It was shown that BDNF delivered this way induced signalling effects in a highly localised region in the brain.[111]

FUS with the presence of microbubbles has been shown to induce transient and local opening of the BBB for the delivery of therapeutic molecules which normally cannot penetrate into the brain. The success of FUS brain-drug delivery relies on its integration with *in vivo* imaging to monitor the kinetic change of therapeutic molecules (image guided drug delivery).

However, it is the area of targeting brain tumours that has attracted most interest in the FUS disrupted BBB.[112] Mei and colleagues investigated the effects of targeted and reversible disruption of the BBB by MRI-guided FUS and delivery of methotrexate to the rabbit brain. The authors recorded that the methotrexate concentration in the sonicated group was notably higher than that in both the control group (intravenous administration) and the internal carotid artery administered group. They observed a greater than 10-fold increase in the drug level compared to internal carotid administration without FUS.[113]

Liu *et al.* investigated the delivery of 1,3-bis(2-chloroethyl)-1-nitrosourea (BCNU) to glioblastomas in rats with induced tumours with the help of FUS. The authors found that FUS significantly enhanced the penetration of BCNU through the BBB in normal and tumour-implanted brains without causing bleeding. Surprisingly, treatment of tumour-implanted rats with FUS alone had no beneficial effect on tumour progression. However, treatment with FUS before BCNU administration controlled tumour progression and improved animal survival relative to untreated controls.[114]

The blood–brain/tumor barrier inhibits the uptake and accumulation of chemotherapeutic drugs, Liu and colleagues recently assessed FUS-mediated delivery of an iron oxide magnetic nanoparticles (MNPs) conjugated to an antineoplastic agent, epirubicin. They used MNPs because of the favourable MRI characteristics, which could facilitate imaging. They demonstrated a substantial accumulation of MNPs, as well as epirubicin, up to 15 times the therapeutic range in the brain when delivered with FUS. They further showed decreased tumour progression in animals with brain tumours that received MNP with epirubicin *via* FUS.[115]

Receptor targeting liposomal nanocarriers have been combined with image guided FUS to treat brain tumours. In a recently presented study it was shown that pulsed HIFU and human atherosclerotic plaque-specific peptide-1 (AP-1)-conjugated liposomes containing doxorubicin (AP-1 Lipo-Dox) acted synergistically in an experimental brain tumour model. Prior to each sonication, AP-1 Lipo-Dox or unconjugated Lipo-Dox were administered intravenously, and the concentration in the brain was quantified. Drug injection with sonication increased the tumour : normal brain doxorubicin ratio of the target tumours by about two fold compared with the control tumours. Moreover, the tumour : normal brain ratio was highest after the injection of AP-1 Lipo-Dox with sonication. The results of this study indicate that combining targeting strategies can substantially enhance delivery of chemotherapy in the brain.[116] In a separate study the authors investigated the pharmacokinetics of $^{111}$I-labeled AP1-Lipo-dox using microSPECT. The authors confirmed that sonication increased liposomal doxorubicin concentrations in tumour areas (murine glioblastoma) and that molecular targeting acts synergistically with FUS.[117]

Targeted gene transfer into central nervous system was investigated using MRI-guided FUS-induced BBB disruption. The results of this study showed that MRI-guided FUS achieved plasmid DNA transfer across the opened BBB furthermore plasmid were endocytosed by neuronal cells presenting heterogeneous distribution. BDNF (and BDNF-EGFP) expressions were markedly enhanced by the combination of ultrasound and plasmid BDNF-EGFP-loaded microbubbles about 20-fold compared to that of the control group. This indicates that this strategy can be used to deliver functional biologicals and gene therapy. The method of using MRI-guided FUS to induce the local BBB disruption could accomplish effective targeted gene therapy in the CNS. In this type of experiment the microbubbles are used as the plasmid carrier. The investigators conjugated plasmids onto the surface of microbubbles and they coated these carriers using polymers in a layer by layer technique.[118]

An exciting application is the delivery of therapeutic stem cells to the brain using FUS to potentially treat neurodegenerative diseases, traumatic brain injury, and stroke. MRI guidance was used to target the ultrasound beam thereby delivering iron-labelled, green fluorescent protein (GFP)-expressing neural stem cells specifically to the striatum and the hippocampus of the rat brain. Immunohistochemical analysis confirmed the presence of GFP-positive cells in the targeted brain regions suggesting that MRIgFUS may be an effective alternative to invasive intracranial surgery for stem cell transplantation.[119]

Contrast-enhanced magnetic resonance imaging (CE-MRI) is used to monitor contrast agent leakage to verify BBB-opening and infer drug deposition. However, despite being administered concurrently, microbubbles, therapeutic agent, and contrast agent have distinct pharmacokinetics and pharmacodynamics, thus complicating the quantification and optimisation of BBB-opening and drug delivery. Multifunctional microbubbles (MB) were loaded with therapeutic agent (doxorubicin; DOX) and conjugated with superparamagnetic iron oxide (SPIO) nanoparticles providing a theranostic tool to treat brain tumours. These DOX-SPIO-MBs were designed to concurrently open the BBB and perform drug delivery upon FUS exposure, act as bimodal MRI and ultrasound contrast agent, and allow magnetic targeting (MT) to achieve enhanced drug delivery (theranostic). Burst-tone FUS was applied after injection of DOX-SPIO-MBs, followed by magnetic targeting with an external magnet attached to the scalp in a rat glioma model. The authors found that BBB-opening and drug delivery were achieved concurrently during the FUS treatment. In addition, MT increased local SPIO deposition in tumour regions by 22.4%. Our findings suggest that DOX-SPIO-MBs with FUS could be an excellent theranostic tool for future image-guided drug delivery to brain tumours.[120]

Magnetic nanoparticles have been suggested in combination with FUS and microbubbles for magnetic hyperthermia. The superparamagnetic properties of the MNPs provide the opportunity to guide them by an externally positioned magnet and also provide contrast for MRI (theranostics). However, their therapeutic effect in treating gliomas, is limited by insufficient local accumulation and retention due to their inability to traverse biological barriers. To overcome this the use of FUS in combination with magnetic targeting was suggested as a mechanism to deliver therapeutic MNPs across the BBB and to enter the brain both passively and actively. MRI was used to monitor and quantify their distribution *in vivo*. Synergistic targeting and image monitoring are powerful techniques for the delivery of macromolecular chemotherapeutic agents into the CNS under the guidance of MRI.[115] This study shows that theranostics and their guided delivery can overcome difficult biological barriers and deliver the therapy at the right site.

## 1.4 Drug Delivery Dosage Forms and FUS Future Perspectives

During recent years there has been an expansion in research in the design of MRgFUS drug delivery. The main dosage forms tested in MRgFUS drug delivery strategy are the thermosensitive liposomes and the lipid based microbubbles that can incorporate drugs or other liposomes on their surface.[118,121]

There is limited research in the area of using other responsive materials or nanocarriers. Rapoport recently discussed the potential of using micelles and FUS[122] for enhanced tissue permeation. Micelles are nanosized carriers able to carry hydrophobic drugs; their combination with FUS could substantially enhance their delivery in tissues. Kostarelos and colleagues suggested the incorporation of thermosensitive peptides onto liposome bilayers to enhance thermo-responsiveness,[123] and the group of Lammers designed polymer based micro-bubbles for US drug release.[124]

A novel type of HIFU triggered active tumour-targeting polymeric micelle was recently prepared and investigated for controlled drug release and enhanced cellular uptake. Amphiphilic hyaluronic acid (HA) conjugates were synthesised to form docetaxel loaded micelles in aqueous conditions with high encapsulation efficiencies of over 80%. It was shown that HIFU enhanced the cellular uptake behaviour by altering the permeability of the cell membrane. It was also able to aid with the extravasation of micelles into the interior of tumours. These micelles can emerge as promising nanocarriers of chemotherapeutic agents for image guided FUS controlled drug release and tumour targeting.[125]

Another recent study suggested the use of a multifunctional hollow mesoporous Prussian blue (HMPBs) theranostic nanoplatform, the hollow structure of which is capable of encapsulating doxorubicin (DOX) and perfluorohexane (HMPBs-DOX/PFH) to be used in combination with FUS. *In vitro* and *in vivo* studies validated that HMPBs-DOX/PFH can be used as an amplifiable dual-mode imaging contrast agent, which can simultaneously enhance US and photoacoustic imaging for guiding and monitoring tumour therapy. When exposed to HIFU, this versatile HMPBs-DOX/PFH agent could increase the cavitation effect and use lower HIFU intensity to achieve coagulative necrosis.[126]

From the above it is evident that already established delivery systems such as different structurally nanocarriers have not been investigated in combination with image guided FUS. It would be interesting to see the effect of FUS on the enhanced permeability of different micelles, polymers (dendrimers) or metal nanoparticles (gold-iron) to tissues. Thermosensitive materials have been hardly explored in this field. Polymers or proteins that respond to small change of temperature could form suitable image guided FUS triggered platforms.

Overall MRgFUS drug delivery is a novel and valuable tool to increase drug targeting and tissue specific drug delivery. It is expected that future studies will prove the clinical efficacy of MRgFUS drug delivery applications.

## References

1. O. Al-Bataineh, J. Jenne and P. Huber, *Cancer Treat. Rev.*, 2012, **38**, 346–353.
2. J. E. Kennedy, G. R. Ter Haar and D. Cranston, *Br. J. Radiol.*, 2003, **76**, 590–599.
3. G. Ter Haar, *Adv. Exp. Med. Biol.*, 2016, **880**, 3–20.
4. R. Deckers and C. T. Moonen, *J. Controlled Release*, 2010, **148**, 25–33.
5. A. S. Mikhail, A. Partanen, P. Yarmolenko, A. M. Venkatesan and B. J. Wood, *Magn. Reson. Imaging Clin. N. Am.*, 2015, **23**, 643–655.
6. J. G. Lynn, R. L. Zwemer, A. J. Chick and A. E. Miller, *J. Gen. Physiol.*, 1942, **26**, 179–193.
7. W. J. Fry and F. J. Fry, *IRE Trans. Med. Electron.*, 1960, **ME-7**, 166–181.
8. J. Haut, J. P. Colliac, L. Falque and Y. Renard, *Ophtalmologie*, 1990, **4**, 138–141.
9. F. A. Jolesz and N. J. McDannold, *Neurol. Clin.*, 2014, **32**, 253–269.
10. D. Weintraub and W. J. Elias, *Mov. Disord.*, 2017, **32**, 20–27.
11. K. Hynynen and N. McDannold, *Int. J. Hyperthermia*, 2004, **20**, 725–737.
12. E. A. Dick and W. M. Gedroyc, *Expert Rev. Med. Devices*, 2010, **7**, 589–597.
13. E. M. Weeks, M. W. Platt and W. Gedroyc, *Eur. Radiol.*, 2012, **22**, 2822–2835.
14. D. Weintraub and W. J. Elias, *Mov. Disord.*, 2016, **32**, 20–27.
15. D. Pfeiffer, J. Berger and A. J. Gross, *BJU Int.*, 2012.
16. H. U. Ahmed, R. G. Hindley, L. Dickinson, A. Freeman, A. P. Kirkham, M. Sahu, R. Scott, C. Allen, J. Van der Meulen and M. Emberton, *Lancet Oncol.*, 2012, **13**, 622–632.
17. A. M. Venkatesan, A. Partanen, T. K. Pulanic, M. R. Dreher, J. Fischer, R. K. Zurawin, R. Muthupillai, S. Sokka, H. J. Nieminen, N. Sinaii, M. Merino, B. J. Wood and P. Stratton, *J. Vasc. Interv. Radiol.*, 2012, **23**, 786–794 e784.

18. D. Jeanmonod, B. Werner, A. Morel, L. Michels, E. Zadicario, G. Schiff and E. Martin, *Neurosurg. Focus*, 2012, **32**, E1.
19. N. McDannold, G. T. Clement, P. Black, F. Jolesz and K. Hynynen, *Neurosurgery*, 2010, **66**, 323–332, discussion 332.
20. A. E. Bond, R. Dallapiazza, D. Huss, A. L. Warren, S. Sperling, R. Gwinn, B. B. Shah and W. J. Elias, *Neurosurgery*, 2016, **63**(suppl. 1), 154.
21. R. Medel, S. J. Monteith, W. J. Elias, M. Eames, J. Snell, J. P. Sheehan, M. Wintermark, F. A. Jolesz and N. F. Kassell, *Neurosurgery*, 2012, **71**, 755–763.
22. M. N. Gallay, D. Moser, F. Rossi, P. Pourtehrani, A. E. Magara, M. Kowalski, A. Arnold and D. Jeanmonod, *J. Ther. Ultrasound*, 2016, **4**, 5.
23. C. R. Hill, J. C. Bamber and G. t. Haar, *Physical Principles of Medical Ultrasonics*, John Wiley & Sons, Chichester, Hoboken, N.J., 2nd edn, 2004.
24. P. Wells, *Biomedical Ultrasound*, Academic Press, Boston, 1st edn, 1977.
25. W. D. O'Brien Jr, *Prog. Biophys. Mol. Biol.*, 2007, **93**, 212–255.
26. K. Hynynen and D. DeYoung, *Int. J. Hyperthermia*, 1988, **4**, 267–279.
27. C. M. Tempany, N. J. McDannold, K. Hynynen and F. A. Jolesz, *Radiology*, 2011, **259**, 39–56.
28. M. Wang and Y. Zhou, *Int. J. Hyperthermia*, 2016, **32**, 569–582.
29. D. Pajek and K. Hynynen, *Phys. Med. Biol.*, 2012, **57**, 4951–4968.
30. X. Fan and K. Hynynen, *Phys. Med. Biol.*, 1996, **41**, 591–608.
31. E. B. Hutchinson and K. Hynynen, *Med. Phys.*, 1998, **25**, 2392–2399.
32. O. Diou, N. Tsapis and E. Fattal, *Expert Opin. Drug Delivery*, 2012, **9**, 1475–1487.
33. S. D. Jo, S. H. Ku, Y. Y. Won, S. H. Kim and I. C. Kwon, *Theranostics*, 2016, **6**, 1362–1377.
34. M. Roy Chowdhury, C. Schumann, D. Bhakta-Guha and G. Guha, *Biomed. Pharmacother*, 2016, **84**, 291–304.
35. R. Zeineldin and J. Syoufjy, *Methods Mol Biol*, 2017, **1530**, 3–12.
36. C. Liang, L. Xu, G. Song and Z. Liu, *Chem. Soc. Rev.*, 2016, **45**, 6250–6269.
37. A. Subramanian, A. Manigandan, P. R. Sivashankari and S. Sethuraman, *Biotechnol. Adv.*, 2015, **33**, 1897–1911.
38. L. Y. Rizzo, B. Theek, G. Storm, F. Kiessling and T. Lammers, *Curr. Opin. Biotechnol.*, 2013, **24**, 1159–1166.
39. W. M. Gedroyc and A. Anstee, *Expert Rev. Med. Devices*, 2007, **4**, 539–547.
40. Y. S. Kim, *Int. J. Hyperthermia*, 2015, **31**, 225–232.
41. B. Quesson, J. A. de Zwart and C. T. Moonen, *J. Magn. Reson. Imaging*, 2000, **12**, 525–533.
42. N. M. Hijnen, A. Elevelt, J. Pikkemaat, C. Bos, L. W. Bartels and H. Grull, *J. Ther. Ultrasound*, 2013, **1**, 8.
43. L. H. Lindner, H. M. Reinl, M. Schlemmer, R. Stahl and M. Peller, *Int. J. Hyperthermia*, 2005, **21**, 575–588.
44. C. Lorenzato, C. Oerlemans, M. van Elk, W. J. Geerts, B. Denis de Senneville, C. Moonen and C. Bos, *Contrast Media Mol. Imaging*, 2016, **11**, 184–194.
45. M. Peller, L. Willerding, S. Limmer, M. Hossann, O. Dietrich, M. Ingrisch, R. Sroka and L. H. Lindner, *J. Controlled Release*, 2016, **237**, 138–146.

46. S. R. Sirsi and M. A. Borden, *Adv. Drug Delivery Rev.*, 2014, **72**, 3–14.
47. G. Kong, G. Anyarambhatla, W. P. Petros, R. D. Braun, O. M. Colvin, D. Needham and M. W. Dewhirst, *Cancer Res.*, 2000, **60**, 6950–6957.
48. K. Kooiman, H. J. Vos, M. Versluis and N. de Jong, *Adv. Drug Delivery Rev.*, 2014, **72**, 28–48.
49. R. Suzuki and A. L. Klibanov, *Adv. Exp. Med. Biol.*, 2016, **880**, 205–220.
50. M. B. Yatvin, J. N. Weinstein, W. H. Dennis and R. Blumenthal, *Science*, 1978, **202**, 1290–1293.
51. G. R. Anyarambhatla and D. Needham, *J. Liposome Res.*, 1999, **9**, 491–506.
52. D. Needham, G. Anyarambhatla, G. Kong and M. W. Dewhirst, *Cancer Res.*, 2000, **60**, 1197–1201.
53. M. L. Hauck, S. M. LaRue, W. P. Petros, J. M. Poulson, D. Yu, I. Spasojevic, A. F. Pruitt, A. Klein, B. Case, D. E. Thrall, D. Needham and M. W. Dewhirst, *Clin. Cancer Res.*, 2006, **12**, 4004–4010.
54. D. Needham and M. W. Dewhirst, *Adv. Drug Delivery Rev.*, 2001, **53**, 285–305.
55. B. J. Wood, R. T. Poon, J. K. Locklin, M. R. Dreher, K. K. Ng, M. Eugeni, G. Seidel, S. Dromi, Z. Neeman, M. Kolf, C. D. Black, R. Prabhakar and S. K. Libutti, *J. Vasc. Interv. Radiol.*, 2012, **23**, 248–255 e247.
56. R. T. Poon and N. Borys, *Future Oncol.*, 2011, **7**, 937–945.
57. C. W. Hong, L. Chow, E. B. Turkbey, R. Lencioni, S. K. Libutti and B. J. Wood, *Cardiovasc. Intervent. Radiol.*, 2016, **39**, 409–416.
58. R. Soloman and A. A. Gabizon, *Clin. Lymphoma Myeloma*, 2008, **8**, 21–32.
59. V. Frenkel, A. Etherington, M. Greene, J. Quijano, J. Xie, F. Hunter, S. Dromi and K. C. Li, *Acad. Radiol.*, 2006, **13**, 469–479.
60. S. Dromi, V. Frenkel, A. Luk, B. Traughber, M. Angstadt, M. Bur, J. Poff, J. Xie, S. K. Libutti, K. C. Li and B. J. Wood, *Clin. Cancer Res.*, 2007, **13**, 2722–2727.
61. A. H. Negussie, P. S. Yarmolenko, A. Partanen, A. Ranjan, G. Jacobs, D. Woods, H. Bryant, D. Thomasson, M. W. Dewhirst, B. J. Wood and M. R. Dreher, *Int. J. Hyperthermia*, 2011, **27**, 140–155.
62. A. Ranjan, G. C. Jacobs, D. L. Woods, A. H. Negussie, A. Partanen, P. S. Yarmolenko, C. E. Gacchina, K. V. Sharma, V. Frenkel, B. J. Wood and M. R. Dreher, *J. Controlled Release*, 2012, **158**, 487–494.
63. A. Partanen, P. S. Yarmolenko, A. Viitala, S. Appanaboyina, D. Haemmerich, A. Ranjan, G. Jacobs, D. Woods, J. Enholm, B. J. Wood and M. R. Dreher, *Int. J. Hyperthermia*, 2012, **28**, 320–336.
64. A. Gasselhuber, M. R. Dreher, A. Partanen, P. S. Yarmolenko, D. Woods, B. J. Wood and D. Haemmerich, *Int. J. Hyperthermia*, 2012, **28**, 337–348.
65. M. de Smet, S. Langereis, S. van den Bosch and H. Grull, *J. Controlled Release*, 2010, **143**, 120–127.
66. E. Evans and D. Needham, *J. Phys. Chem.*, 1987, **91**, 4219–4228.
67. T. Tagami, M. J. Ernsting and S. D. Li, *J. Controlled Release*, 2011, **154**, 290–297.
68. T. Tagami, W. D. Foltz, M. J. Ernsting, C. M. Lee, I. F. Tannock, J. P. May and S. D. Li, *Biomaterials*, 2011, **32**, 6570–6578.

69. K. Kono, S. Nakashima, D. Kokuryo, I. Aoki, H. Shimomoto, S. Aoshima, K. Maruyama, E. Yuba, C. Kojima, A. Harada and Y. Ishizaka, *Biomaterials*, 2011, **32**, 1387–1395.
70. C. Y. Lin, J. R. Li, H. C. Tseng, M. F. Wu and W. L. Lin, *Nanomedicine*, 2012, **8**, 900–907.
71. T. Yu, Y. Zhang, H. He, S. Zhou, Y. Liu and P. Huang, *Mol. Pharm.*, 2011, **8**, 1408–1415.
72. N. Rapoport, *Int. J. Pharm.*, 2004, **277**, 155–162.
73. Z. Deng, Y. Xiao, M. Pan, F. Li, W. Duan, L. Meng, X. Liu, F. Yan and H. Zheng, *J. Controlled Release*, 2016, **243**, 333–341.
74. E. V. Rosca, M. Wright, R. Gonitel, W. Gedroyc, A. D. Miller and M. Thanou, *Mol. Pharm.*, 2015, **12**, 1335–1346.
75. N. Kamaly, J. A. Pugh, T. L. Kalber, J. Bunch, A. D. Miller, C. W. McLeod and J. D. Bell, *Mol. Imaging Biol.*, 2010, **12**, 361–366.
76. N. Kamaly, T. Kalber, M. Thanou, J. D. Bell and A. D. Miller, *Bioconjugate Chem.*, 2009, **20**, 648–655.
77. N. Frazier, A. Payne, C. Dillon, N. Subrahmanyam and H. Ghandehari, *Nanomedicine*, 2017, **13**, 1235–1243.
78. N. Frazier, A. Payne, J. de Bever, C. Dillon, A. Panda, N. Subrahmanyam and H. Ghandehari, *J. Controlled Release*, 2016, **241**, 186–193.
79. M. N. Centelles, M. Wright, W. Gedroyc and M. Thanou, *Pharmacol. Res.*, 2016, **114**, 144–151.
80. J. VanOsdol, K. Ektate, S. Ramasamy, D. Maples, W. Collins, J. Malayer and A. Ranjan, *J. Controlled Release*, 2017, **247**, 55–63.
81. A. Yudina and C. Moonen, *Int. J. Hyperthermia*, 2012, **28**, 311–319.
82. V. Frenkel and K. C. Li, *Future Oncol.*, 2006, **2**, 111–119.
83. C. W. Song, H. J. Park, C. K. Lee and R. Griffin, *Int. J. Hyperthermia*, 2005, **21**, 761–767.
84. S. Wang, V. Zderic and V. Frenkel, *Future Oncol.*, 2010, **6**, 1497–1511.
85. G. Kong, R. D. Braun and M. W. Dewhirst, *Cancer Res.*, 2001, **61**, 3027–3032.
86. J. Friedl, E. Turner and H. R. Alexander Jr, *Int. J. Oncol.*, 2003, **23**, 611–616.
87. H. L. Liu, Y. Y. Wai, W. S. Chen, J. C. Chen, P. H. Hsu, X. Y. Wu, W. C. Huang, T. C. Yen and J. J. Wang, *Ultrasound Med. Biol.*, 2008, **34**, 598–606.
88. V. Frenkel, *Adv. Drug Delivery Rev.*, 2008, **60**, 1193–1208.
89. R. E. Apfel, *Br. J. Cancer, Suppl.*, 1982, **5**, 140–146.
90. S. Mo, C. C. Coussios, L. Seymour and R. Carlisle, *Expert Opin. Drug Delivery*, 2012, **9**, 1525–1538.
91. M. W. Miller, T. A. Sherman and A. A. Brayman, *Ultrasound Med. Biol.*, 2000, **26**, 1317–1326.
92. C. Y. Lai, C. H. Wu, C. C. Chen and P. C. Li, *Ultrasound Med. Biol.*, 2006, **32**, 1931–1941.
93. J. J. Kwan, S. Graham, R. Myers, R. Carlisle, E. Stride and C. C. Coussios, *Phys. Rev. E: Stat., Nonlinear, Soft Matter Phys.*, 2015, **92**, 023019.
94. J. L. Mestas, R. A. Fowler, T. J. Evjen, L. Somaglino, A. Moussatov, J. Ngo, S. Chesnais, S. Rognvaldsson, S. L. Fossheim, E. A. Nilssen and C. Lafon, *J. Drug Targeting*, 2014, **22**, 688–697.

95. S. M. Graham, R. Carlisle, J. J. Choi, M. Stevenson, A. R. Shah, R. S. Myers, K. Fisher, M. B. Peregrino, L. Seymour and C. C. Coussios, *J. Controlled Release*, 2014, **178**, 101–107.
96. M. Gyongy and C. C. Coussios, *IEEE Trans. Biomed. Eng.*, 2010, **57**, 48–56.
97. C. Poon, D. McMahon and K. Hynynen, *Neuropharmacology*, 2017, **120**, 20–37.
98. A. H. Mesiwala, L. Farrell, H. J. Wenzel, D. L. Silbergeld, L. A. Crum, H. R. Winn and P. D. Mourad, *Ultrasound Med. Biol.*, 2002, **28**, 389–400.
99. N. Sheikov, N. McDannold, S. Sharma and K. Hynynen, *Ultrasound Med. Biol.*, 2008, **34**, 1093–1104.
100. J. Deng, Q. Huang, F. Wang, Y. Liu, Z. Wang, Q. Zhang, B. Lei and Y. Cheng, *J. Mol. Neurosci.*, 2012, **46**, 677–687.
101. K. Hynynen, N. McDannold, N. Vykhodtseva and F. A. Jolesz, *Acta Neurochir. Suppl.*, 2003, **86**, 555–558.
102. F. A. Jolesz, K. Hynynen, N. McDannold and C. Tempany, *Magn. Reson. Imaging Clin. N. Am.*, 2005, **13**, 545–560.
103. N. McDannold, N. Vykhodtseva, S. Raymond, F. A. Jolesz and K. Hynynen, *Ultrasound Med. Biol.*, 2005, **31**, 1527–1537.
104. M. Kinoshita, N. McDannold, F. A. Jolesz and K. Hynynen, *Proc. Natl. Acad. Sci. U. S. A.*, 2006, **103**, 11719–11723.
105. K. Hynynen, *Expert Opin. Drug Delivery*, 2007, **4**, 27–35.
106. M. Kinoshita, N. McDannold, F. A. Jolesz and K. Hynynen, *Biochem. Biophys. Res. Commun.*, 2006, **340**, 1085–1090.
107. L. H. Treat, N. McDannold, N. Vykhodtseva, Y. Zhang, K. Tam and K. Hynynen, *Int. J. Cancer*, 2007, **121**, 901–907.
108. F. Vlachos, Y. S. Tung and E. E. Konofagou, *Phys. Med. Biol.*, 2010, **55**, 5451–5466.
109. J. J. Choi, J. A. Feshitan, B. Baseri, S. Wang, Y. S. Tung, M. A. Borden and E. E. Konofagou, *IEEE Trans. Biomed. Eng.*, 2010, **57**, 145–154.
110. M. A. O'Reilly, A. C. Waspe, M. Ganguly and K. Hynynen, *Ultrasound Med. Biol.*, 2011, **37**, 587–594.
111. B. Baseri, J. J. Choi, T. Deffieux, G. Samiotaki, Y. S. Tung, O. Olumolade, S. A. Small, B. Morrison and E. E. Konofagou, *Phys. Med. Biol.*, 2012, **57**, N65–N81.
112. A. B. Etame, R. J. Diaz, C. A. Smith, T. G. Mainprize, K. Hynynen and J. T. Rutka, *Neurosurg. Focus*, 2012, **32**, E3.
113. J. Mei, Y. Cheng, Y. Song, Y. Yang, F. Wang, Y. Liu and Z. Wang, *J. Ultrasound Med.*, 2009, **28**, 871–880.
114. H. L. Liu, M. Y. Hua, P. Y. Chen, P. C. Chu, C. H. Pan, H. W. Yang, C. Y. Huang, J. J. Wang, T. C. Yen and K. C. Wei, *Radiology*, 2010, **255**, 415–425.
115. H. L. Liu, M. Y. Hua, H. W. Yang, C. Y. Huang, P. C. Chu, J. S. Wu, I. C. Tseng, J. J. Wang, T. C. Yen, P. Y. Chen and K. C. Wei, *Proc. Natl. Acad. Sci. U. S. A.*, 2010, **107**, 15205–15210.
116. F. Y. Yang, T. T. Wong, M. C. Teng, R. S. Liu, M. Lu, H. F. Liang and M. C. Wei, *J. Controlled Release*, 2012, **160**, 652–658.

117. F. Y. Yang, H. E. Wang, R. S. Liu, M. C. Teng, J. J. Li, M. Lu, M. C. Wei and T. T. Wong, *PLoS One*, 2012, **7**, e45468.
118. Q. Huang, J. Deng, F. Wang, S. Chen, Y. Liu, Z. Wang and Y. Cheng, *Exp. Neurol.*, 2012, **233**, 350–356.
119. A. Burgess, C. A. Ayala-Grosso, M. Ganguly, J. F. Jordao, I. Aubert and K. Hynynen, *PLoS One*, 2011, **6**, e27877.
120. C. H. Fan, C. Y. Ting, H. J. Lin, C. H. Wang, H. L. Liu, T. C. Yen and C. K. Yeh, *Biomaterials*, 2013, **34**, 3706–3715.
121. I. Lentacker, B. Geers, J. Demeester, S. C. De Smedt and N. N. Sanders, *Mol. Ther.*, 2010, **18**, 101–108.
122. N. Rapoport, *Int. J. Hyperthermia*, 2012, **28**, 374–385.
123. Z. S. Al-Ahmady, W. T. Al-Jamal, J. V. Bossche, T. T. Bui, A. F. Drake, A. J. Mason and K. Kostarelos, *ACS Nano*, 2012, **6**, 9335–9346.
124. S. Fokong, B. Theek, Z. Wu, P. Koczera, L. Appold, S. Jorge, U. Resch-Genger, M. van Zandvoort, G. Storm, F. Kiessling and T. Lammers, *J. Controlled Release*, 2012, **163**, 75–81.
125. S. Zheng, Z. Jin, J. Han, S. Cho, V. D. Nguyen, S. Y. Ko, J. O. Park and S. Park, *Colloids Surf., B*, 2016, **143**, 27–36.
126. N. Zhang, X. Cai, W. Gao, R. Wang, C. Xu, Y. Yao, L. Hao, D. Sheng, H. Chen, Z. Wang and Y. Zheng, *Theranostics*, 2016, **6**, 404–417.

CHAPTER 2

# *Image-guided Drug Delivery Systems Based on NIR-absorbing Nanocarriers for Photothermal-chemotherapy of Cancer*

JINPING WANG, XIAOXIAO TAN, XIAOJUAN PANG, QING YOU, LI LIU, QI SUN, FENGPING TAN AND NAN LI*

School of Pharmaceutical Science and Technology, Tianjin University, 300072, Tianjin, PR China
*E-mail: linan19850115@163.com

## 2.1 Introduction

Cancer is one of the most devastating malignant diseases threatening human health.[1] Traditional cancer therapies mainly include chemotherapies, surgery, and radiotherapies. While surgery in many circumstances could not completely remove all tumor cells in the human body, radiotherapy and chemotherapy can have serious adverse effects to normal tissues and limited specificities to tumor cells.[2] Photothermal therapy (PTT) using light-absorbing agents to convert near-infrared (NIR) light into heat to burn cancer cells has attracted great attention in recent years as

Drug Discovery Series No. 63
Theranostics and Image Guided Drug Delivery
Edited by Maya Thanou

Published by the Royal Society of Chemistry, www.rsc.org

a promising approach for cancer treatment.[3–5] However, cancer therapy currently relying on a single PTT treatment still remains unsatisfying. Strategies of combining photothermal therapy with chemotherapy have been considered as a promising anti-tumor approach for those aggressive tumors with low overall survival rates.[6] Apart from being directly utilized for photothermal ablation of cancer cells, the photothermal effect of NIR-absorbing nanomaterials has also been exploited for remotely controlled drug release. This photothermal-controlled drug delivery system provides promising approaches to reverse multidrug resistance, for which the poor cellular uptake and insufficient intracellular drug release remain rate-limiting steps for reaching the drug concentration level within the therapeutic window.

In addition, for effective, safe and personalized photothermal/chemical combination treatment, it is crucial to identify the location and size of the tumors before therapy, to monitor the *in vivo* distribution of the photothermal or drug agents during therapy and to evaluate effectiveness after therapy with appropriate imaging technologies.[7–9] Therefore, imaging-guided combination therapy is desired for cancer therapy and those photothermal agents with dual imaging and therapeutic functions have attracted intensive research interest.[10–15] With the rapid development of nanotechnology in recent years, image guided photothermal-chemo therapy based on nanomaterials have attracted tremendously increasing interest (Figure 2.1).

Recently, many nanostructures have been employed as image guided chemo-photothermal combination therapies for enhancing the efficacy of tumor therapy, which can be divided into two major categories: inorganic and organic based nanomaterials.[16,17] Inorganic nanomaterials including different gold nanostructures,[18,19] carbon nanomaterials,[20] copper sulfide nanoparticles,[21] and a few other newly reported ones,[22,23] as well as organic nanoparticles such

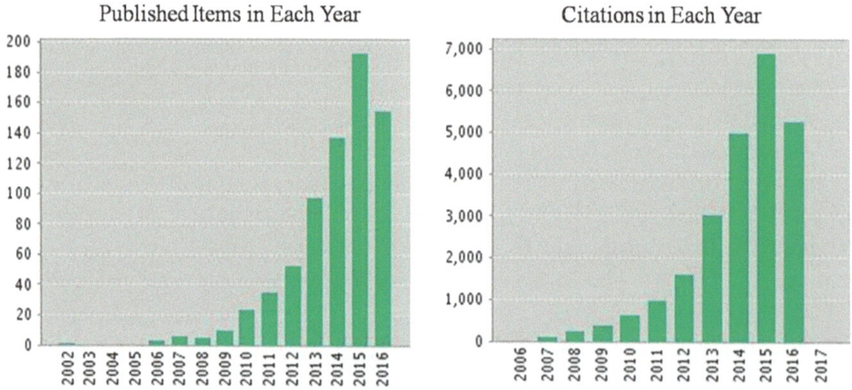


**Figure 2.1** Statistics of publications and their citations with topics including "photothermal", "drug delivery", and "imaging", based on Web of Science searching conducted on September 29th, 2016.

as NIR-absorbing conjugated polymers,[24,25] porphysomes,[26] and nanoparticles encapsulated NIR dyes,[27] have been widely explored by many research groups including ours for photothermal-controlled drug delivery for killing cancer cells *in vitro* and *in vivo*.

In this chapter, we comprehensively summarize recent progress regarding the development of photothermal-responsive drug delivery nanocarriers for cancer treatment using functional nanomaterials. Various inorganic and organic NIR-absorbing nanomaterials explored for combined photothermal-chemo therapy cancer treatment will be reviewed.

## 2.2 Inorganic Nanocarriers Used as Photothermal-controlled Drug Delivery Systems

### 2.2.1 Metallic Nanocarriers

#### *2.2.1.1 Gold Nanocarriers*

Gold nanocarriers have become one of the most superior metallic materials applied to treating cancer due to their unique surface plasmon resonance (SPR) and remarkable biocompatibility.[28–30] Gold nanocarriers can also be used as a contrast agent for photoacoustic (PA) imaging in cancer therapy due to their large optical adsorption coefficients in the near-infrared region.[31–33]

Gold nanopaticles (AuNPs) have attracted great research interest in sensors, imaging diagnostic agents and drug delivery. Depending on their unique SPR character, AuNPs could control some drugs' fluorescence properties from quenching to being excited to realize imaging. In the therapeutic system, which contained a chemotherapy and imaging agent, reported by Chen *et al.*,[34] doxorubicin (Dox) was conjugated to AuNPs *via* a thiole–Au bond by using a peptide substrate, CPLGLAGG, as the linker. When carriers arrived at tumor sites, the peptide substrate could be hydrolyzed by the overexpressed matrix metalloproteinases-2 (MMP-2) protease, then the pronounced fluorescence of Dox recovered the quenching to an exciting state due to the nano-surface energy transfer effect induced by AuNPs. This fluorescence could be used for tumor imaging. At the same time the released Dox could induce the apoptosis of tumor cells to achieve tumor inhibition. The functionalized AuNPs carriers with tumor-triggered drug release could achieve both cancer therapy and diagnosis. Our group[35] has reported the stepwise therapeutic strategy of a pH (low) insertion peptide (pHLIP) modified hollow gold nanospheres (HAuNs) carried Chlorin e6 (Ce6) system (HAuNS-pHLIP-Ce6) (Figure 2.2). We constructed an efficient smart-responsive delivery system that actively targeted and directly translocated into tumor cells under lower pH value, subsequently releasing and dequenching its payload by PTT effect induced by HAuNs. The dequenched Ce6 not only contributed to the photodynamic therapy (PDT) effect, but also acted as an imaging agent to indicate tumor regions.

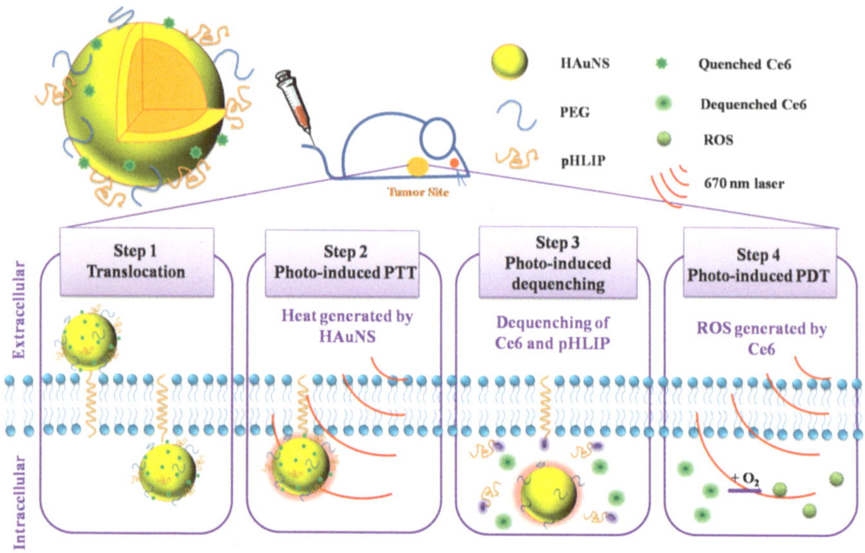


**Figure 2.2** Schematic representation of HAuNS-pHLIP-Ce6 for synergistic PTT/PDT therapeutics based on 4 steps. Step 1: Upon entering the low pH of the tumor microenvironment, pHLIP spontaneously formed helical structures and swiftly translocated the nanocarriers into the cell. Step 2: Under exposure to 670 nm laser irradiation, HAuNS converted light-to-heat to perform PTT effect. Step 3: Followed by hyperthermia on gold surface, Ce6 and pHLIP were unloaded and dequenched from vehicles. Step 4: Ce6 was induced to produce ROS upon irradiation of 670 nm laser. Reproduced from *Biomaterials*, **79**, M. Yu, F. Guo, J. Wang, F. Tan and N. Li, A pH-Driven and photoresponsive nanocarrier: Remotely-controlled by near-infrared light for stepwise antitumor treatment, 25–35. Copyright 2016 with permission from Elsevier.

In addition, gold nanoclusters (AuNCs) with high luminescent ability have been widely used as the nanocarrier for image guided cancer therapy. Compared with other luminescent agents, such as organic dyes or quantum dots (QDs), AuNCs have excellent photostability, good biocompatibility, and low toxicity.[36,37] Chen *et al.*[38] designed a multifunctional nanocomposite that provided selective targeting, *in-situ* imaging and anticancer therapy functionalities by encapsulating a hydrophobic drug (paclitaxel) in a luminescent gold nanocluster (L-AuNC) conjugated to a pH-labile targeting polymeric carrier. The nanocomposite was water-stable and biocompatible. In addition, this nanocomposite could release its drug payload at a weakly acidic pathological site *via* a polymer degradation processes, while the L-AuNCs provided positioning information and the hydrophobic drug provided the therapeutic action at the targeted cells. Both *in vitro* and *in vivo* studies confirmed the efficacy of this theranostic composite for cancer cell imaging and targeted treatment.

In 2013, Ma and co-workers developed a novel multifunctional drug-delivery platform based on cholesteryl succinyl silane (CSS) nanomicelles loaded with doxorubicin, $Fe_3O_4$ magnetic nanoparticles, and gold nanoshells (CDF-Au-shell nanomicelles) to combine magnetic resonance (MR) imaging, magnetic-targeted drug delivery, light-triggered drug release, and photothermal therapy.[39] Their CDF-Au-shell nanomicelles displayed enhanced $T_2$-weighted MR imaging and surface plasmon absorbance in the NIR region, thus exhibiting an NIR-induced temperature elevation and an NIR light-triggered and stepwise release behavior of doxorubicin. The synergistic effect in killing cancer cells was found by the combined photothermal therapy and the magnetic-field-guided drug delivery. The "all-in-one" drug-delivery system, which combines the functional units into one nanoparticle, provided unprecedented opportunities for developing advanced imaging-guided drug-delivery systems.

### 2.2.1.2 *Two-dimensional Nanosheets*

In recent years, transition metal dichalcogenides (TMDCs), have attracted tremendous interest in a range of fields, such as electrictronics, energy storage, catalysis, and the biomedical field due to the novel physical and chemical properties.[40–42] Recent studies have demonstrated that certain members of TMDCs could be used as a photothermal-controlled drug nanocarrier due to their low cytotoxicity and high photothermal-conversion efficiency.[43,44] In addition, TMDCs with high-M element can be adopted for contrast-enhanced X-ray computed tomography (CT) imaging.[43] Therefore, TMDCs represent a new 2D nanosystem for efficient tumor hyperthermia and other specific theranostic biomedical applications.

Single-layer $MoS_2$ nanosheets are the most representative and explored TMDCs showing high performance in broad range of applications, mostly in electronic devices, energy storage and catalysis.[45–47] Besidethe use in biomedical applications, Chou *et al.* first demonstrated the possibility of using single-layer $MoS_2$ sheets as a novel NIR absorbing agent, which showed higher absorbance in the NIR region than that of both graphene and gold nanorods.[48] Therefore, $MoS_2$ sheets could also be endowed with great potential for a NIR-triggered drug delivery system in the biomedical field due to their advantages including strong NIR photothermal absorption and unique 2D structure. Zhao and co-workers prepared chitosan (CS) coated $MoS_2$ nanosheets with controllable sizes from commercial $MoS_2$ flakes *via* a safe, yet simple oleum treatment exfoliation method (Figure 2.3).[49] *Via* π–π stacking and hydrophobic interactions, the DOX was absorbed onto the surface of the $MoS_2$ nanosheets. Highly effective *in vivo* photothermal therapy plus chemotherapy for tumors under NIR laser was then realized, after intravenous injection with a moderate dose of this formulation into KB cells and Panc-1 cells, as well as bearing mice. Additionally, they found that the Mo-based $MoS_2$ nanosheets could be used as a contrast agent in CT imaging. Therefore, this 2D $MoS_2$-based on-demand drug delivery system has implicated increased opportunities for simultaneous imaging diagnosis and efficient therapy in the biomedical field.

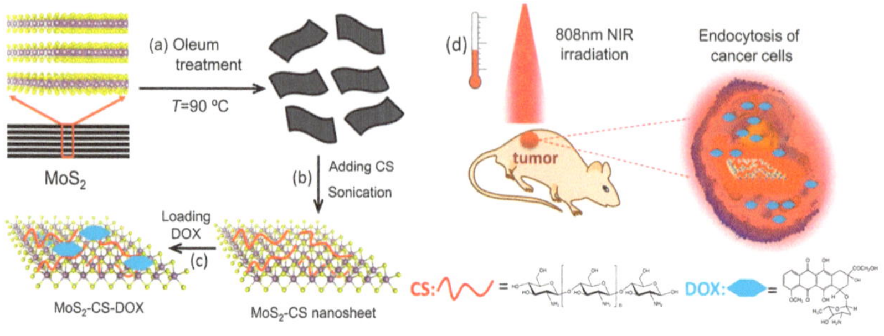


**Figure 2.3** Schematic illustration of the high-throughput synthesis of $MoS_2$–CS nanosheets as a NIR photothermal triggered drug delivery system for efficient cancer therapy. (a,b) oleum treatment exfoliation process to produce single layer $MoS_2$ nanosheets and then modified with CS, (c) DOX loading process, and (d) NIR photothermal-triggered drug of the $MoS_2$ nanosheets to the tumor site. Reproduced with permission from ref. 49. Copyright 2014 The American Chemical Society.

In addition, $WS_2$ nanosheets were reported to be used as a dual-modal CT/photoacoustic imaging guided photothermal therapy of tumors by Liu's group.[43] Due to the unique 2D structure, the $WS_2$ nanosheets could also be employed as the drug nanocarrier. In 2015, Yang and co-workers fabricated mesoporous silica coated, iron oxide decorated $WS_2$ nanosheets ($WS_2$-IO@MS-PEG) as a theranostic platform for modal imaging guided cancer combination therapy.[50] With excellent stability and low cytotoxicity, $WS_2$-IO@MS-PEG nanomaterials could act as a drug carrier for efficient loading of chemotherapeutics such as DOX, whose release on the other hand may be triggered by external NIR light *via* the photothermal effect. Utilizing inherent physical properties of their composite nanoparticles, three modes of imaging, fluorescence, MR and CT were carried out on tumor-bearing mice, evidencing efficient retention of nanomaterials in the tumor after systemic administration. In addition, effective inhibition of tumor growth was also realized after the combined photothermal and chemotherapy delivered by $WS_2$-IO@MS-PEG/DOX.

### *2.2.1.3 CuS Nanostructures*

Copper sulfide nanoparticles (CuS NPs), a rising new star in NIR resonant materials, have drawn widespread attention in biomedical applications and especially in cancer theranostics due to their low cost, biodegradability,[51] and high photothermal conversion efficiency,[52] as well as synthetic simplicity.[53] Recently, with special physical properties, CuS nanoparticles are gradually emerging as a promising platform for photothermal therapy,[54] *in vivo* imaging,[55] drug delivery,[56] as well as multifunctional agents that can integrate both imaging and therapy.[57]

Semiconductor CuS nanoparticles (CuS NPs), as a new class of photothermal nanoparticles can provide an alternative for gold analogues.

Unlike the optical absorption in gold nanostructures based on the SPR, the NIR light absorption of CuS NPs derives from the d–d transition of $Cu^{2+}$ ions,[58] so CuS NPs are more stable under laser irradiation and cannot easily be altered by the solvent or the surrounding environment compared with traditional gold nanostructures when they are formulated or delivered *in vivo*. What's more, the absorption wavelength for d–d transitions peaks at ~900 nm, which is in the NIR range and suitable for *in vivo* applications. This eliminates the need for specifically designed and manufactured CuS nanoparticles which can require special and sometimes complicated procedures.

When irradiated with a NIR laser, CuS NPs perform a photothermal effect and then generate hyperthermia to destroy the tumor cells. Nowadays, more and more researchers have reported the PTT effect of the CuS and verified this *in vitro* and *in vivo*. Zha and co-workers have successfully synthesized copper sulphide nanoparticles and the *in vitro* tests showed the CuS nanoparticles had a good PTT effect.[59] Other hollow nanospheres,[60] CuS nanorods,[61] tubular structures,[61] nanowires[62] and so on all exhibited a satisfactory PTT effect.

In recent years, CuS nanoparticles were reported as a novel class of contrast agents for photoacoustic imaging to guide phototherapy (including PTT and PDT). Ding *et al.* reported PEGylated copper sulfide nanoparticles and further measured the *in vivo* photoacoustic imaging of tumors with the nanoparticles.[63] Ku *et al.* also synthesized a kind of copper sulfide nanoparticle and a 1064 nm NIR laser source was chosen to be used with these NP due to its low absorption and coefficient in normal tissues. Optical photoacoustic imaging was used for the detection of brain and lymph nodes in mouse models. Noticeably, even implanted at 5 cm depth of a chicken breast, the CuS nanoparticles could be imaged.[64]

Apart from the ability to perform the photothermal effect and photoacoustic imaging, CuS nanoparticles can also be used as drug delivery vehicles to load antitumor drugs like doxorubicin for photothermal-chemotherapy multiple therapies. Feng and co-workers reported a hyaluronic acid modified hollow copper sulfide for carrying doxorubicin to achieve image guided photothermal-chemo combination therapy.[65] Thus, CuS nanoparticles have been popular in the development of the drug delivery methods for cancer therapy. On one hand, CuS nanoparticles have a relative PTT effect in the NIR region and they can also act as desired drug delivery vehicles. On the other hand, CuS nanoparticles also show potential for photoacoustic imaging which can be used as a guide for the whole platform when they are used for delivery in the living body.

### 2.2.2 Nanocarbons

Nanocarbon materials, such as carbon nanotubes and nanographene, are one of the most promising cancer theranostic vehicles owing to their numerous advantages, such as high drug-loading capacities, strong NIR region absorption and excellent cell penetration ability.[66–68]

#### 2.2.2.1 Carbon Nanotubes

Carbon nanotubes (CNTs) specifically single-wall carbon nanotubes (SWCNTs), exhibit many advantages as drug delivery vehicles, such as high surface area for carrying drug cargoes, excellent photo-thermal converting capacities, remarkable cell membrane penetrability and strong Raman spectroscopy signals. It has been confirmed that SWCNTs would be an ideal cancer theranostic nanocarrier with appropriate surface modifications.[69–71]

Xie *et al.*[72] have prepared a long circulating SWCNTs complex with the assistance of a dispersion agent (Evans blue, EB). This complex was endowed with fluorescent/photoacoustic image-guided enhanced synergistic PDT and PTT abilities by self-assembly loaded with an albumin attached photosensitizer, Chlorine e6 (Ce6) *via* the high affinity between albumin and EB. With the assistance of EB, the formulations were well dispersed and functionally long-circulating. What's more, compared to sole PDT or PTT treatment, the simultaneous PDT and PTT were achieved under the guidance of fluorescence and photoacoustic imaging results and managed to destroy tumors and prevent recurrence. Generally, their studies showed that the SWCNTs based theranostic systems were capable of image-guidance and simultaneous PDT/PTT therapy as well as detecting tumors. The modifications of EB onto SWCNTs functionalization can be easily applied to other nanomaterials for improving their *in vivo* stability and prolonging circulation time (Figure 2.4).

Hou *et al.*[73] have described a functional redox-sensitive system based on SWCNTs, targeting agent hyaluronic acid (HA), anticancer drug doxorubicin (DOX), and MRI contrast agents gadolinium (Gd). The resulting Gd/SWCNTs-HA-ss-DOX was tested for magnetic resonance imaging guided chemo-photothermal therapy abilities. *In vitro* release studies revealed that this SWCNTs image guided system achieved much faster drug release under reducing conditions. Confocal laser Scanning Microscopy (CLSM) studies demonstrated that this system was capable of simultaneously delivering DOX and SWCNTs into HA receptor overexpressed cells *via* HA receptor-mediated endocytosis due to the existence of HA. *In vivo* studies demonstrated that this Gd/SWCNTs-HA-ss-DOX system could effectively accumulate in tumor sites and act as a contrast agent for MRI to identify tumor margins. Under the localized NIR irradiation, tumor cell killing effectiveness of Gd/SWCNTs-HA-ss-DOX was significantly strengthened, demonstrating the synergistic chemo-photothermal therapy abilities. These results demonstrated that Gd/SWCNTs-HA-ss-DOX system has a great potential for MRI guiding synergistic chemo-photothermal therapy and redox-responsive releasing function in future clinical anticancer applications.

#### 2.2.2.2 Nanographene

The distinctive physicochemical properties have made nanographene and its derivatives, such as graphene oxide (GO) and reduced graphene oxide (rGO), novel materials in nanocarbons research for PTT-based cancer treatment including enhancement of image guided PTT, as well as chemotherapy.[74,75]

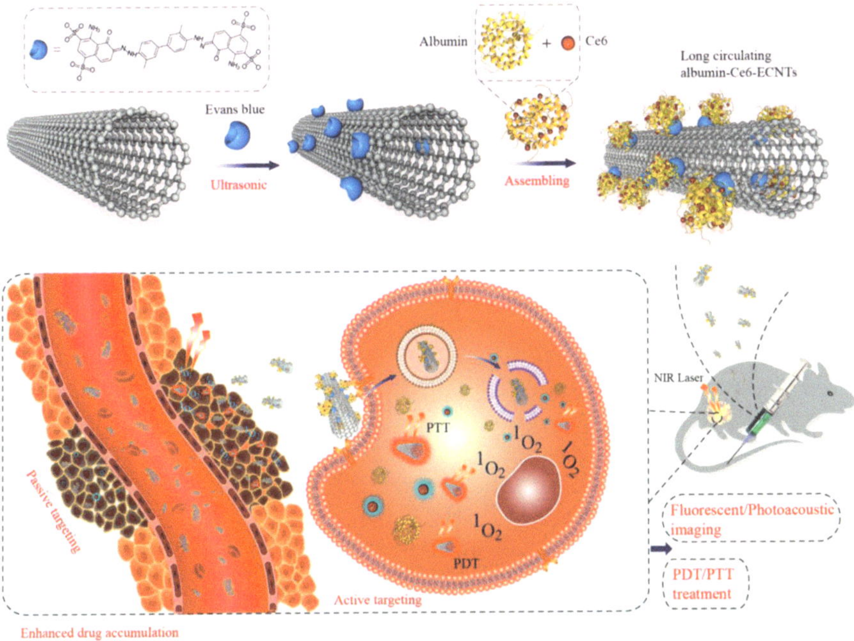


**Figure 2.4** Preparation and application of albumin/Ce6 fabricated EB/carbon nanotube-based delivery system. Reproduced from *Biomaterials*, **103**, L. Xie, G. Wang, H. Zhou, F. Zhang, Z. Guo, C. Liu, X. Zhang and L. Zhu, Functional long circulating single walled carbon nanotubes for fluorescent/photoacoustic imaging-guided enhanced phototherapy, 219–228. Copyright 2016 with permission from Elsevier.

To date, considerable efforts have been made to investigate graphene-based nanostructures in PTT treatment, particularly in combination therapy for cancer treatment.

Guo *et al.*[76] have prepared an MRI guided GO complex for prostate cancer targeting and efficient drug delivery. MRI contrast agent gadolinium diethylene triamine pentaacetate (Gd-DTPA), anti-cancer drug DOX and prostate stem cell antigen (PSCA) targeting monoclonal antibody (mAb) was conjugated with GO nanosheets through amino group capped dendrimers (DEN). The as-prepared GO-DEN (Gd-DTPA)-mAb/DOX can specifically target PSCA over-expressed cancer cells and transport drugs to malignant prostate tumor sites. In addition, GO-DEN (Gd-DTPA)-mAb/DOX displayed promising *in vivo* MRI results for targeting and indicating of prostate tumors. Thus, the GO composite with combination of numerous functionalities may be a promising nanoplatform for applications in MRI guided synergistic tumor therapy of prostate cancers.

Guo and co-workers[77] have developed a brain-targeting transferrin conjugated GO complex for brain-targeting simultaneous chemo-photothermal therapy. Rhodamine B (Rho B) was applied to label the nanocarriers by physical adsorption and chemotherapeutic agent DOX was loaded onto the transferrin-conjugated PEGylated nanoscale graphene oxide (TPG). The as prepared TPG-DOX (TPGD) was proven to deliver drugs across the blood brain barrier and improve drug accumulation both *in vitro* and *in vivo*, thus performing dual functional photothermal-chemotherapy of the tumor.

### 2.2.3 Other Inorganic Nanocarriers

Besides metallic nanocarriers and carbon nanomaterials, which have been extensively explored in imaging guided chemo-photothermal combination cancer therapy, a number of other inorganic nanomaterials with strong NIR absorbance have also shown promise as effective photothermal-controlled drug delivery nanosystems. Recently, Li and co-workers successfully prepared a photothermal controlled drug-delivery nanoagent based on $Bi_2Se_3$ nanomaterials capable of simultaneous real-time imaging and thermo-chemotherapy (Figure 2.5). The obtained nanoparticles displayed high stability, safety, biocompatibility, and bioimaging both *in vitro* and *in vivo*.[78]

Another inorganic photothermal-responsive nanocarrier was Prussian blue (PB) nanoparticles.[79] Prussian blue (PB) is a clinical medicine for the treatment of radioactive exposure, which has been approved by the USA Food and Drug Administration. In addition, PB nanoparticles have also been developed as a new generation of photothermal-controlled drug delivery carriers due to their high absorption in the NIR region. A Mn-containing Prussian blue analogue (MnPBA) was easily coated onto both the outer surface and the inner mesoporous channels of Prussian blue nanoparticles (HMPBs) to form the core–shell hollow-structured HMPB@MnPBA.[80] Furthermore, the hollow

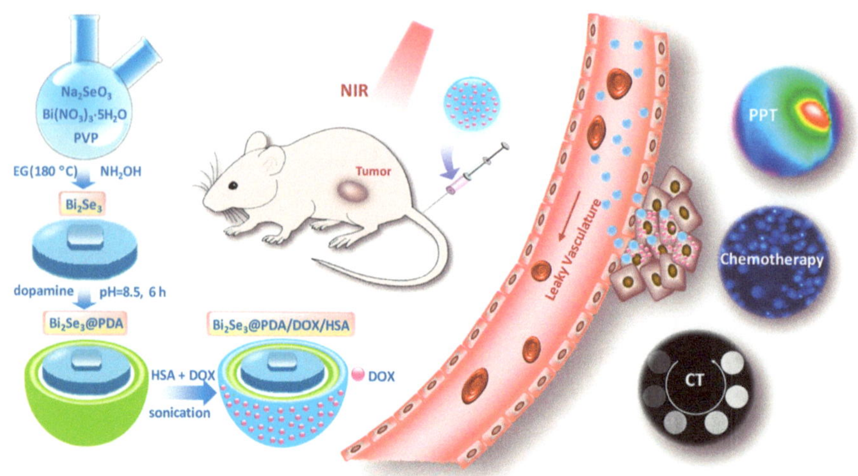


**Figure 2.5** Schematic Illustration of the synthesis and multifunction of $Bi_2Se_3$@PDA/DOX/HSA NPs. Reproduced with permission from ref. 78. Copyright 2016 The American Chemical Society.

mesoporous structure and high surface area endowed HMPB-Mn with high drug loading capacity of doxorubicin (DOX) for tumor chemotherapy. Those DOX loaded HMPB@MnPBA nanoparticles had a powerful positive correlation between pH-responsive MRI intensity and DOX release, which can achieve efficient drug release monitoring by MRI. In 2015, Wu and co-workers designed doxorubicin loaded PB nanocages, which served as dual-mode therapeutic agents for hepatocellular carcinoma chemo-photothermal treatment.[81] All of those results demonstrated that PB nanoparticles could act as the promising photothermal-controlled drug delivery for chemo-photothermal treatment of cancers.

In summary, various inorganic nanomaterials with high absorbance in the NIR region have been extensively used in photothermal-controlled drug delivery as functional carriers, showing encouraging therapeutic results both *in vitro* and *in vivo*. However, most of these inorganic nanocarriers are nonbiodegradable and usually would retain in the body for long periods, raising concerns regarding their potential long-term toxicity. Before these therapeutic strategies are tested in cancer patients, much more effort should be focused on the *in vivo* behaviors of those inorganic nanocarriers, including their pharmacokinetics, pharmacodynamics, and potential long-term toxicity.

## 2.3 Organic Nanocarriers Used as Photothermal-controlled Drug Delivery Systems

### 2.3.1 Conjugated Polymer Nanocarriers

#### *2.3.1.1 Polyaniline Nanoparticles*

Among conjugated polymers, polyaniline (PANI) was the first one to be utilized as a photothermal agent for cancer ablation.[82] PANI nanomaterials have attracted tremendous interest in bioelectronics and biomedical application due to their excellent conductivity, good biocompatibility, mechanical flexibility, low cost and simple synthesis with mild conditions.[83–85] Our group[86] designed a multifunctional rapamycin/DiR loaded lipid-PANI hybrid nanoparticle system, RDLPNPs (Figure 2.6). Notably, PANI as the appropriate model receptor of fluorescence resonance energy transfer (FRET) could effectively convert the NIR fluorescence energy from the donor DiR resulting in enhanced PTT efficacy. In addition, the rapamycin release was subsequently triggered by the enhanced PTT effect after laser irradiation, which could cut down the tumor vessels and inhibit tumor growth. After intravenous injection of RDLPNPs, fluorescent (from DiR) and enhanced photoacoustic (from DLPNPs) signals could be monitored over time and reached a peak at the 6 h time point. These results showed that RDLPNPs was a promising imaging guided photothermal drug delivery nanoplatform for chemo-photothermal therapy.

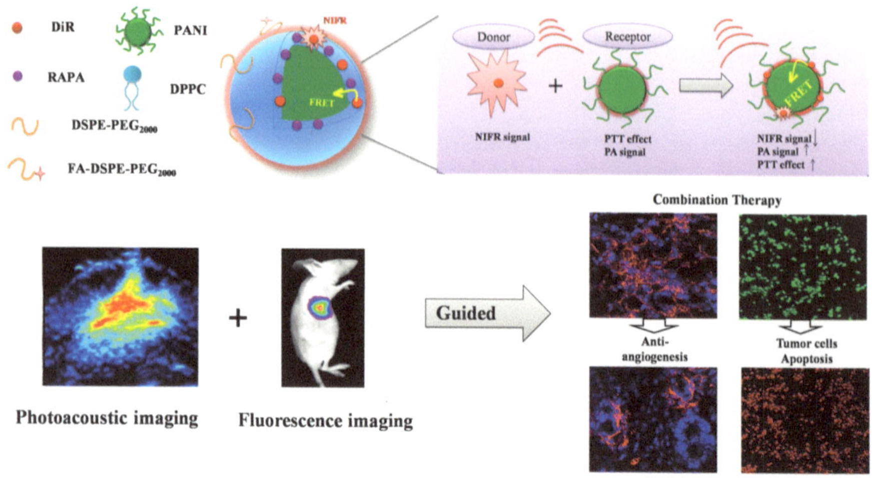


**Figure 2.6** Schematic illustration of RDLPNPs exhibiting photoacoustic and fluorescence imaging guided combination therapy of antiangiogenesis and enhanced photothermal performance mediated by FRET. Reproduced from *J. Control. Release*, **237**, J. Wang, F. Guo, M. Yu, L. Liu, F. Tan, R. Yan and N. Li, Rapamycin/DiR loaded lipid-polyaniline nanoparticles for dual-modal imaging guided enhanced photothermal and anti-angiogenic combination therapy, 23–34. Copyright 2016 with permission from Elsevier.

Recently, Chen *et al.* successfully fabricated graphene oxide and AuNP core PANI shell (GO-Au@PANI) nanocomposites for synchronous surface-enhanced Raman scattering (SERS) imaging and chemo-photothermal cancer ablation.[87] PANI was decorated onto AuNPs by facile *in situ* oxidative polymerization; then the Au@PANI core–shell nanoparticles were deposited on the graphene oxide (GO) sheet *via* π–π stacking and electrostatic interaction. GO-Au@PANI also revealed ultrahigh drug-loading capability with the doxorubicin (DOX)-loading efficiency of 189.2% and sensitive NIR/pH-responsive drug release behavior. The real-time monitoring of intracellular DOX release dynamics from the nanocomposites was carried out using SERS-fluorescence dual-mode imaging. Finally, the GO-Au@PANI/DOX nanocomposites exhibited effectively *in vitro* and *in vivo* imaging guided chemo-photothermal ablation of cancer cells as a high-performance chemo-photothermal therapeutic nanoagent.

### *2.3.1.2 Polypyrrole Nanoparticles*

Polypyrrole (PPy) has been extensively utilized in biosensing and tissue engineering applications due to its high conductivity, remarkable stability and outstanding biocompatibility.[88] PPy nanomaterials are generally prepared by *in situ* polymerization of a pyrrole monomer. Furthermore, with strong absorbance in the NIR region, PPy has also been explored by many groups not only as a promising photothermal agent for effective cancer ablation but also as a novel polymer-based optical contrast agent for enhanced PA imaging.[89] Considering the advantages and disadvantages of different therapeutic strategies, multifunctional nanoplatforms with good patient compliance and multiple therapeutic functions together with imaging capabilities have received great attention in the development of novel tumor theranostic approaches. For instance, Wang *et al.*[90] developed a multifunctional nanocomposite based on PEGylated $Fe_3O_4$@PPy loaded with aromatic chemotherapeutic drug doxorubicin (DOX) for image-guided, remotely controlled combination photo-chemotherapy (Figure 2.7). In this design, the $Fe_3O_4$ core was used for magnetically controlled drug delivery as well as a MR imaging contrast. On the other hand, the PPy shell exhibited a strong photothermal effect, which was useful to kill cancer cells directly by hyperthermia or indirectly by intracellularly promoted uptake and triggered release of DOX to enhance chemotherapeutic efficiency. This photothermal-controlled drug delivery system was then applied to MR imaging guided combined photothermal-chemotherapies both *in vitro* and *in vivo*, demonstrating outstanding and promising synergistic therapeutic efficacy. Meanwhile, Liang *et al.*[91] successfully constructed PEGylated PPy nanoparticles conjugating gadolinium chelates (Gd-PEG-PPy NPs) with good physiological stability, strong NIR absorption, low dark toxicity and high magnetization for dual-modal MR and PA imaging guided photothermal therapy with NIR laser irradiation.

### 2.3.1.3 *Polydopamine Nanoparticles*

Polydopamine nanoparticles are effective PTT agents with high photothermal conversion efficiency and good biodegradability.[92] The photothermal heating effect can not only directly kill cancer cell by hyperthermia, but also be expected to trigger the release of loaded drug, and consequently improve the efficacy of therapy.[93] Recently, Lu and co-workers developed a biodegradable coordination polymer (CP) encapsulated PDAs nanocomplex (PDAs@$CP_3$).[94] This resultant PDAs@$CP_3$ could combine the property of PDAs and CP together, and serve as a longitudinal ($T_1$)/transversal ($T_2$) dual mode contrast agent for MRI. Upon irradiation with a 808 nm NIR laser, the DOX loaded PDAs@$CP_3$ nanocomplex (denoted as PDAs@$CP_3$-DOX) exhibited high photothermal conversion efficiency and the remote laser-triggered release behavior, which resulted in prominent chemo-photothermal synergistic cancer cell killing effect both *in vitro* and *in vivo*. Their work helped the development of PDAs-based theranostic agents, which enabled $T_1/T_2$ dual mode MRI-guided chemo-photothermal synergistic therapy in an effective and safe way.

### 2.3.1.4 *Other Conjugated Polymer Nanocarriers*

Other conjugated polymer nanocarriers, including poly(3,4-ethylenedioxythiophene) : poly(4-styrenesulfonate) (PEDOT : PSS) and donor–acceptor (D-A) conjugated polymers (PBIBDF-BT), have also been used for the

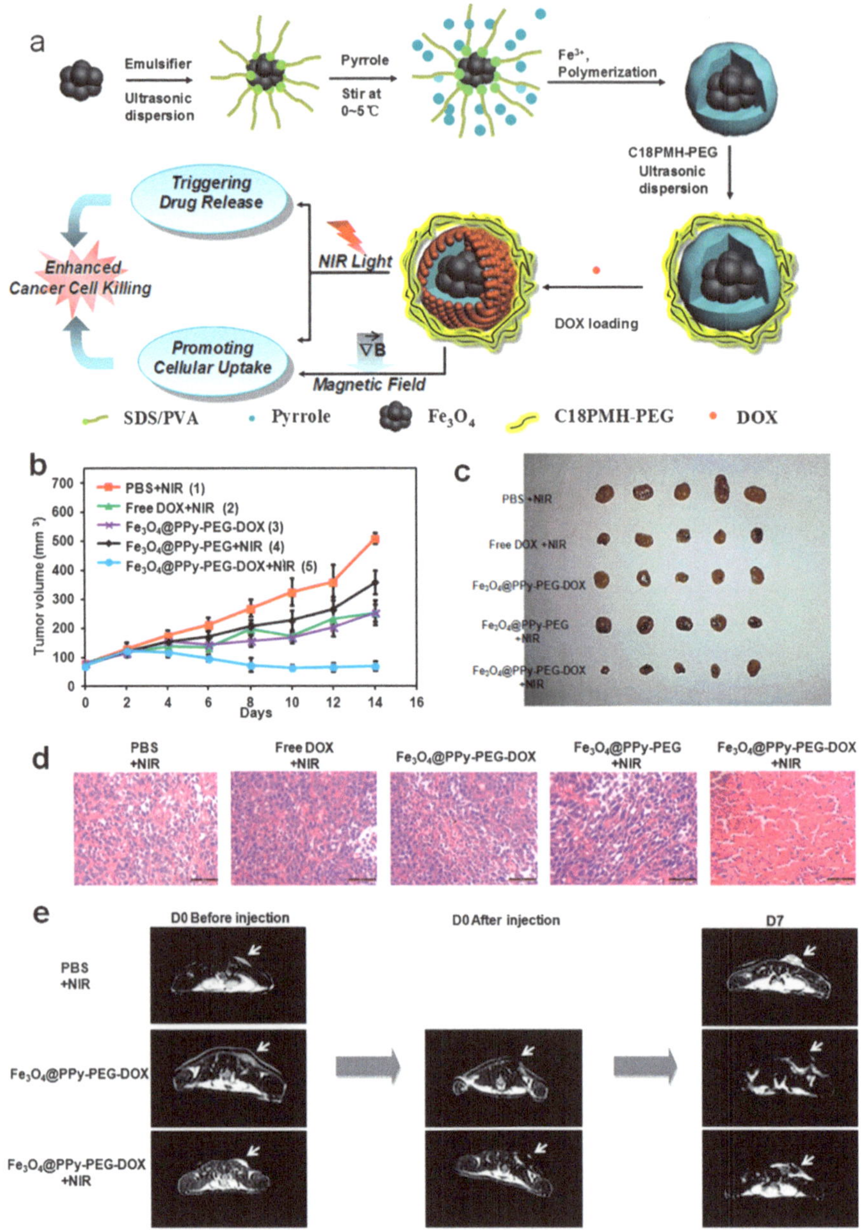


**Figure 2.7** Multifunctional $Fe_3O_4$@PPy-PEG composite nanoparticles for imaging-guided combination cancer therapy. (a) The scheme shows the preparation of $Fe_3O_4$@PPy-PEG nanoparticles, the subsequent DOX loading, and remotely controlled cancer therapy triggered by both NIR light and magnetic field. (b–e) *In vivo* combination cancer therapy: (b) tumor growth curves of different groups of mice after various treatments as indicated (five mice per group). (c) Photos of the tumors collected from different groups of mice at the end of treatment (day 14).

imaging guided photothermal-responsive drug delivery.[95,96] In 2013, Gong *et al.* reported PEDOT:PSS nanoparticles,[97] which were used as a drug carrier to load various types of therapeutic molecules. It was found that organic nanoparticles based on polyethylene glycol (PEG) coated PEDOT:PSS, a conjugated polymer with strong NIR absorbance, showed rather high passive tumor accumulation owing to the enhanced permeability and retention (EPR) effect of cancerous tumors, and could serve as a powerful photothermal agent with excellent tumor ablation effect under NIR laser irradiation, in mouse tumor model experiments. Importantly, PEDOT:PSS-PEG nanoparticle could be used as the drug carrier for combined cancer therapy. In their work, DOX, a photosensitizer chlorine 6(Ce6), and a water-insoluble drug SN38, were effectively loaded on PEDOT:PSS-PEG nanoparticles with good stability and high loading capacity. Loading of those therapeutics on PEDOT:PSS-PEG showed a number of unique benefits, including improving water-solubility for the water-insoluble drug, accelerating cellular uptake of the photosensitizer for enhanced efficacy in photodynamic therapy, as well as enabling combined photothermal- and chemotherapy for synergistic cancer cell killing.

Yang and co-workers developed DOX loaded PBIBDF-BT-based nanoparticles decorated with amphiphilic copolymer mPEG-*b*-PHEP through a single emulsion method.[96] The obtained PBIBDF-BT@$NP_{PPE}$ with a size of 50 nm showed strong and sharp NIR absorption and had a maximal absorption peak at approximately 811 nm. By intratumoral injection plus NIR irradiation, PBIBDF-BT@$NP_{PPE}$ successfully inhibited tumor growth, exhibiting superior anticancer efficacy. More importantly, PBIBDF-BT@$NP_{PPE}$ efficiently encapsulated a chemotherapy drug and achieved NIR-triggered intracellular release and synergistic therapeutic effects of thermo-chemotherapy.

## 2.3.2 Near-infrared Cyanine Dyes

### *2.3.2.1 Indocyanine Green Nanoparticles*

Indocyanine green (ICG) is an amphiphilic, inert and non-toxic compound, which has a molecular weight of 751.4 Da and a hydrodynamic diameter of 1.2 nm. ICG, the only nearinfrared (NIR) dye approved by the FDA, has been widely applied for determining hepatic function, liver blood flow, and cardiac output, as well as in ophthalmic angiography.[98] What's more, as ICG

---

(d) H&E-stained tumor slices collected from different groups of mice post-treatments (scale bar = 50 μm). (e) Representative MR images of mice from three different groups obtained at day zero before and after injection and day seven post-treatment (tumors marked by the white arrow). All the above data confirmed that the combined photothermal and chemotherapy ($Fe_3O_4$@PPy-PEG-DOX + NIR) offered an obvious synergistic tumor killing effect. Adapted with permission from ref. 25. Copyright 2013 The American Chemical Society.

has high NIR light absorption characteristics ($\lambda$ = 650–900 nm), and generates high fluorescence and PA signals,[99] providing it with a wide range of potential applications in fluorescent imaging/PA and laser-mediated localized phototheratherapy.[100]

The functions of ICG for tumor treatments include different methods. Recently, the use of ICG in PDT and PTT has drawn widely increased attention due to its strong absorption band that causes deeper tissue penetration generating significant heating.[101] For the PTT effect, photoabsorbing agents (PTA), ICG, can generate hyperthermia from absorbed laser energy to cause cancer cell death. In another way, PDT provides toxicities to tumor cells by the photosensitizers (PS) transferring light energy to the surrounding oxygen to generate cytotoxic singlet oxygen. In terms of the mechanism of ICG cytotoxicity, it appears that its decomposition products are linked immediately to the singlet oxygen.[102] On account of these good properties in the contexts, several research groups (Lu Han *et al.*, Rui Chen *et al.*, Ismail Ocsoy *et al.*, Vaibhav Mundra *et al.*)[103–106] have proposed the use of ICG for PDT and PTT effect.

In 2009, Ishizawa *et al.* first introduced the imaging application of ICG in hepatic surgery for the visualization of colorectal liver metastases and hepatocellular carcinomas.[107] Despite its several advantages, the employment of ICG in clinic is still limited. On one hand, ICG is unstable in an aqueous solution and easy to bind to plasma proteins after intravenous injection, resulting in a very short half-life of only about 2–4 min.[108] On the other hand, ICG cannot actively target tumor sites. In order to address the several issues of instability, the encapsulation of the ICG into carriers has been proposed. Experimental studies demonstrated that this approach has successfully improved ICG stability, prolonged its circulation in the blood and targeted the dye to a specific site, making ICG an ideal fluorescence marker. Polymeric nanocarriers,[109] liposomes,[110] lipid-polymer nanoparticles[111] and inorganic particles[112] have all been used to make nanocarriers for ICG to form imaging guided combined phototherapy and chemotherapy systems.

#### 2.3.2.2 Other Near-infrared Cyanine Dye Nanoparticles

With a similar structure to ICG, heptamethine indocyanine dye has also been used as an NIR imaging probe.[113–117] Peng *et al.* synthesized a multifunctional micelle containing an $^{188}$Re-labeled radionucleotide for single-photon emission computed tomography (SPECT) imaging and an IR-780 iodide NIR dye for fluorescent imaging and PTT.[113] Li *et al.* reported IR825/DOX loaded NIR-absorptive stealth liposomes for imaging guided combinational cancer treatment.[115] Both *in vitro* and *in vivo* results confirmed that the IR825/DOX loaded liposomes, along with concurrent NIR treatment, could lead to better tumoricidal efficacy than either liposomal DOX medication or sole photothermal ablation. Huang *et al.* developed a ferritin (DFRT) nanocage, loaded with the near-infrared dye IR820, that showed strong absorbance in the NIR region for simultaneous fluorescence/photoacoustic/photothermal multimodal imaging-guided enhanced PTT.[117]

### 2.3.3 Other Organic Nanocarriers

Some other organic nanomaterials have also been reported in recent years as photothermal-controlled drug delivery methods. Porphyrin-based micelle developed by Su and co-workers have also shown great potential in photothermal-chemo combination cancer therapy recently[118] (Figure 2.8). In addition, porphysomes, which were organic nanoparticles self-assembled from phospholipid-porphyrin conjugates, have also been developed due to their

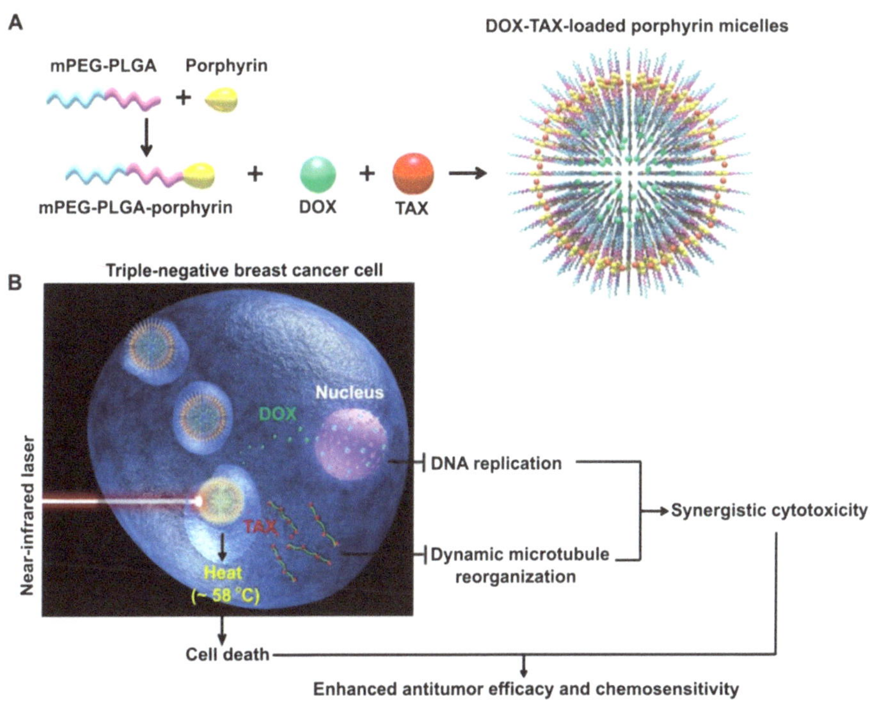


**Figure 2.8** Schematic illustration of the assembly process and the antitumor mechanism of DOX-TAX-loaded porphyrin micelles. (A) Construction of the DOX-TAX-loaded porphyrin micelles from the amphiphilic polymer mPEG-PLGAeporphyrin. (B) The therapeutic mechanism of DOX-TAX-loaded porphyrin micelles for combination photothermal-chemotherapy. Reproduced from *Biomaterials*, **80**, S. Su, Y. Ding, Y. Li, Y. Wu, G. Nie, Integration of photothermal therapy and synergistic chemotherapy by a porphyrin self-assembled micelle confers chemosennsitivity in triple-negative breast cancer, 169–178. Copyright 2016 with permission from Elsevier.

liposome-like structure, high porphyrin loading capacity, strong absorption in the NIR region, and excellent biocompatibility. Recently, Ng *et al.* synthesized activatable porphyrin nanodiscs by complexing apolipoproteins to porphyrin-lipid using the detergent dialysis method.[119] The ability of those nanodiscs to incorporate large amounts of photosensitizer molecules into such a compact structure allowed for phototherapeutic action, fluorescence diagnostic applications, and the potential to effectively deliver photosensitizers deep into poorly permeable tumors.

Overall, organic nanomaterial based drug delivery systems are encouraging alternatives to inorganic nanomaterials for applications in chemo-photothermal combination therapy. Different from inorganic nanomaterials based drug delivery systems, which would retain in the body for a long time, small organic dyes should be able to be easily cleared out from the body with little long-term retention. Porphysomes solely containing biocompatible and biodegradable molecules are also safe agents that are promising for clinical use. However, other organic nanomaterials, such as conjugated polymers, still need further study to understand their potential biodegradation behaviors *in vivo*.

## 2.4 Conclusions and Outlook

In this chapter, we have systematically reviewed the recent research in chemo-phototherapies of cancer using functional inorganic/organic nanocarriers. Due to the various advantages involving non-invasiveness, high drug loading efficiency, and great therapeutic efficacy, imaging guided photothermal-responsive drug delivery systems have been proposed as attractive cancer treatment techniques in recent years.

However, despite the tremendous amount of exciting results reported in the past few years in this field, there are still many challenges ahead towards further clinical applications of those functional nanomaterials in image guided chemo-photothermal therapy of cancer. (1) One of the most important issues is the potential long-term safety concerns of the nanomaterials, especially those inorganic ones that are not biodegradable and would remain in the body for a long time after administration. (2) Another major challenge in photothermal therapy of tumors is the limited light penetration depth. The effective penetration depth of NIR light is still usually limited to be no deeper than 1 cm. For some types of cancers, such as skin cancers, esophageal cancer and oral cancer, light can be induced to locally irradiate the tumors with the help of certain facilities. For other types of cancers with tumors located deep inside the body, effective phototherapy would require the appropriate design of medical devices (*e.g.*, with optical fibers) that can deliver light into those deep lesions. Thus, the future development of imaging-guided chemo-photothermal therapy based multifunctional nanocarriers may indeed have great potential in our fight against cancer.

## Acknowledgements

This work was partially supported by the National Natural Science Foundation of China (81503016), the Application Foundation and Cutting-edge Technologies Research Project of Tianjin (Young Program) (15JCQNJC13800), and the National Basic Research Project (973 Program) of China (2014CB932200).

## References

1. M. Ferrari, *Nat. Rev. Cancer*, 2005, **5**, 161–171.
2. D. Peer, J. M. Karp, S. Hong, O. C. Farokhzad, R. Margalit and R. Langer, *Nat. Nanotechnol.*, 2007, **2**, 751–760.
3. S. Lal, S. E. Clare and N. J. Halas, *Acc. Chem. Res.*, 2008, **41**, 1842–1851.
4. M. P. Melancon, M. I. N. Zhou and C. Li, *Acc. Chem. Res.*, 2011, **44**, 947–956.
5. X. Huang, P. K. Jain, I. H. El-Sayed and M. A. El-Sayed, *Lasers Med. Sci.*, 2008, **23**, 217–228.
6. J. O. You, P. Guo and D. T. Auguste, *Angew. Chem., Int. Ed.*, 2013, **125**, 4235–4240.
7. M. K. Yu, D. Kim, I. H. Lee, J. S. So, Y. Jeong and S. Jon, *Small*, 2011, **7**, 2241–2249.
8. C. L. Peng, Y. H. Shih, P. C. Lee, T. M. H. Hsieh, T. Y. Luo and M. J. Shieh, *ACS Nano*, 2011, **5**, 5594–5607.
9. K. Yang, L. Hu, X. Ma, S. Ye, L. Cheng, X. Shi and Z. Liu, *Adv. Mater.*, 2012, **24**, 1868–1872.
10. A. M. Gobin, M. H. Lee, N. J. Halas, W. D. James, R. A. Drezek and J. L. West, *Nano Lett.*, 2007, **7**, 1929–1934.
11. J. W. Kim, E. I. Galanzha, E. V. Shashkov, H. M. Moon and V. P. Zharov, *Nat. Nanotechnol.*, 2009, **4**, 688–694.
12. M. Zhou, R. Zhang, M. Huang, W. Lu, S. Song, M. P. Melancon, M. Tian, D. Liang and C. Li, *J. Am. Chem. Soc.*, 2010, **132**, 15351–15358.
13. H. Ke, J. Wang, Z. Dai, Y. Jin, E. Qu, Z. Xing, C. Guo, X. Yue and J. Liu, *Angew. Chem., Int. Ed.*, 2011, **50**, 3017–3021.
14. T. D. MacDonald, T. W. Liu and G. Zheng, *Angew. Chem., Int. Ed.*, 2014, **27**, 7076–7079.
15. J. Yu, C. Yang, J. Li, Y. Ding, L. Zhang, M. Z. Yousaf and F. Sheng, *Adv. Mater.*, 2014, **26**, 4114–4120.
16. L. Cheng, C. Yuan, S. Shen, X. Yi, H. Gong, K. Yang and Z. Liu, *ACS Nano*, 2015, **9**, 11090–11101.
17. K. Yang, H. Xu and L. Cheng, *Adv. Mater.*, 2012, **24**, 5586–5592.
18. M. Yu, F. Guo and J. Wang, *Biomaterials*, 2016, **79**, 25–35.
19. Y. Xia, W. Li, C. Cobley, J. Chen, X. Xia, Q. Zhang, M. Yang, E. C. Cho and P. K. Brown, *Acc. Chem. Res.*, 2011, **44**, 914–924.
20. J. Ge, Q. Jia, W. Liu, L. Guo, Q. Liu, M. Lan and P. Wang, *Adv. Mater.*, 2015, **27**, 4169–4177.

21. G. Ku, M. Zhou, S. Song, Q. Huang, J. Hazle and C. Li, *ACS Nano*, 2012, **6**, 7489–7496.
22. Z. Li, Y. Hu, K. A. Howard, T. Jiang, X. Fan, Z. Miao, Y. Sun, F. Besenbacher and M. Yu, *ACS Nano*, 2015, **10**, 984–997.
23. R. Lv, P. Yang, F. He, S. Gai, C. Li, Y. Dai, G. Yang and J. Li, *ACS Nano*, 2015, **9**, 1630–1647.
24. J. Wang, F. Guo, M. Yu, L. Liu, F. Tan, R. Yan and N. Li, *J. Controlled Release*, 2016, **237**, 23–34.
25. C. Wang, H. Xu, C. Liang, Y. Liu, Z. Li, G. Yang, L. Cheng, Y. Li and Z. Liu, *ACS Nano*, 2013, **7**, 6782–6795.
26. J. F. Lovell, C. S. Jin, E. Huynh, H. Jin, C. Kim, J. L. Rubinstein, W. C. W. Chan, W. Cao, L. V. Wang and G. Zheng, *Nat. Mater.*, 2011, **10**, 324–332.
27. S. Su, Y. Tian, Y. Li, Y. Ding, T. Ji, M. Wu, Y. Wu and G. Nie, *ACS Nano*, 2015, **9**, 1367–1378.
28. Z. Z. J. Lim, J. E. J. Li, C. T. Ng, L. Y. L. Yung and B. H. Bay, *Acta Pharmacol. Sin.*, 2011, **32**, 983–990.
29. X. Huang and M. A. El-Sayed, *J. Adv. Res.*, 2010, **1**, 13–28.
30. K. Saha, S. S. Agasti, C. Kim, X. Li and V. M. Rotello, *Chem. Rev.*, 2012, **112**, 2739–2779.
31. P. Huang, J. Lin, W. Li, P. Rong, Z. Wang, S. Wang, X. Wang, X. Sun, M. Aronova, G. Niu, R. D. Leapman, Z. Nie and X. Chen, *Angew. Chem.*, 2013, **125**, 14208–14214.
32. T. A. El-Brolossy, T. Abdallah, M. B. Mohamed, S. Abdallah, K. Easawi, S. Negm and H. Talaat, *Eur. Phys. J.: Spec. Top.*, 2008, **153**, 361–364.
33. P.-C. Li, C.-R. C. Wang, D.-B. Shieh, C.-W. Wei, C.-K. Liao, C. Poe, S. Jhan, A.-A. Ding and Y.-N. Wu, *Opt. Express*, 2008, **16**, 18605–18615.
34. W. Chen, X. Xu, H. Jia, Q. Lei, G. Luo, S. Cheng, R. Zhuo and X. Zhang, *Biomaterials*, 2013, **34**, 8798–8807.
35. M. Yu, F. Guo, J. Wang, F. Tan and N. Li, *Biomaterials*, 2016, **79**, 25–35.
36. H. Liu, X. Zhang, X. Wu, L. Jiang, C. Burda and J. J. Zhu, *Chem. Commun.*, 2011, **47**, 4237–4239.
37. T. Zhou, Y. Huang, W. Li, Z. Cai, F. Luo, C. J. Yang and X. Chen, *Nanoscale*, 2012, **4**, 5312–5315.
38. D. Chen, Z. Luo, N. Li, D. Chen, Z. Luo, N. Li, J. Y. Lee, J. Xie and J. Lu, *Adv. Funct. Mater.*, 2013, **23**, 4324–4331.
39. Y. Ma, X. Liang, S. Tong, G. Bao, Q. Ren and Z. Dai, *Adv. Funct. Mater.*, 2013, **23**, 815–822.
40. Q. Wang, K. Kalantar-Zadeh, A. Kis, J. N. Coleman and M. S. Strano, *Nat. Nanotechnol.*, 2012, **7**, 699–712.
41. D. Voiry, A. Goswami, R. Kappera, C. de C. C. e. Silva, D. Kaplan, T. Fujita, M. Chen, T. Asefa and M. Chhowalla, *Nat. Chem.*, 2015, **7**, 45–49.
42. W. S. Yun, S. W. Han, S. C. Hong, I. G. Kim and J. D. Lee, *Phys. Rev. B*, 2012, **85**, 033305.
43. L. Cheng, J. Liu, X. Gu, H. Gong, X. Shi, T. Liu and W. Bu, *Adv. Mater.*, 2014, **26**, 1886–1893.

44. L. Cheng, J. Liu, X. Gu, H. Gong, X. Shi, T. Liu and W. Bu, *Chem. Soc. Rev.*, 2015, **44**, 2713–2731.
45. M. Chhowalla, H. S. Shin, G. Eda, L. J. Li, K. P. Loh and H. Zhang, *Nat. Chem.*, 2013, **5**, 263–275.
46. X. Huang, Z. Zeng and H. Zhang, *Chem. Soc. Rev.*, 2013, **42**, 1934–1946.
47. H. Peng, W. Dang, J. Cao, Y. Chen, D. Wu, W. Zheng and Z. Liu, *Nat. Chem.*, 2012, **4**, 281–286.
48. S. S. Chou, B. Kaehr, J. Kim, B. M. Foley, M. De, P. E. Hopkins, J. Huang, C. J. Brinker and V. P. Dravid, *Angew. Chem., Int. Ed.*, 2013, **52**, 4160–4164.
49. W. Yin, L. Yan, J. Yu, G. Tian, L. Zhou, X. Zheng, X. Zhang, Y. Yong, J. Li, Z. Gu and Y. Zhao, *ACS Nano*, 2014, **8**, 6922–6933.
50. G. Yang, H. Gong, T. Liu, X. Sun, L. Cheng and Z. Liu, *Biomaterials*, 2015, **60**, 62–71.
51. L. Guo, I. Panderi, D. D. Yan, K. Szulak, Y. Li, Y. T. Chen, H. Ma, D. B. Niese, N. Seeram, A. Ahmed, B. Yan, D. Pantazatos and W. Lu, *ACS Nano*, 2013, **7**, 8780–8793.
52. Q. Tian, M. Tang, Y. Sun, R. Zou, Z. Chen, M. Zhu, S. Yang, J. Wang, J. Wang and J. Hu, *Adv. Mater.*, 2011, **23**, 3542–3547.
53. M. Zhou, R. Zhang, M. Huang, W. Lu, S. Song, M. P. Melancon, M. Tian, D. Liang and C. Li, *J. Am. Chem. Soc.*, 2010, **132**, 15351–15358.
54. S. B. Lakshmanan, X. Zou, M. Hossu, L. Ma, C. Yang and W. Chen, *J. Biomed. Nanotechnol.*, 2012, **8**, 883–890.
55. G. Ku, M. Zhou, S. Song, Q. Huang, J. Hazle and C. Li, *ACS Nano*, 2012, **6**, 7489–7496.
56. S. Ramadan, L. Guo, Y. Li, B. Yan and W. Lu, *Small*, 2012, **8**, 3143–3150.
57. S. Goel, F. Chen and W. Cai, *Small*, 2014, **10**, 631–645.
58. Y. Li, W. Lu, Q. Huang, M. Huang, C. Li and W. Chen, *Nanomedicine*, 2010, **5**, 1161–1171.
59. Z. Zha, S. Zhang, Z. Deng, Y. Li, C. Li and Z. Dai, *Chem. Commun.*, 2013, **49**, 3455–3457.
60. X. Liu, G. Xi, Y. Liu, S. Xiong, L. Chai and Y. Qian, *J. Nanosci. Nanotechnol.*, 2007, **7**, 4501–4507.
61. X. Zhang, G. Wang, A. Gu, Y. Wei and B. Fang, *Chem. Commun.*, 2008, 5945–5947.
62. X.-H. Liao, N.-Y. Chen, S. Xu, S.-B. Yang and J.-J. Zhu, *J. Cryst. Growth*, 2003, **252**, 593.
63. K. Ding, J. Zeng, L. Jing, R. Qiao, C. Liu, M. Jiao, Z. Li and M. Gao, *Nanoscale*, 2015, **7**, 11075–11081.
64. G. Ku, M. Zhou, S. Song, Q. Huang, J. Hazle and C. Li, *ACS Nano*, 2010, **6**, 7489–7496.
65. Q. Feng, Y. Zhang, W. Zhang, X. Shan, Y. Yuan, H. Zhang, L. Hou and Z. Zhang, *Acta Biomater.*, 2016, **38**, 129–142.
66. J. T. Robinson, S. M. Tabakman, Y. Liang, H. Wang, H. S. Casalongue, D. Vinh and H. Dai, *J. Am. Chem. Soc.*, 2011, **133**, 6825–6831.
67. B. Tian, C. Wang, S. Zhang, L. Feng and Z. Liu, *ACS Nano*, 2011, **5**, 7000–7009.

68. Z. Liu and X. J. Liang, *Theranostics*, 2012, **2**, 235–237.
69. L. Meng, X. Zhang, Q. Lu, Z. Fei and P. J. Dyson, *Biomaterials*, 2012, **33**, 1689–1698.
70. A. M. A. Elhissi, W. Ahmed, I. U. Hassan, V. R. Dhanak and A. D'Emanuele, *J. Drug Delivery*, 2011, **2012**, 1–10.
71. S. K. Vashist, D. Zheng, G. Pastorin, K. Al-Rubeaan, J. H. T. Luong and F. S. Sheu, *Carbon*, 2011, **49**, 4077–4097.
72. L. Xie, G. Wang, H. Zhou, F. Zhang, Z. Guo, C. Liu, X. Zhang and L. Zhu, *Biomaterials*, 2016, **103**, 219–228.
73. L. Hou, X. Yang, J. Ren, Y. Wang, H. Zhang, Q. Feng, Y. Shi, X. Shan, Y. Yuan and Z. Zhang, *Int. J. Nanomed.*, 2016, **11**, 607–624.
74. B. Tian, C. Wang, S. Zhang, L. Feng and Z. Liu, *ACS Nano*, 2011, **5**, 7000–7009.
75. X. Ma, H. Tao, K. Yang, L. Feng, L. Cheng, X. Shi, Y. Li, L. Guo and Z. Liu, *Nano Res.*, 2012, **5**, 199–212.
76. L. Guo, H. Shi, H. Wu, Y. Zhang, X. Wang, D. Wu, L. An and S. Yang, *Carbon*, 2016, **107**, 87–99.
77. H. Dong, M. Jin, Z. Liu, H. Xiong, X. Qiu, W. Zhang and Z. Guo, *Lasers Med. Sci.*, 2016, **31**, 1123–1131.
78. Z. Li, Y. Hu, K. A. Howard, T. Jiang, X. Fan, Z. Miao, Y. Sun, F. Besenbacher and M. Yu, *ACS Nano*, 2016, **10**, 984–997.
79. G. Fu, W. Liu, S. Feng and X. Yue, *Chem. Commun.*, 2012, **48**, 11567–11569.
80. X. Cai, W. Gao, M. Ma, M. Wu, L. Zhang, Y. Zheng, H. Chen and J. Shi, *Adv. Mater.*, 2015, **27**, 6382–6389.
81. M. Wu, Q. Wang, X. Liu and J. Liu, *RSC Adv.*, 2015, **5**, 30970–30980.
82. J. Yang, J. Choi, D. Bang, E. Kim, E.-K. Lim, H. Park, J.-S. Suh, K. Lee, K.-H. Yoo, E.-K. Kim, Y.-M. Huh and S. Haam, *Angew. Chem., Int. Ed.*, 2011, **50**, 441–444.
83. P. R. Bidez, S. Li, A. G. MacDiarmid, E. C. Venancio, Y. Wei and P. I. Lelkes, *J. Biomater. Sci., Polym. Ed.*, 2006, **17**, 199–212.
84. E. T. Kang, K. G. Neoh and K. L. Tan, *Prog. Polym. Sci.*, 1998, **23**, 277–324.
85. Y. Hong, S. Hwang, D. Heo, B. Kim, M. Ku, E. Lee, S. Haam, D. S. Yoon, J. Yang and J.-S. Suh, *Nanoscale*, 2015, **7**, 1661–1666.
86. J. Wang, F. Guo, M. Yu, L. Liu, F. Tan, R. Yan and N. Li, *J. Controlled Release*, 2016, **237**, 23–34.
87. H. Chen, Z. Liu, S. Li, C. Su, X. Qiu, H. Zhong and Z. Guo, *Theranostics*, 2016, **6**, 1096–1104.
88. X. Song, Q. Chen and Z. Liu, *Nano Res.*, 2015, **8**, 340–354.
89. Z. Zha, Z. Deng, Y. Li, C. Li, J. Wang, S. Wang, E. Qu and Z. Dai, *Nanoscale*, 2013, **5**, 4462–4467.
90. C. Wang, H. Xu, C. Liang, Y. Liu, Z. Li, G. Yang, L. Cheng, Y. Li and Z. Liu, *ACS Nano*, 2013, **7**, 6782–6795.
91. X. Liang, Y. Li, X. Li, L. Jing, Z. Deng, X. Yue, C. Li and Z. Dai, *Adv. Funct. Mater.*, 2015, **25**, 1451–1462.
92. Y. Liu, K. Ai, J. Liu, M. Deng, Y. He and L. Lu, *Adv. Mater.*, 2013, **25**, 1353–1359.

93. S. P. Sherlock, S. M. Tabakman, L. Xie and H. Dai, *ACS Nano*, 2011, **5**, 1505–1512.
94. Y. Chen, K. Ai, J. Liu, X. Ren, C. Jiang and L. Lu, *Biomaterials*, 2016, **77**, 198–206.
95. L. Xu, L. Cheng, C. Wang, R. Peng and Z. Liu, *Polym. Chem.*, 2014, **5**, 1573–1580.
96. D. D. Li, J. X. Wang, Y. Ma, H. S. Qian, D. Wang, L. Wang, G. Zhang, L. Qiu, Y. C. Wang and X. Z. Yang, *ACS Appl. Mater. Interfaces*, 2016, **8**, 19312–19320.
97. H. Gong, L. Cheng, J. Xiang, H. Xu, L. Feng, X. Shi and Z. Liu, *Adv. Funct. Mater.*, 2013, **23**, 6059–6067.
98. J. T. Alander, I. Kaartinen, A. Laakso, T. Patil, T. Spillmann, V. V. Tuchin, M. Venermo and P. Valisuo, *Int. J. Biomed. Imaging*, 2012, **7**, 1–26.
99. M. Ogawa, N. Kosaka, P. L. Choyke and H. Kobayashi, *Cancer Res.*, 2009, **69**, 1268–1272.
100. J. Yu, D. Javier, M. A. Yaseen, N. Nitin, R. Richards-Kortum, B. Anvari and M. S. Wong, *J. Am. Chem. Soc.*, 2010, **132**, 1929–1938.
101. V. Saxena, M. Sadoqi and J. Shao, *Int. J. Pharm.*, 2006, **308**, 200–204.
102. E. Engel, R. Schraml, T. Maisch, K. Kobuch, B. König, R. M. Szeimies, J. Hillenkamp, W. Bäumler and R. Vasold, *Invest. Ophthalmol. Visual Sci.*, 2008, **49**, 1777–1783.
103. L. Han, Y. Zhang, X. W. Chen, Y. Shu and J. H. Wang, *J. Mater. Chem. B*, 2016, **4**, 105–112.
104. R. Chen, X. Wang, X. Yao, X. Zheng, J. Wang and X. Jiang, *Biomaterials*, 2013, **34**, 8314–8322.
105. I. Ocsoy, N. Isiklan, S. Cansiz, N. Ozdemir and W. Tan, *RSC Adv.*, 2016, **6**, 30285–30292.
106. V. Mundra, Y. Peng, S. Rana, A. Natarajan and R. I. Mahato, *J. Controlled Release*, 2015, **220**, 130–140.
107. T. Ishizawa, N. Fukushima, J. Shibahara, K. Masuda, S. Tamura, T. Aoki, K. Hasegawa, Y. Beck, M. Fukayama and N. Kokudo, *Cancer*, 2009, **115**, 2491–2504.
108. A. K. Kirchherr, A. Briel and K. Mader, *Mol. Pharm.*, 2009, **6**, 480–491.
109. J. R. Vorst, B. E. Schaafsma, M. Hutteman, F. P. R. Verbeek, G. J. Liefers, H. H. Hartgrink, V. T. H. B. M. Smit, C. W. G. M. Lowik, C. J. H. Velde, J. V. Frangioni and A. L. Vahrmeijer, *Cancer*, 2013, **119**, 3411–3418.
110. A. Akbarzadeh, R. Rezaei-Sadabady, S. Davaran, S. W. Joo, Y. Hanifephour, M. Samiei, M. Kouhi and K. Nejati-Koshki, *Nanoscale Res. Lett.*, 2013, **8**, 102–110.
111. S. Krishnamurthy, R. Vaiyapuri, L. Zhang and J. M. Chan, *Biomater. Sci.*, 2015, **3**, 923–936.
112. K. Deng, Z. Hou, X. Deng, P. Yang, C. Li and J. Lin, *Adv. Funct. Mater.*, 2015, **25**, 7280–7290.
113. C. L. Peng, Y. H. Shih, P. C. Lee, T. M. H. Hsieh, T. Y. Luo and M. J. Shieh, *ACS Nano*, 2011, **5**, 5594–5607.
114. F. Guo, M. Yu, J. Wang, F. Tan and N. Li, *ACS Appl. Mater. Interfaces*, 2015, **7**, 20556–20567.

115. M. Li, C. The, C. Y. Ang, S. Y. Tan, Z. Luo, Q. Qu, Y. Zhang, V. Korzh and Y. Zhao, *Adv. Funct. Mater.*, 2015, **25**, 5602–5610.
116. X. Song, R. Zhang, C. Liang, Q. Chen, H. Gong and Z. Liu, *Biomaterials*, 2015, **57**, 84–92.
117. P. Huang, P. Rong, A. Jin, X. Yan, M. G. Zhang, J. Lin, H. Hu, Z. Wang, X. Yue, W. Li, G. Niu, W. Zeng, W. Wang, K. Zhou and X. Chen, *Adv. Mater.*, 2014, **26**, 6401–6408.
118. S. Su, Y. Ding, Y. Li, Y. Wu and G. Nie, *Biomaterials*, 2016, **80**, 169–178.
119. K. K. Ng, J. F. Lovell, A. Vedadi, T. Hajian and G. Zheng, *ACS Nano*, 2013, **7**, 3484–3490.

CHAPTER 3

# Applications of Magnetic Nanoparticles in Multi-modal Imaging

J. T.-W. WANG, J. BAI AND K. T. AL-JAMAL*

Institute of Pharmaceutical Science, Faculty of Life Sciences & Medicine, King's College London, Franklin-Wilkins Building, London SE19 NH, United Kingdom
*E-mail: khuloud.al-jamal@kcl.ac.uk

## 3.1 Nanoparticles and Magnetic Nanoparticles

### 3.1.1 Application of Nanoparticles in Biomedicine

Due to the extensive development in nanotechnology, nanocarriers are gradually playing substantial roles in biomedical fields. There are many types of nanocarriers, including lipid and polymer-based nanoparticles (NPs), carbon-based nanomaterials, inorganic NPs and silica porous NPs *etc.* Table 3.1 provides some examples of a range of nanotherapeutic and/or nanodiagnostic approaches that are currently employed and/or have been approved in clinical trials. Lipid and polymer-based NPs dominate clinical trials. Liposomes are formed by the hydrophilic and hydrophobic interaction of lipid–lipid and lipid–water molecules and characterised as a bilayer structure composed of lipid molecules with an aqueous compartment in the centre.[1] Polymeric NPs are another major platform with the capacity of carrying drugs either

Drug Discovery Series No. 63
Theranostics and Image Guided Drug Delivery
Edited by Maya Thanou

Published by the Royal Society of Chemistry, www.rsc.org

**Table 3.1** Representative nanoparticles in clinical trials.[a]

| Nanocarriers | Active Ingredient | Brand Name | Indication | Status |
|---|---|---|---|---|
| Silica nanoparticles | Cyanine 5.5 | Crgdy-peg-c dots | Imaging in head and neck cancer, melanoma, breast cancer, cervical cancer, uterine cancer | I |
| Polymeric Nanoparticles | Camptothecin | Crlx101 | Solid tumours, small cell lung carcinoma, non-small cell lung carcinoma, lung neoplasms | I/II |
| | Docetaxel | Docetaxel-pnp | Solid tumours | I |
| | Docetaxel | BIND-014 | KRAS positive or squamous cell non-small cell lung cancer, metastatic castration-resistant prostate cancer | II |
| | Docetaxel | Cripec® docetaxel | Metastatic cancer, solid tumours | I |
| | Rapamycin | Sel-110 | Gout | I |
| | Paclitaxel | Genexol-pm | Locally advanced head and neck squamous cell carcinoma | II |
| Liposome | Doxorubicin | Caelyx®/doxil® | Breast cancer, multiple myeloma, ovarian cancer, sarcoma, kaposi | Approved |
| | Doxorubicin | Myocet® | Breast cancer | Approved |
| | Daunorubicin | Daunoxome® | Acute myeloid leukaemia | Approved |
| | Vincristine | Onco-tcs® | Acute lymphoblastic leukaemia | Approved |
| | Cisplatin | Lipoplatin | Pancreatic cancer | Approved |
| | Fus1 protein | Dotap:chol-fus1 | Lung cancer | I/II |
| | P53 dna | Sgt-53 | Glioblastoma | II |
| | Vincristine | Marqibo® | Leukaemia | II |
| | Sirna | Dcr-myc | Hepatocellular carcinoma | I/II |
| Lipid nanoparticle | Amphotericin b | Camb/mat2203 | Candidiasis, chronic mucocutaneous | II |
| | Paclitaxel | Abraxane ® | Breast cancer, pancreatic cancer | Approved |
| Albumin nanoparticles | Rapamycin | Nab-rapamycin | Stage IV uterine corpus cancer | II |
| | Hafnium oxide | Nbtxr3 (pep503) | Head and neck cancer, rectal cancer | I/II |
| Inorganic nanoparticles | Curcumin | Theracurmin | Schizophrenia, cognition, psychosis | I/II |
| Polysaccharide nanoparticles | Rsv fusion protein | Rsv f vaccine | Respiratory syncytial virus infections | III |
| Virus nanoparticles | Mir-16 microrna | Targomirs | Malignant pleural mesothelioma, non-small cell lung cancer | I |
| Bacterial nanoparticles | | | | |

[a]Results were obtained from clinicaltrials.gov website on 05/06/2016.

embedded/encapsulated within or conjugated to the surface. Both platforms are biocompatible and biodegradable. Other types of NPs already approved for clinical use are albumin-based and metallic NPs, such as iron oxide NPs for magnetic resonance imaging.

Their applications in cancer therapy for instance, takes advantage of the enhanced permeability and retention effect (EPR), a unique physiological feature in tumour tissues (Figure 3.1).[2] Tumour cells proliferate extremely rapidly; therefore, they have a high demand for blood supply to gain sufficient oxygen and nutrients. Their abnormal vascular network is known to be dilated, tortuous and lacking conventional hierarchy of blood vessels. One consequence of the vascular abnormalities is the heterogeneity of vessel distribution, which may cause hypoxia and necrosis in tumour tissue.[3] In addition, the rapid tumour growth also results in structural abnormalities. The tumour vasculature tends to be leaky and permeable due to endothelia defects and poor supporting of the adherent pericytes.[4] The high permeability of blood vessels enables nanocarriers to reach tumours from the bloodstream. The impaired lymphatic drainage system also prevents the clearance of those nanocarriers from the tumour, resulting in enhanced retention[5] (Figure 3.2a).

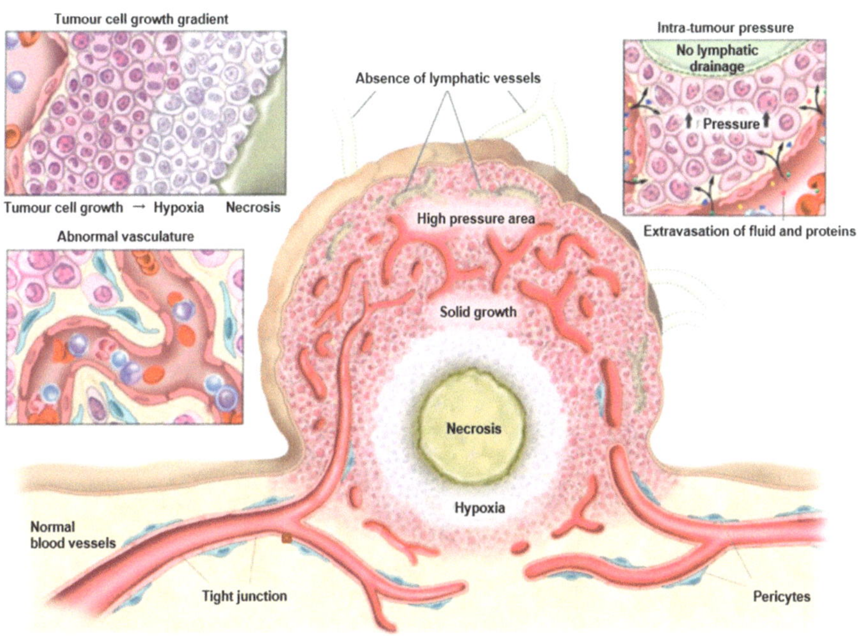


**Figure 3.1** Physiological characteristics of a solid tumour. Heterogeneous vasculature distribution leads to hypoxia and necrosis in tumour tissues. Extravasation of extracellular fluid and plasma protein results in high tumour interstitial pressure. Figure adapted with permission from Kobayashi *et al.*, *Theranostics*, 2014, **4**(1), 81–89, DOI:10.7150/thno.7193,[2] © Ivyspring International Publisher.

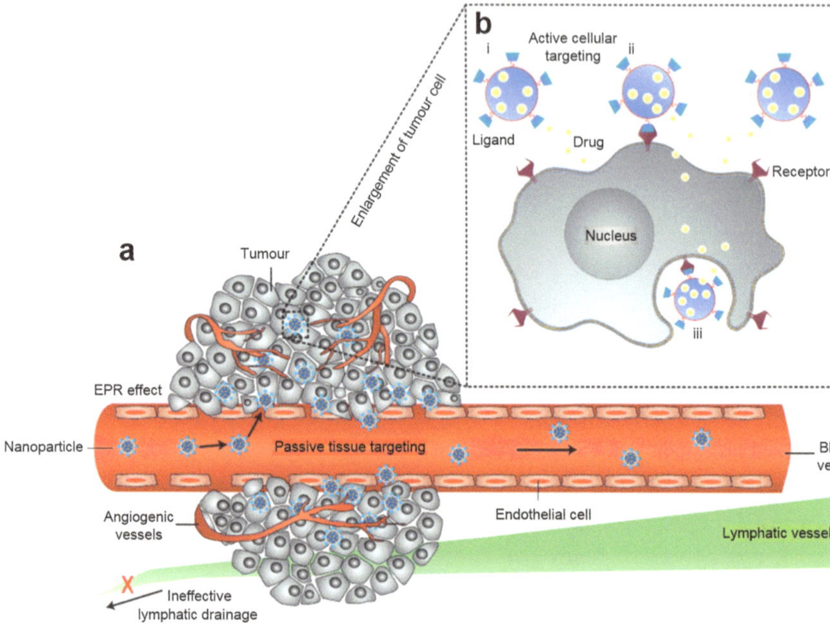


**Figure 3.2** Schematic illustration of nanocarrier accumulation in tumour *via* different targeting mechanisms. (a) Passive targeting of nanocarriers *via* EPR effect. (b) Active cellular targeting *via* ligand-receptor binding. Nanocarriers can: (i) release the drug in close proximity to the target cells; (ii) attach to the cell membrane and act as an extracellular sustained release drug depot; or (iii) internalise into the cell. Reprinted by permission from Macmillan Publishers Ltd: *Nature Nanotechnology* (ref. 6), Copyright 2007.

Moreover, the large functional surface area of nanocarriers can be chemically conjugated with targeting ligands that have high affinity to some receptors overexpressed on certain types of cancer. The active ligand-receptor interaction will consequently improve intracellular internalisation of these nanocarriers in tumours (Figure 3.2b). In addition to passive and active targeting, nanocarriers offer many advantages over free anti-cancer drugs. They can serve as a drug reservoir with increased payload and protect the drug from premature degradation or interaction with the biological environment.[6] Drugs encapsulated within the nanocarriers normally show sustained release patterns or stimuli-responsive (pH, temperature, ultrasound and magnetic field, *etc.*) release profiles.[7]

One of the drawbacks of early-developed formulations is the fast clearance from the blood circulation by the reticuloendothelial system, thereby leading to low tumour accumulation of the therapeutic agents. Several strategies have been investigated and one of the most commonly used approaches to prolong blood circulation is incorporating a synthetic polymer polyethylene glycol (PEG) to improve passive uptake in tumours. The most significant difference of the formulations in current clinical trials, compared to those have been approved clinically, is that most of them are with targeting moieties. These targeting ligands allow the specific accumulation of the nanocarriers/drugs in the target tissues *via* an active targeting approach. An example is BIND-014 (ClinicalTrials.gov Identifier NCT02283320), a polymeric NP formulation containing a chemotherapeutic agent, docetaxel (DTX). This formulation can target prostate-specific membrane antigens which express on prostate cancer cells and on the vasculature of many non-prostate solid tumours, including non-small cell lung cancer. The clinical trial data demonstrated higher accumulations in tumours and better efficacies than Taxotere®, a commercial Docetaxel-Tween 80 formulation.[8]

## 3.1.2 Magnetic Nanoparticles

### *3.1.2.1 Magnetic Properties of Magnetic Nanoparticles*

Magnetic nanoparticles (MNPs) have been frequently used in biomedical and biotechnology studies recently, including magnetic drug targeting, magnetic hyperthermia therapy, magnetic resonance imaging contrast enhancement, biosensors and magnetic cell separation.[9]

Naked MNPs are not stable in water at neutral pH or in physiological fluids and they tend to agglomerate and precipitate quickly. Therefore to stabilise the MNP either sterically or electrostatically, coating agents with good biocompatibility such as polysaccharides, polymers and lipids are commonly used.[10] In addition, these particles can be further functionalised by attaching carboxyl groups, amines, biotin, streptavidin, antibodies *etc.* for targeting purposes.[11] Meanwhile, the size and surface properties, *e.g.*, charge and hydrophobicity can be altered and consequently affect the MNPs' behaviour, such as colloidal stability and pharmacokinetic performance.

The underlying physics of biomedical applications of MNPs is based on the magnetisation concept. There are several forms of magnetism in nature. When the magnetic material is subjected to a cycle of positive and negative magnetic fields, the magnetisation shows a non-linear behaviour. The magnetisation can be described by a hysteresis loop which consists of three main parameters, namely saturation magnetisation (Ms), remnant magnetisation (Mr) and coercive field ($H_c$). Ms represents the limit value to which the curve ends within the high field, and is reached when all the magnetic moments in the material are aligned with the external field. The remnant magnetisation refers to the retained magnetisation at zero field from the saturation state. The coercive field represents the field needed in order to completely demagnetise the sample.[12] Based on these three parameters, the magnetism of magnetic materials can be classified into different types (Figure 3.3).

Iron oxide particles are widely used due to their good chemical stability, magnetic responsiveness and biocompatibility.[9] The most frequently used MNPs in biomedical applications are the superparamagnetic iron

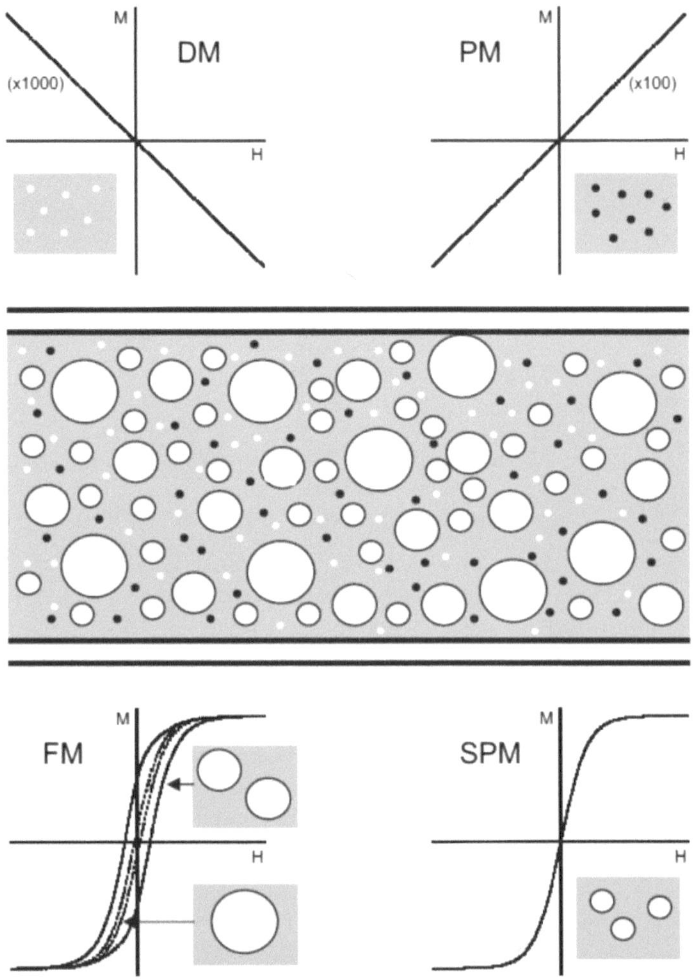


**Figure 3.3** Magnetic responses of different classes of magnetic material. M–H curves are shown for diamagnetic (DM) and paramagnetic (PM) biomaterials in the blood vessel, and for the ferromagnetic (FM) injected particles, where the response can be either multi-domain (- - - - in FM diagram), single-domain (—— in FM diagram) or superparamagnetic (SPM), depending on the size of the particle. Figure adapted from Pankhurst, *et al.*, *Journal of Physics D: Applied Physics,* 2003.[12] © IOP Publishing. Reused with permission. All rights reserved.

oxide nanoparticles (SPIONs).[13] The magnetic moment of the iron oxide nanoparticles usually has two orientations due to the magnetic anisotropy. At finite temperature, there is a finite probability for the magnetisation to flip and reverse its direction. The flipping frequency increases as the particle size increases.[12] The size required to achieve superparamagnetism can vary with core materials, which is typically with a size threshold below 30 nm for iron-based nanoparticles.[14] In the absence of an external magnetic field, the magnetic moments of individual crystallites are randomly oriented and compensate for each other resulting in zero net magnetisation with negligible remanence magnetisation and coercivity (Figure 3.4a).[15] When an external magnetic field is applied, the SPIONs behave similarly to the paramagnetic particles which will align along the magnetic field, leading to strong and rapid magnetic response to the magnetic field (Figure 3.4b). This superparamagnetic property is ideal for *in vivo* use since they can literally drag drug molecules to their target site in the body under the influence of an applied magnet field, with zero remnant magnetisation retained after the removal of the magnetic field. This is advantageous since the residual magnetism results in agglomeration and clearance from the circulation by phagocytic cells and may also pose greater risks for vascular embolism.[13]

The most commonly used SPIONs are magnetite ($Fe_3O_4$) and maghemite ($\gamma$-$Fe_2O_3$). They have attracted extensive interests due to many reasons: Firstly, magnetite and maghemite can be synthesised with controllable sizes ranging from a few nanometres up to tens of nanometres.[15] Secondly, the surface of MNPs is usually modified with coating agents which can be used for bioconjugation with biological molecules, to improve their biocompatibility and binding to biological entities.[13] Thirdly, these nanoparticles are highly responsive to a magnetic field. They can either be manipulated by a magnetic field gradient to a target region of the body such as the tumour, or oscillate in an alternative magnetic field and generate thermal energy, thus

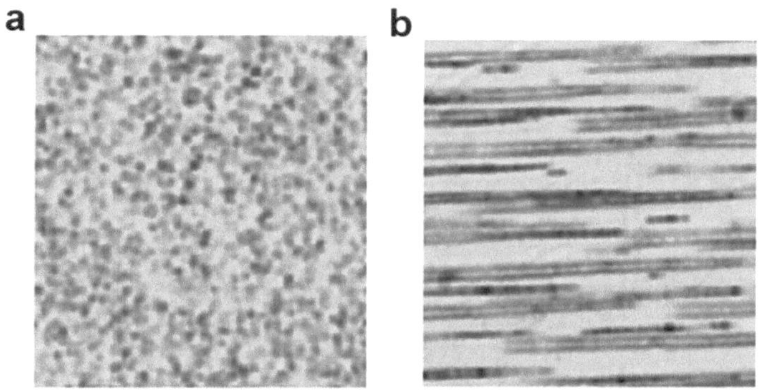


**Figure 3.4** Optical microscope images of the superparamagnetic behaviour of MNPs. SPIONs observed under an optical microscope in the absence and in the presence of a magnetic field, respectively. When no magnetic field is applied (a), SPIONs exhibit the same properties as non-magnetic ones. In contrast, in the presence of a magnetic field (b), a magnetic dipole is induced at the level of each droplet. Reprinted from *Journal of Magnetism and Magnetic Materials*, **250**, Preparation and characterization of narrow sized (o/w) magnetic emulsion, F. Montagne, O. Mondain-Monval, C. Pichot, H. Mozzanega and A. Elaıssari, 302–312, Copyright 2002, with permission from Elsevier.

used as hyperthermia agents. All these magnetic characteristics have made the MNPs promising possibilities in biomedicine.

### 3.1.2.2 *Classification of Magnetic Nanocarriers*

Depending on the synthesis and preparation procedure, the magnetic nano-formulations can be divided into MNPs and magnetic nanocarriers (MNCs). Initially, drugs are covalently attached to the SPION surface or adsorbed within the coating materials used to stabilise the SPION cores, *i.e.* starch, dextran or hydrophilic polymers.[16,17] Although these surface modifications improve the water solubility of MNPs, limited targeting efficiency has been shown due to fast MNP clearance *in vivo*. Another problem for direct use of MNPs in cancer theranostics is that MNPs have limited drug loading capacity. This is because the drug loading onto MNPs is by physical entrapment into the polymer coating matrix on the MNPs. If the polymer coating onto the inorganic particle itself is not stable, which is often the case, the drug can desorb once encountering the biological environment.[18–20] This premature release of the drug into the blood is a key issue.

By contrast, MNCs comprise a magnetic reservoir with the drug confined to an aqueous or oily compartment. Examples of MNCs are magnetoliposomes,[21,22] micelles,[23] polymeric nanospheres,[24] fullerenes,[25] Au shell nanoparticles,[26] chitosan nanoparticles,[27] lipo-polymersomes,[28] and 'nanoworms'.[29] Those MNCs either incorporated with hydrophilic (coated with hydrophilic polymers such as starch, dextran, and citric acid) or hydrophobic MNPs (normally coated with oleic acid or oleylamine), depending on the nature of the MNCs. Due to the stringent regulatory approval requirements, only a few MNP formulations have been approved for biomedical use, and not yet for therapeutic purposes but only as imaging agents.

One of the most studied is the magnetic liposome. Magnetic cationic liposomes (MCL) have proven successful by electrostatic interaction of their positive surface charge with an oppositely charged cell surface (particularly in the Kupffer cells in the liver), making them potential mediators for successful passive drug targeting.[30] Their positive surface also allows them to electronically interact with negatively charged nucleic acid, thereby improving gene delivery. MCLs also show the potential to generate heat at a locally targeted site when an alternating magnetic field is applied, which further outlines the benefits of MCLs in cancer hyperthermia therapy.[31]

Other widely explored types of nanocarriers are those based on polymer-associated magnetic nanoparticles where MNPs are either embedded or encapsulated in the polymers. The polymeric nanoparticles have been extensively studied due to the emergence of biodegradable and biocompatible polymers. A number of SPION-encapsulated polymeric MNCs have been reported by using a range of biocompatible and biodegradable polymers such as poly-e-caprolactone (PCL), poly(lactide) (PLA) or poly(lactide-*co*-glycolide)

(PLGA). Hydrophobic SPIONs are typically used due to the same hydrophobic nature as the polymers. These formulations encapsulate anti-cancer agents such as docetaxel,[32] paclitaxel[33] and cisplatin,[34] and have shown potential uses in MRI and targeted cancer therapy.

Encapsulation of MNPs into nanocarriers acts as an elegant way to modify the surface properties of MNPs and their pharmacokinetic profiles *in vivo*.

## 3.2 Applications of MNPs in Biomedical Imaging

### 3.2.1 Imaging MNPs

Various ways of imaging MNP-based agents can be performed by utilising the intrinsic magnetic properties of MNPs which respond differently when different forms of the magnetic fields are applied. These imaging modalities include magnetic resonance imaging (MRI) with a static magnetic field, magnetic particle imaging (MPI) with an alternating magnetic field or magneto-motive ultrasound imaging (MMUS) and magneto-photoacoustic imaging (MPA) with a pulsed magnetic field (Figure 3.5). Multi-modal imaging can also be performed when MNPs are combined with secondary imaging agents such as optical probes or radioisotopes.

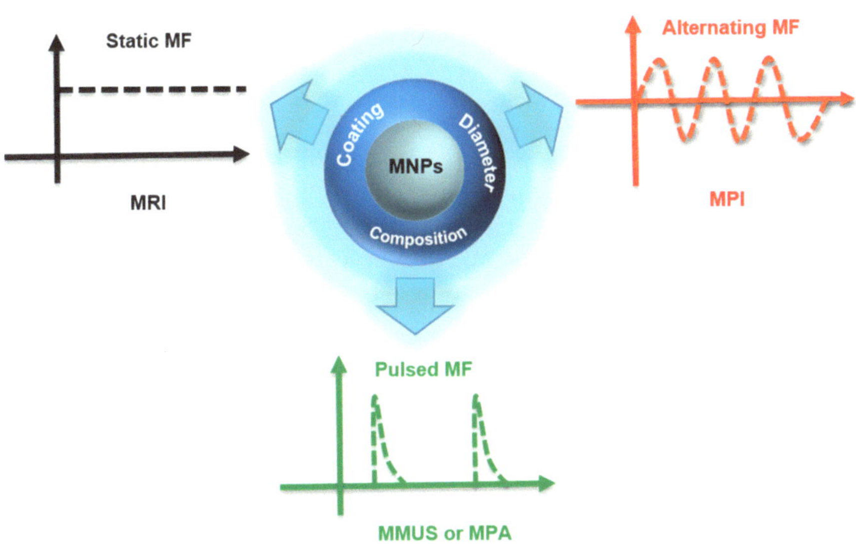


**Figure 3.5** The versatility of MNPs as a platform for various imaging modalities utilising their intrinsic magnetic properties. MNPs can respond differently under a static, alternating or pulsed magnetic field (MF) to be used for magnetic resonance imaging (MRI), magnetic particle imaging (MPI), or magneto-motive ultrasound imaging (MMUS) and magneto-photoacoustic imaging (MPA).

### 3.2.1.1 Magnetic Resonance Imaging

Magnetic resonance imaging is a widely used tool in pharmaceutical research and the clinical field due to its excellent soft tissue contrast property that provides three-dimensional (3D) anatomic images with high spatial resolution.[35] There is no limit on volume or depth of the analysed target. MRI is based on the magnetism property of protons that align themselves in a very large magnetic field. These protons originate from water molecules present in the body tissue. A radiofrequency generated at a particular frequency, known as the "resonance frequency," can flip the spin of a proton. When the electromagnetic field is turned off, the proton flips back to the original state, generating a radiofrequency signal (Figure 3.6). This process is called "relaxation". There are two relaxation processes: longitudinal relaxation ($T_1$) and transverse relaxation ($T_2$), which can be monitored to generate MR images. Both $T_1$ and $T_2$ can be shortened by the use of a magnetic contrast agent. The most commonly used contrast agents are gadolinium (Gd)-based nanoparticles which can enhance the positive signal on $T_1$-weighted images, and MNP-based contrast agents which decrease the signal intensity on $T_2$-weighted images.[12,35,36]

MNP-based contrast agents have been extensively investigated as MRI contrast agents with low toxicity and more potent contrast properties, and as carriers for targeted drug delivery in theranostic medicine studies. With the iron concentration increasing in the imaged tissue, the signal intensity gradually decreases. The MRI contrast properties of a contrast agent are usually characterised by its relaxivity ($r_1$ and $r_2$) which is proportional to the measured rate of relaxation ($T_1$ and $T_1$, respectively) over a range of contrast agent concentrations.[37–39] The MRI contrast relies on the differential uptake of MNPs by different tissues, which is mainly affected by the size of MNPs. MNPs with a size larger than ~30 nm usually rapidly accumulate in the liver.[12,40] Particles smaller than 10 nm can be used for vasculature imaging as they are observed to have a long blood circulation and end up with the collection of reticuloendothelial cells in the body, including lymph modes and bone marrow.[12]

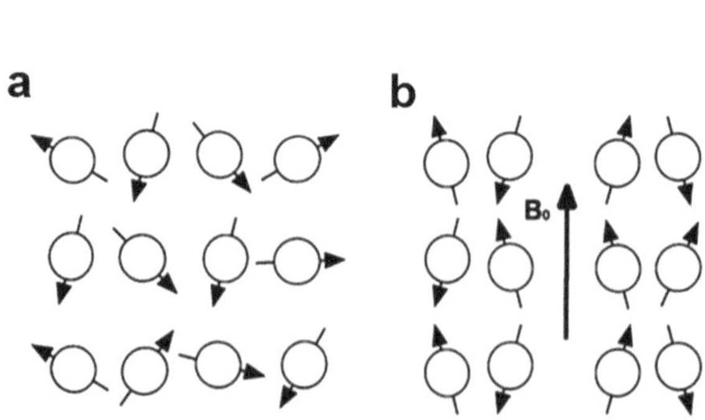


**Figure 3.6** Effect of an external magnetic field on the orientation of nuclear spin of water protons. Spinning protons in the absence (a) or presence (b) of an externally applied magnetic field $B_0$.

**Table 3.2** Clinically tested MNP-based $T_2$ MRI Contrast Agents.

| Brand name | Coating agents | Size (nm) | Application | References |
|---|---|---|---|---|
| Endorem® (Europe)/ Feridex® (US) | Dextran | 120–180 | Liver | 43 |
| Resovist® | Carboxydextran | 60 | Liver | 43 |
| Sinerem® | Dextran | 15–30 | Lymph node metastasis | 43 |
| Clariscan® | Starch | 11.9 | Liver and vasculature | 44 |
| Rienso® | Carboxylmethyl-dextran | 30 | Vasculature | 45 |
| Lumirem® (Europe)/ GastroMARK® (US) | Silicon | 300 | Liver, bowel | 46 |

Many $T_2$ contrast agents with different coating materials have been designed and clinically tested (Table 3.2). It should be noted that Clariscan® did not receive approval and many others have been withdrawn from the market.[41] Resovist® is commercially available but limited to Japan.[42]

Magnetic resonance imaging, without a contrast agent, is known to have low sensitivity since the image signals can be confused with other endogenous components such as calcification, air, haemorrhage and blood clots. Extensive research has been focused on developing the MRI probes with higher sensitivity by modifying their relaxation rates. Factors that can influence the relaxation rate of MNP-based MRI contrast agents include the particle size and composition.

**3.2.1.1.1 MNPs as $T_2$ Contrast Agents.** Particle size can greatly affect the magnetic properties of nanoparticles. Both $r_1$ and $r_2$ values decrease with size reduction of MNPs. Since $r_2$ values are proportional to the magnetisation of the MNPs, they are more strongly affected by size. A study conducted by Jun *et al.* investigated nanoscale size effects on the magnetism and induced MR signals using a series of synthetically controlled magnetic $Fe_3O_4$ nanocrystal with size regime from 4 to 12 nm.[47] The magnetisation values were strengthened when the particle size increased from 4 nm to 12 nm. This also induced a significant enhancement in MR $T_2$ signals. In addition, it has been shown that the SPION clustering can also enhance the $r_2$ values. In theory, the cluster itself may be considered as a large magnetised sphere and this helps improve the overall magnetic moment of SPIONs.[48] For example, several-fold higher $r_2$ relaxivity than Feridex® was observed when SPIONs were clustered as seen under the transmission electron microscopy (TEM).[22]

The other factor that may alter the MRI contrast properties of MNPs is their chemical composition. As mentioned earlier, the $r_2$ values are dependent on the Ms of MNPs. Replacing iron with other metals can be another approach to modulate MRI $T_2$ contrast. For example, the $T_2$ MR contrast properties have been tested on a series of metal-doped iron oxide nanoparticles, including manganese (Mn), cobalt (Co) and nickel (Ni). The results suggested

Mn-doped $MnFe_2O_4$ demonstrated higher $r_2$ relaxivity (by 27%) compared to $Fe_3O_4$, whereas the $T_2$ MR contrast reduced by 12% and 65% when Fe was replaced by Co ($CoFe_2O_4$) and Ni ($NiFe_2O_4$), respectively.[49] In another case, the effect of Zn-doping level in SPIONs on $T_2$ contrast was investigated and the highest $r_2$ relaxivity achieved ($Zn_{0.4}Fe_{2.6}O_4$) was 8-fold higher compared to Feridex®.[50]

**3.2.1.1.2 MNPs as $T_1$ Contrast Agents.** MNPs show both longitudinal and transverse relaxation processes. However, their influence on $T_2$ relaxation is much greater than on $T_1$ relaxation. The $T_1$ contrast effect can be relatively improved by reducing the nanoparticle size. For example, a new classification of SPIONs, known as ultrasmall superparamagnetic iron oxide nanoparticles (USPIOs) was reported and they are much smaller than most of the clinically tested SPIONs such as Feridex®. They exhibited enhanced $T_1$ contrast properties and relatively weaker $T_2$ contrast. Reducing the particle size down to 5 nm helped with $T_1$ shortening, but limited the $T_2$ contrast effect due to the low magnetisation. When these USPIOs are coated with hydrophilic polymers such as phosphate-functionalised PEG and PEG polymer containing bisphosphonate, which can effectively prevent the particles from aggregation, their $r_1$ relaxivity could be approximately 2-fold higher than that observed with the commercially available Magnevist® (Gd-DTPA). Their long blood circulation times allow their use as blood pool contrast agents to visualise blood vessels and vascularised organs with a high spatial definition.[51,52]

In addition, a number of studies also investigated the embedding of Gd into SPIONs near the surface of SPIONs in order to enhance the $T_1$ contrast effect. The embedded Gd species in SPIONs (~5 nm) led to great enhancements in $T_1$ contrast, with 28% and 138% increase compared to SPIONs alone (~5 nm) and Magnevist® (Gd-DTPA), respectively.[53]

### *3.2.1.2 Magnetic Particle Imaging*

The concept of MPI was firstly introduced by Gleich *et al.* in 2005.[54] MPI contrast agents, mainly SPIONs, rely on the non-linearity of the magnetisation curves of magnetic material and the fact that the particle magnetisation saturates at some magnetic field strength (Figure 3.7).[55] MPI is scalable and is potentially capable of full-body imaging in both small animals and human. It offers greater sensitivity, lower cost and quantitative detection compared to MRI.[56]

A gold standard MPI contrast agent is Resovist® and it is considered as a benchmark by which all potential tracers are evaluated.[57] However, it is suggested that less than 3% of particles actually contribute to the MPI.[54] Many studies have been focused on investigating the nanoparticle size effects on improving MPI sensitivity and resolution. In a study by Ferguson *et al.*, two commercial (Resovist® and Feridex®) and one synthesised SPIONs (coated with amphiphilic polymers) with a particle size of ~5 nm (both Resovist® and

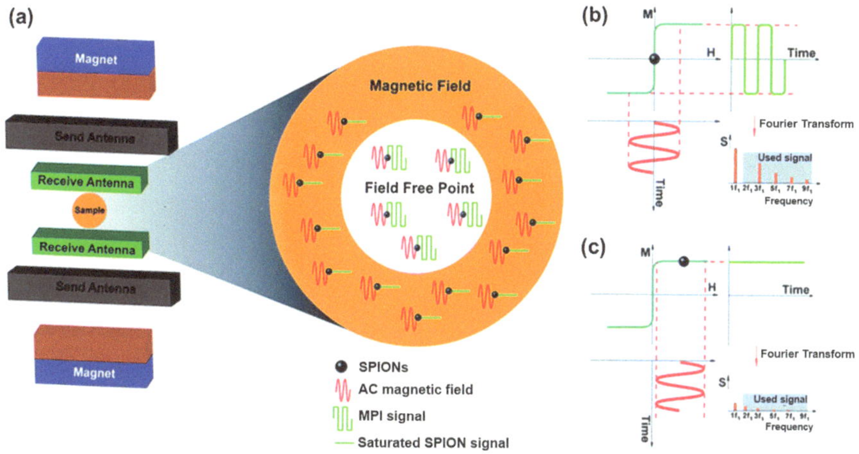


**Figure 3.7** Basic principle of magnetic particle imaging (MPI). (a) Schematic illustration of the MPI instrument and the MPI signal generation. (b) and (c) Response of magnetic particles to an external magnetic field. Adapted from ref. 55 with permission from The Royal Society of Chemistry.

Feridex®) and ~21 nm, respectively, were compared.[58] A size-dependence in MPI response was observed. MPI sensitivity and spatial resolution of SPIONs with large size were improved by 4 and 1.2 fold, respectively, compared to Resovist®.

Gleich and his co-workers further expanded the use of MPI for 3D high-resolution real-time imaging and applied this technique in imaging magnetic nanoparticle blood flow in a cardiovascular system *in vivo*.[59] The 3D real-time imaging offers higher spatial resolution and therefore fast dynamic blood flow information. This may be potentially used to measure micro-vascularisation and monitor blood supply for tumour staging and therapy assessment.[60]

### 3.2.1.3 *Magneto-motive Ultrasound Imaging*

Magneto-motive ultrasound imaging has been introduced recently as an ultrasound-based imaging modality capable of using SPIONs as contrast agents to expand the application of ultrasound (US) imaging.[61] MMUS benefits from the notable advantages of US imaging and it is considered to be a real-time, non-ionising, deeply penetrating and cost efficient method.[62] It also overcomes some limitations of US imaging such as the lack of ideal imaging contrast agents (*e.g.* small size, long blood circulation time and good stability).[63] In MMUS, an external high-strength pulsed magnetic field is applied to induce motion (displacement) within SPION-accumulated tissues and ultrasound is used to detect the induced magnetic displacement.[64] Since most tissue components are weakly diamagnetic materials, they do not significantly interact with the applied magnetic field. In contrast, SPIONs can be strongly magnetised due to their high magnetic susceptibility and move towards lower magnetic potential (Figure 3.8).[65] As listed in Table 3.3, the magnetic susceptibility of most iron oxide based nanoparticles such as Feridex® (*ca.* 53 emu $g^{-1}$) is at least seven orders of magnitude higher than the tissue components (*ca.* $-10^{-6}$ emu $g^{-1}$); therefore, the latter causes very low noise level, providing MMUS high sensitivity.[65]

Since higher magnetic susceptibility generates larger magneto-motive forces and consequently greater displacement and better signal-to-noise ratio, a number of studies have been carried out on the investigation of improving the magnetic properties of SPIONs for *in vivo* MMUS.[65] For example, Mehrmohammadi *et al.* reported that the magnetic susceptibility of SPIONs increased from 50 emu $g^{-1}$ to 161 emu $g^{-1}$ when iron oxide was doped with zinc (Zn).[63] This caused signal increments of 225–277% compared to

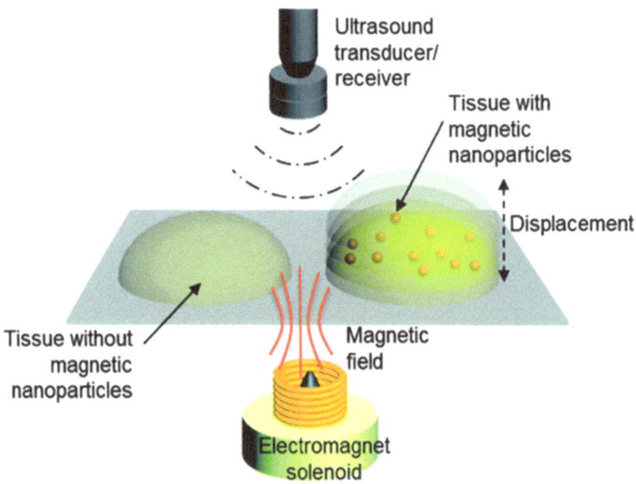


**Figure 3.8** Schematic illustration of magneto-motive ultrasound (MMUS). Reproduced from ref. 63 with permission from The Royal Society of Chemistry.

**Table 3.3** Mass susceptibility of various tissue constituents and MNPs.

| Materials | Magnetic susceptibility (emu $g^{-1}$) |
|---|---|
| Water (37 °C) | $-9.05 \times 10^{-6}$ |
| Red blood cell (deoxygenated) | $-5.96 \times 10^{-6}$ |
| Haemoglobin (deoxygenated) | $0.11 \times 10^{-6}$ |
| Whole blood (deoxygenated) | $-7.47 \times 10^{-6}$ |
| Air | $300 \times 10^{-6}$ |
| Human tissues | $-(11–7) \times 10^{-6}$ |
| Liver (severe iron overload) | 0 |
| Feridex® | 53.43 |
| 20 nm carbon-coated cobalt nanoparticles | 131.95 |
| Zinc-doped MNPs | 161 |

Feridex®. Owing to the high sensitivity, the developed Zn–Fe composite magneto-active imaging contrast agent was used to identify tumour lesions in nude mice bearing xenograft tumours.

#### *3.2.1.4 Magneto-photoacoustic Imaging*

Photoacoustic imaging (PA) is based on the absorption of light energy provided by short-pulsed lasers.[66] Imaging contrast is determined by optical absorption while its spatial resolution is determined by ultrasound (US).[66,67] However, background signals from endogenous chromophores such as haemoglobin, melanin and lipids can strongly interfere with the imaging and result in a high noise level. To address the limitation, PA imaging can be combined with MNPs as a new hybrid imaging technique, MPA, by integrating US, PA and MMUS imaging. Under the influence of a pulsed external magnetic field, the MNPs create a vibrating motion which distinguishes the PA signals from the background noise since most endogenous photo-absorbers are non-magnetic and remain still (Figure 3.9).[64] As a result, this highly improves the imaging contrast of the traditional PA imaging. For example, one study showed that Endorem® can be used as an MPA contrast agent in melanoma imaging. Due to enhanced contrast by the magneto-motion, Endorem® was able to overcome the interference from the melanin and photoacoustically detect the melanoma metastases in lymph nodes. The detection may allow for resection of lymph nodes in cancer surgery, thereby saving time and preventing the recall of patients for additional

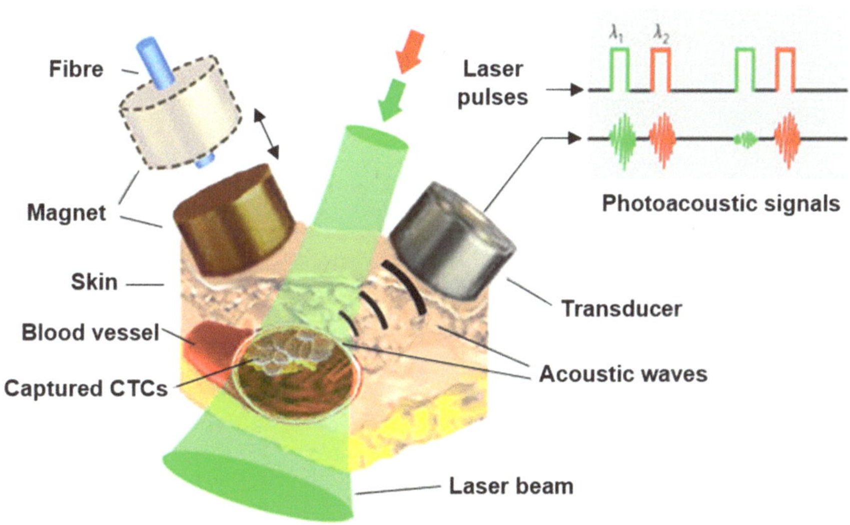


**Figure 3.9** Schematic illustration of Magneto-photoacoustic imaging (MPA). Reprinted by permission from Macmillan Publishers Ltd: *Nature Nanotechnology* (ref. 69), Copyright 2009.

operations.[68] In addition, similar to the MRI $T_2$ contrast, MNP clustering can also enhance PA signals and improve MPA imaging. Galanzha and co-workers used MNPs conjugated with an amino-terminal fragment of the urokinase plasminogen activator as a MPA imaging agent, utilising the intrinsic absorption of their $Fe_2O_3$ core to bind, capture and visualise circulating breast cancer cells under a magnet. They demonstrated that the PA signals from MNP-labelled cells were significantly amplified due to the MNP clustering in cells.[69]

#### 3.2.1.5 *Magnetic Nanoparticles in Multi-modal Imaging*

A number of other imaging techniques have been developed for cancer imaging, providing anatomic and physiological information.[70] Some of the representative imaging modalities in either preclinical research or clinical settings are MRI, positron emission tomography (PET), computed tomography (CT), SPECT, near infrared fluorescence (NIRF) imaging, US imaging.[64,70] The emerging field of multi-modal imaging using multi-labelled imaging probes promises to detect the same nanocarriers with multiple imaging techniques. It is well-known that each imaging technique has its own unique advantages and intrinsic limitations, such as insufficient sensitivity or spatial resolution, which makes it difficult to obtain accurate and reliable information at the disease site. The advantages and disadvantages of each imaging technique are summarised in Table 3.4.

Multi-modal imaging systems therefore are proposed to provide more accurate and reliable data than with a single imaging modality alone.[35,71,72]

**Table 3.4** Comparison and characteristics of different imaging modalities.

| Imaging modality | Spatial resolution (SR) | Temporal resolution (TR) | Sensitivity (mol $L^{-1}$) | Advantages | Disadvantages |
|---|---|---|---|---|---|
| NIRF | 3–5 mm | seconds to minutes | $10^{-9}$–$10^{-12}$ | Low cost, quick scanning, high sensitivity | Low penetration depth, low SR |
| MRI | 20–100 μm | minutes to hours | $10^{-3}$–$10^{-5}$ | High SR, high penetration depth | Low sensitivity, long requisition time |
| PET<br>SPECT | 1–2 mm | seconds to minutes | $10^{-11}$–$10^{-12}$<br>$10^{-10}$–$10^{-11}$ | High sensitivity, no limit of penetration depth | Low SR, ionising radiation, high cost |
| CT | 50–200 μm | minutes | - | High SR | Low contrast, ionising radiation |
| US | 50–100 μm | seconds to minutes | - | Low cost, high SR | Low contrast, low penetration depth |

Some representative multi-modal imaging studies using MNP-based nanoparticles are summarised in Table 3.5 and more details are discussed below.

**3.2.1.5.1 MNPs Combined with Fluorescence Probes for MRI-optical Imaging.** Optical imaging is the mostly used imaging modality for preclinical studies at the current stage. It allows the microscopic visualisation of the distribution of fluorescent labels at the subcellular scale in animals by using fluorescence or confocal microscopic techniques.[80,81] So far, the U.S. Food and Drug Administration (FDA) approved fluorophores for biomedical applications are indocyanine green (ICG) and fluorescein.

To combine MNPs with optical imaging, these organic fluorescent molecules are usually conjugated to the MNPs (Figure 3.10). The MNP surface may be further functionalised with targeting moieties and used as diagnostic agents that can specifically target cancer. Developing molecular imaging probes, such as antibodies, to target specific cancers has been a main focus of. However, antibodies can be immunogenic and their relatively large size usually cause slow excretion and thereby increase background signals.[82,83] A possible solution is to use the antibody derivatives such as peptides or single chain antibodies with smaller molecular weight as diagnostic agents, conjugated with near infrared (NIR)-fluorophores.[84] For example, Moore *et al.* a engineered cross-linked iron oxide (CLIO) nanoparticle with the functionalisation of a peptide (CDR3 V(H) sequence of ASM2 monoclonal antibody) and an NIR dye, cyanine 5.5. The CLIO nanoparticles were able to specifically bind to the underglycosylated mucin-1 (uMUC-1) tumour antigen due to the presence of a targeting peptide on the CLIO surface.[77,85] Meanwhile, dual-modal fluorescence/MR imaging was performed to detect and follow up the early progression of cancer.

Compared with the direct conjugation of fluorescent dyes to SPIONs, encapsulation of SPIONs and fluorescence dyes in a nanocarrier is more advantageous (Figure 3.10). Firstly, chemical conjugates, particularly amide bonds, are susceptible to cleavage *in vivo*, thereby hindering the long-term *in vivo* imaging.[86] Multiple agents can compete for the same functional groups on SPION surface and this, in turn, may affect the efficacy of both agents.[87] Secondly, the encapsulation protects fluorescent dyes from photochemical destruction and improves their stability. Saxena *et al.* reported that the photo-degradation, thermal-degradation and aqueous-instability of ICG can be significantly enhanced when it was encapsulated into polymeric nanoparticles. It was hypothesised that the polymeric envelope protected the ICG by isolating it from the surrounding environment.[88] As a result, encapsulation of the organic dyes in a nanocarrier can be a strategy to facilitate the long-term biodistribution of the optical probe *in vivo*. Foy *et al.* developed a nanocarrier consisting of hydrophobic oleic acid (OA) layered SPIONs which were further coated with a Pluronic block copolymer. Hydrophobic NIR dyes could partition into the OA layer while being protected by the Pluronic coating. This design allowed *in vivo* optical imaging up to 11 days and consequently enabled the assessment of biodistribution,

**Table 3.5** Representative studies of MNP-based multi-modal imaging. Abbreviations: UCNPs: upconversion nanoparticles; PASA: polyaspartic acid; HSA: human serum albumin; PLGA-PEG: poly (ethylene glycol)-b-poly(lactide-*co*-glycolide); MWCNTs: multiwalled carbon nanotubes; BPs: bisphosphonates; DOTA: 1,4,7,10-tetraazacyclododecane-1,4,7,10-tetraacetic acid; ICG: indocyanine green.

| Formulation | Imaging modality | Imaging probe | Finding | Ref. |
|---|---|---|---|---|
| Chitosan coated MNPs | $T_2$-weighted MRI<br>$T_1$-weighted MRI | SPIONs<br>Gd | $r_1 = 53.7$; $r_2 = 75.5$ mM$^{-1}$ s$^{-1}$; potential to be used as a $T_1$–$T_2$ dual-modality contrast agent | 73 |
| Carbon nanotubes | $T_2$-weighted MRI<br>SPECT | SPIONs<br>$^{99m}$Tc | $r_2 = 425$ mM$^{-1}$ s$^{-1}$; MWCNTs were radiolabelled with $^{99m}$Tc through BPs. Dual labelled MWCNTs enabled dual MRI and SPECT imaging *in vivo* | 74 |
| PEG coated MNPs | $T_1$-weighted MRI<br>SPECT | SPIONs<br>$^{99m}$Tc | $r_1 = 9.5$ mM$^{-1}$ s$^{-1}$; $r_2/r_1 = 2.97$; PEG-BP-SPIONs were radiolabelled with $^{99m}$Tc through BPs for quantitative MRI-SPECT imaging of blood vessels and vascular organs | 51 |
| UCNPs-MNPs | $T_2$-weighted MRI<br>Upconversion luminescence<br>Fluorescence | SPIONs<br>Lanthanide ions<br>Squaraine dye | $r_2 = 84$ mM$^{-1}$ s$^{-1}$; enabled magnetic targeted drug delivery and *in vivo* cancer cell tracking | 75 |
| PASP coated MNPs | $T_2$-weighted MRI<br>PET | SPIONs<br>$^{64}$Cu | $r_2 = 105.5$ mM$^{-1}$ s$^{-1}$; MNPs with integrin $\alpha v\beta 3$ and DOTA to allow simultaneous dual PET and MRI of tumour integrin expression | 76 |
| Crossed-linked MNPs | $T_2$-weighted MRI<br>Fluorescence | SPIONs<br>Cyanine 5.5 | $r_2 = 53.4$ mM$^{-1}$ s$^{-1}$; MNPs with EPPT1 peptide and used for specific uMUC-1 positive tumour imaging | 77 |
| HSA nanoparticles | $T_2$-weighted MRI<br>PET<br>Fluorescence | SPIONs<br>$^{64}$Cu<br>Cyanine 5.5 | $r_2 = 314$ mM$^{-1}$ s$^{-1}$; complementary information of quantification and distribution of MNPs in U87MG tumour-bearing mice | 78 |
| PLGA-PEG nanocapsules | $T_2$-weighted MRI<br>SPECT<br>Fluorescence | SPIONs<br>$^{111}$In<br>ICG | $r_2 = 404$ mM$^{-1}$ s$^{-1}$; complementary information provided on the spatial distribution of the nanocarriers in tumour with magnetic targeting | 79 |

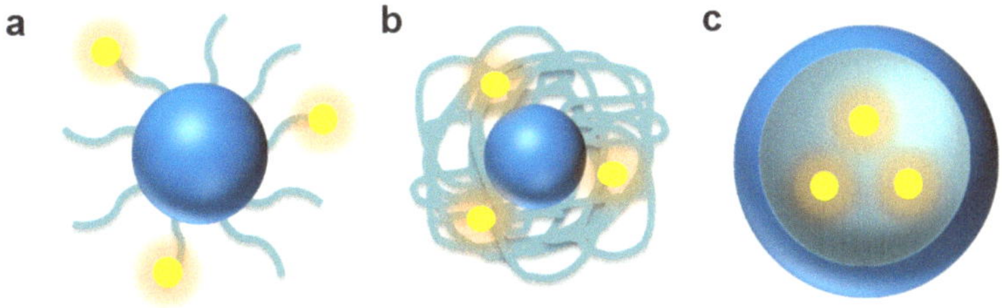


**Figure 3.10** Fluorescence labelling of MNPs. Optical probes can be combined with MNPs by chemical conjugation (a), physical embedding (b) and encapsulation (c).

tumour retention of the SPIONs and clearance when exposed to an externally applied magnetic field.[89]

In addition to these organic fluorophores, there has been intensive research carried out on investigating the combination of MNPs with inorganic optical imaging materials. Quantum dots (QDs) and upconversion nanoparticles (UCNPs) are good examples. QDs are inorganic fluorescent semiconductor nanocrystals.[90] They have drawn interest in biomedical research due to their unique optical properties including high quantum yield, high extinction coefficients, broad absorption with narrow symmetric photoluminescence spectra and high resistance to photo-bleaching and photo/chemical degradation.[91] To combine MNPs with QDs, Ang *et al.* developed a magnetic bifunctional system consisting of CdSe-QDs. SPIONs and CdSe-QDs were assembled as a single particle. The initially formed SPIONs served as seeds for CdSe-QD's nucleation and growth. CdSe-QDs started to grow as a single-component particle when the nucleation sites in MNP seeds were fully occupied. The emission peaks became broader with an increase in growth time due to the particle aggregation or heterodimer formation. The tuneable optical properties made CdSe-QDs a key component of versatile multimodal probes for specific biological applications with magnetic targeting.[92] Nevertheless, serious concerns have been raised in previous studies of QDs about the potential toxicity in biomedical imaging due to the use of heavy metal elements.[93] For example, an *in vitro* experiment has shown that QDs can internalise into cells and release free cadmium, thereby causing toxicity by generating free radical oxygen species.[94] An alternative to CdSe-QDs is the silicon-based QDs (Si-QDs). Silicon has better inherent biocompatibility and may ultimately be used to develop a image-guided magnetic drug delivery system for cancer while minimising collateral damage to normal tissues.[95]

UCNPs constitute a new class of optical imaging materials. Upconversion refers to the process that converts low-energy sources, generally in the NIR region, to higher energy light, usually in the ultra-violent and visible region, through sequential absorption of multiple photons. The process is different from non-linear multiphoton absorption in organic dyes and QDs, which involves simultaneous absorption of two or more photons through a virtual state.[93] This unique frequency conversion capability is usually unavailable for any endogenous and exogenous fluorophores in the body, thereby making UCNPs promising alternatives to the conventional optical probes. The virtually negligible auto-fluorescence background and strong tissue penetration ability of UCNPs provide high signal-to-noise ratio in biomedical imaging. In addition, their high resistance to photo-bleaching and photo-blinking is also a requisite for long-term repetitive imaging. UCNPs and MNPs can be incorporated together by encapsulating them into the polymer matrix. Fluorescent dyes and drugs can be further loaded inside the nanocomposite offering triple-modal imaging (upconversion luminescence/down-conversion fluorescence/magnetic resonance) and cancer therapy.[75] These nanocomposites also have the potential to be used for the delivery of NIR-light induced UCNP-based photodynamic therapy in combination with chemotherapy, providing a better therapeutic response.[26]

**3.2.1.5.2 MNPs Combined with Radioisotopes for MRI-PET/SPECT Imaging.** Nuclear imaging of cancer has been the main emphasis of nuclear medicine research and practice for decades. One of the main merits is its very low interference from the biological environment, which increases the imaging sensitivity and significantly reduces the radioisotope dose required, in a few orders of magnitude lower than optical probes. The selection of radioisotopes and approaches of radiolabelling of nanoparticles can vary remarkably depending on the physicochemical properties of nanoparticles. To justify the choice of radioisotope, several physical factors are discussed including the mode of decay, physical half-life, ease of labelling, cost, and availability.[96] Table 3.6 lists the clinically useful radioisotopes along with their nuclear characteristics. For diagnostic purpose, imaging agents are preferred to have relatively low cytotoxicity but with appropriate decay time. Radioisotopes used in SPECT imaging should have gamma energy emission in the range of 100–250 keV. Those that decay with gamma energies lower than this range produce too much scattered radiation, whereas gamma energies greater than 250 keV are difficult to collimate, resulting in poor imaging quality.[96] Short physical half-life is desirable for imaging to minimise the radiation exposure to subjects.[97] However, half-life of the employed radioisotope must match the time of imaging cycle to allow sufficient time for clinical image acquisition. Among those listed in Table 3.6, more than 80% of radiotracers used in diagnostic nuclear medicine are $^{99m}$Tc-based imaging agents due to the relatively low cost. Other commonly used gamma-emitters are $^{18}$F, $^{68}$Ga and $^{111}$In for PET or SPECT imaging.[98]

Nanoparticles can be radiolabelled through several approaches, dependent on the type of nanoparticles and radioisotopes. These approaches can be classified into three main categories: direct, chelator mediated and physical encapsulation labelling (Figure 3.11). $^{99m}$Tc, for instance, can directly interact with several functional groups on the nanoparticle surface.[104] These groups include hydroxyl (–OH), carboxylic (–COOH), and amino (–$NH_2$) groups, which serve as donor atoms/groups that bind $^{99m}$Tc.[96,105,106]

**Table 3.6** Commonly used radioisotopes for PET/SPECT imaging.

| Radioisotopes | Half-life (h) | Emission type | Energy (keV) | Imaging modality | Ref. |
|---|---|---|---|---|---|
| $^{18}F$ | 1.8 | β | 634 | PET | 99 |
| $^{64}Cu$ | 12.7 | β | 579, 653 | PET | 100 |
| $^{68}Ga$ | 1.1 | β | 770, 1890 | PET | 101 |
| $^{124}I$ | 100.3 | β, γ | 820, 1543, 2146 | PET | 102 |
| $^{99m}Tc$ | 6.0 | γ | 141 | SPECT | 103 |
| $^{111}In$ | 67.2 | γ 81.2%, β | 172, 245 | SPECT | 101 |
| $^{67}Ga$ | 79.2 | γ | 93, 185, 296 | SPECT | 101 |

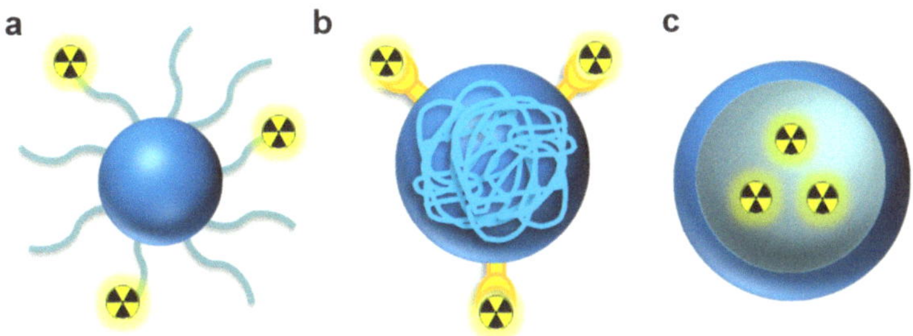


**Figure 3.11** Radiolabelling approaches of nanocarriers. Radioisotopes can be labelled to nanocarriers by direct labelling on the surface (a), labelling with bifunctional chelating agents (b) and encapsulation (c).

However, most of radioisotopes cannot be conjugated to nanoparticles directly. Therefore, bifunctional chelating agents (BFCAs), permitting covalent attachment and chelation, are required to facilitate the radio-complex formation.[107] In some cases, the radioisotopes can also be converted into lipophilic forms to be encapsulated into the nanoparticle cavity. $^{111}In$-oxine,[108] $^{99m}Tc$-hexamethylpropyleneamine oxime[109] and $^{111}In$-acetylacetonate[110] were encapsulated into the liposomes either by passing through the outer layer and being trapped inside or getting encapsulated within the lipid bilayers, respectively.

Dual-modality nuclear/MR imaging allows dynamic imaging of structure and function simultaneously, which directly provides information on pharmacokinetics and metabolism of the therapeutic agents and facilitates non-invasive monitoring of treatment.[111] Radioisotopes are usually attached to MNPs by chelating agents. Small molecules that simultaneously bind to radioisotopes and the surface of nanoparticles are normally used to avoid *trans*-metalation and *trans*-chelation induced by endogenous divalent ions and proteins, respectively. For example, Zolata *et al.* used the pentadeca-1(15),11,13-triene-3,6,9, -triacetic acid (PCTA) as a BFCA for radiolabelling of SPIONs with indium-111. Maleimide groups were introduced on SPIONs to the thiolated PCTA, allowing both MRI and SPECT imaging, to gain high resolution and quantitative biodistribution information.[112]

Another recent approach to substitute BFCAs was explored to radiolabel SPIONs using bisphosphonates (BPs). BPs are used to treat osteoporosis or cancer but they can also strongly bind the surface of metal or metal/oxide such as SPIONs for at least a month at pH 7. An ion chelating agent is usually conjugated to the SPIONs to bind radioisotopes. Examples of the radiolabelled BPs conjugates are $^{64}$Cu-dithiocarbamate-BP and $^{99m}$Tc-D-Penicillamine (DPA)-BP for PET and SPECT imaging, respectively.[113,114] Sandiford *et al.* synthesised USPIOs with long blood circulation time and low $r_2/r_1$ ratio. The radiolabelled BPs displaced molecules of the coating or bonded to gaps on the inorganic core. The results suggested that the changes were so small (quantitatively and qualitatively) that their colloidal or magnetic properties were not affected. As a result, the USPIOs-$^{99m}$Tc-DPA-BP were used as both a $T_1$ MR contrast agent for the visualisation of blood vessels/vascularised organs, and a SPECT imaging agent for quantitative biodistribution assessments.[51]

## 3.3 Applications of MNPs in Drug Delivery

### 3.3.1 Biocompatibility of MNPs

As mentioned earlier, SPIONs are favourable in biomedical application since they are less likely to cause any agglomeration in the body when the externally applied magnetic field is removed. Studies have shown that the clearance of SPIONs from the body was a slow process. Hanini *et al.* reported that the clearance of SPIONs was slow and the decline in serum iron level was observed at one week after SPION administration.[115] This could be due to the fact that iron can be incorporated into the body's iron storage and progressively into the red blood cells (haemoglobin).[116] Numerous studies have proven the safety of SPIONs when used as imaging agents or drug delivery systems *in vivo*. For example, no measurable median lethal dose ($LD_{50}$) of SPIONs was observed and no acute or sub-acute toxicity in rats and beagle dogs was found following an intravenous administration of Feridex® at 150 times higher than the recommended MR imaging dosage.[117]

Despite the overall agreement on SPION's safety use, the toxicity and eliminating rate of SPIONs vary depending on numerous parameters, such as the structural properties, dosage and administration routes of SPIONs.[118] For example, the clearance of two dextran-coated SPIONs with different sizes, Feridex® (120–180 nm) and Sinerem® (15–30 nm), differed from 16–21% after 84 days to 20% after 14 days, respectively, but none of them showed any obvious toxicities.[116,119] An *in vitro* study conducted by Berry *et al.* demonstrated that their synthesised dextran-coated SPIONs (10–15 nm), on the other hand, could alter cell morphology and even cause cell apoptosis due to their high cell uptake compared to uncoated SPIONs.[120] As a result, it is difficult to predict the toxicity of different types of SPIONs. Extensive experiments regarding the safety of SPIONs both *in vitro* and *in vivo* are still required at the current stage.

### 3.3.2 Obstacles and Challenges of MNPs in Drug Delivery Applications

Although many advances in magnetically targeted drug delivery have been achieved, the challenges faced by using MNPs still remain. Extending MNP blood circulation time has been recognised as one of the primary challenges in the development of MNPs.[121] To achieve long blood circulation, it is crucial to evade uptake by the RES organs. The physiochemical properties of MNPs including size, surface charge, shape and surface chemistry can determine the pharmacokinetic profiles, biodistribution, tumour accumulation and toxicity of MNPs. MNPs coated with hydrophilic polymers such as PEG or encapsulated into sophisticated nanocarriers such as liposomes and polymeric nanoparticles can efficiently improve their blood circulation time and alter their biodistribution profiles. Further conjugation of molecular targeting ligands to the surface of the MNPs or magnetic nanocarriers can be used in conjunction with magnetic targeting to improve cell internalisation of MNPs.[122–124]

Another limitation is related to the small size of MNPs, which is a prerequisite for superparamagnetism. The superparamagnetic property is preferable in biomedical application since it ensures no agglomeration of MNPs after the removal of the external magnetic field. However, the small size of MNPs leads to low magnetic response and weak magnetic forces. This, in turn, makes MNPs difficult to be guided by the magnetic field in the bloodstream while withstanding the drag of blood flow.[12] To solve this problem, one suggested method is to improve the magnetisation of the SPIONs.[125] This helps the SPIONs generate strong magnetic forces and overcome the blood convection. The other approach is to have a SPION reservoir within the nanocarriers and this will allow the clustering of SPIONs. The clusters of SPIONs can be considered as a large magnetic sphere and help improve the overall magnetic moment while maintaining the superparamagnetic properties of each individual SPION.[48,79,126–128]

In addition, the magnetic targeting studies have been limited by an insufficient understanding of particle transport in the body and a lack of real-time MNP sensing capabilities. Advanced multi-modal imaging techniques for image-guided magnetic drug delivery are required. Currently, the efficacy of magnetic drug targeting is usually assessed by MRI due to the intrinsic MR contrast property of MNPs.[129–131] This technique provides relatively high spatial resolution and loco-regional MNP distribution in the tumour. However, its high signal-to-noise ratio makes it a less accurate tool for quantification. To improve this, a few studies also used optical imaging in combination with MRI to obtain complementary information and only one study been has carried out on characterising magnetic drug targeting using triple-modal imaging (two optical imaging modalities and MRI).[75,79,89,122] Investigating the behaviours of MNPs within tumour vasculature is challenging since visualisation techniques with satisfactory resolution at vascular level are currently limited. Pioneering work has been performed to observe the dynamic magnetic targeting of MNPs in tumour vasculature.[126] Reversible changes in distribution of fluorescent MNPs was visualised upon application and removal of a magnetic field using fibered confocal fluorescence microscopy. This study provided important information to help elucidate how magnetic drug targeting overcomes blood convection.

Currently, many promising results have confirmed the magnetic targeting in pre-clinical animal studies. The next step, the extrapolation from animal models to human, is also challenging because of limited studies on MNP characterisation and pharmacokinetic profile, therapeutic efficacy and toxicity (direct and their degradation product-induced responses) assessment.[132] To scale up from small animals to man, many physiological parameters should also be taken into account, ranging from differences in body weight, blood hydrodynamics such as blood flow velocity and vessel diameter, tumour vessel density, permeability, tumour volume/location and tissue diffusion *etc.*[12,133] With the aid of the retrospective design of experiment (DoE) approach, it has been found that key attributes of tumour heterogeneity such as tumour microvessel density, permeability and diameter, play different roles but all have great influence on passive and magnetic targeting of MNPs towards tumour.[128] Another study has demonstrated that magnetic targeting efficiency can be mathematically extrapolated to predict the success in pre-clinical or clinical settings by accounting the factors such as the diameters of MNPs, tumour-to-magnet distance and blood vessel characteristics *etc.*[127] More work is required for better clinical translation from experimental studies in the field of magnetic drug targeting.

## Acknowledgements

The authors would like to thank funding from the Biotechnology and Biological Sciences Research Council (BB/J008656/1) and Worldwide Cancer Research (12-1054). J.B. acknowledges funding from the King's-China Scholarship Council (CSC).

## References

1. M. R. Mozafari, Liposomes: an overview of manufacturing techniques, *Cell. Mol. Biol. Lett.*, 2005, **10**(4), 711.
2. H. Kobayashi, R. Watanabe and P. L. Choyke, Improving conventional enhanced permeability and retention (EPR) effects; what is the appropriate target? *Theranostics*, 2014, **4**(1), 81.
3. M. Konerding, A. Miodonski and A. Lametschwandtner, Microvascular corrosion casting in the study of tumor vascularity: a review, *Scanning Microsc.*, 1994, **9**(4), 1233–1243, discussion 43–44.
4. S. K. Hobbs, W. L. Monsky, F. Yuan, W. G. Roberts, L. Griffith and V. P. Torchilin, *et al.*, Regulation of transport pathways in tumor vessels: role of tumor type and microenvironment, *Proc. Natl. Acad. Sci.*, 1998, **95**(8), 4607–4612.

5. H. Maeda, The enhanced permeability and retention (EPR) effect in tumor vasculature: the key role of tumor-selective macromolecular drug targeting, *Adv. Enzyme Regul.*, 2001, **41**(1), 189–207.
6. D. Peer, J. M. Karp, S. Hong, O. C. Farokhzad, R. Margalit and R. Langer, Nanocarriers as an emerging platform for cancer therapy, *Nat. Nanotechnol.*, 2007, **2**(12), 751–760.
7. T. Sun, Y. S. Zhang, B. Pang, D. C. Hyun, M. Yang and Y. Xia, Engineered nanoparticles for drug delivery in cancer therapy, *Angew. Chem., Int. Ed.*, 2014, **53**(46), 12320–12364.
8. R. Natale, M. Socinski, L. Hart, O. Lipatov, D. Spigel and B. Gershenhorn, *et al.*, 41 Clinical activity of BIND-014 (docetaxel nanoparticles for injectable suspension) as second-line therapy in patients (pts) with stage III/IV non-small cell lung cancer, *Eur. J. Cancer*, 2014, **50**, 19.
9. R. Jurgons, C. Seliger, A. Hilpert, L. Trahms, S. Odenbach and C. Alexiou, Drug loaded magnetic nanoparticles for cancer therapy, *J. Phys.: Condens. Matter*, 2006, **18**(38), S2893.
10. W. Wu, Q. He and C. Jiang, Magnetic iron oxide nanoparticles: synthesis and surface functionalization strategies, *Nanoscale Res. Lett.*, 2008, **3**(11), 397–415.
11. J. Dobson, Magnetic nanoparticles for drug delivery, *Drug Dev. Res.*, 2006, **67**(1), 55–60.
12. Q. A. Pankhurst, J. Connolly, S. Jones and J. Dobson, Applications of magnetic nanoparticles in biomedicine, *J. Phys. D: Appl. Phys.*, 2003, **36**(13), R167.
13. M. Arruebo, R. Fernández-Pacheco, M. R. Ibarra and J. Santamaría, Magnetic nanoparticles for drug delivery, *Nano Today*, 2007, **2**(3), 22–32.
14. S. A. Wahajuddin, Superparamagnetic iron oxide nanoparticles: magnetic nanoplatforms as drug carriers, *Int. J. Nanomed.*, 2012, **7**, 3445.
15. F. Montagne, O. Mondain-Monval, C. Pichot, H. Mozzanega and A. Elaıssari, Preparation and characterization of narrow sized (o/w) magnetic emulsion, *J. Magn. Magn. Mater.*, 2002, **250**, 302–312.
16. A. J. Cole, A. E. David, J. Wang, C. J. Galbán and V. C. Yang, Magnetic brain tumor targeting and biodistribution of long-circulating PEG-modified, cross-linked starch-coated iron oxide nanoparticles, *Biomaterials*, 2011, **32**(26), 6291–6301.
17. U. Häfeli, *Scientific and Clinical Applications of Magnetic Carriers*, Springer, 1997.
18. R. Panda, N. Gajbhiye and G. Balaji, Magnetic properties of interacting single domain $Fe_3O_4$ particles, *J. Alloys Compd.*, 2001, **326**(1), 50–53.
19. L. Wang, K.-G. Neoh, E.-T. Kang, B. Shuter and S.-C. Wang, Biodegradable magnetic-fluorescent magnetite/poly (dl-lactic acid-co-α,β-malic acid) composite nanoparticles for stem cell labeling, *Biomaterials*, 2010, **31**(13), 3502–3511.
20. P. A. Jarzyna, T. Skajaa, A. Gianella, D. P. Cormode, D. D. Samber and S. D. Dickson, *et al.*, Iron oxide core oil-in-water emulsions as a multifunctional nanoparticle platform for tumor targeting and imaging, *Biomaterials*, 2009, **30**(36), 6947–6954.

21. M.-S. Martina, J.-P. Fortin, L. Fournier, C. Ménager, F. Gazeau and O. Clément, *et al.*, Magnetic targeting of rhodamine-labeled superparamagnetic liposomes to solid tumors: *in vivo* tracking by fibered confocal fluorescence microscopy, *Mol. Imaging*, 2006, **6**(2), 140–146.
22. G. Mikhaylov, U. Mikac, A. A. Magaeva, V. I. Itin, E. P. Naiden and I. Psakhye, *et al.*, Ferri-liposomes as an MRI-visible drug-delivery system for targeting tumours and their microenvironment, *Nat. Nanotechnol.*, 2011, **6**(9), 594–602.
23. C. Huang, Z. Tang, Y. Zhou, X. Zhou, Y. Jin and D. Li, *et al.*, Magnetic micelles as a potential platform for dual targeted drug delivery in cancer therapy, *Int. J. Pharm.*, 2012, **429**(1), 113–122.
24. M. Licciardi, C. Scialabba, C. Fiorica, G. Cavallaro, G. Cassata and G. Giammona, Polymeric nano-carriers for magnetic targeted drug delivery: preparation, characterization and *in vitro* and *in vivo* evaluation, *Mol. Pharm.*, 2013.
25. J. Shi, X. Yu, L. Wang, Y. Liu, J. Gao and J. Zhang, *et al.*, PEGylated Fullerene/Iron Oxide nanocomposites for photodynamic therapy, targeted drug delivery and MR imaging, *Biomaterials*, 2013, **34**(37), 9666–9677.
26. L. Cheng, K. Yang, Y. Li, X. Zeng, M. Shao and S.-T. Lee, *et al.*, Multifunctional nanoparticles for upconversion luminescence/MR multimodal imaging and magnetically targeted photothermal therapy, *Biomaterials*, 2012, **33**(7), 2215–2222.
27. Y. Sun, Z.-l. Chen, X.-x. Yang, P. Huang, X.-p. Zhou and X.-x. Du, Magnetic chitosan nanoparticles as a drug delivery system for targeting photodynamic therapy, *Nanotechnology*, 2009, **20**(13), 135102.
28. S. H. Hu, B. J. Liao, C. S. Chiang, P. J. Chen, I. W. Chen and S. Y. Chen, Core-shell nanocapsules stabilized by single-component polymer and nanoparticles for magneto-chemotherapy/hyperthermia with multiple drugs, *Adv. Mater.*, 2012, **24**(27), 3627–3632.
29. J. H. Park, G. von Maltzahn, L. Zhang, M. P. Schwartz, E. Ruoslahti and S. N. Bhatia, *et al.*, Magnetic iron oxide nanoworms for tumor targeting and imaging, *Adv. Mater.*, 2008, **20**(9), 1630–1635.
30. J.-P. Fortin-Ripoche, M. S. Martina, F. Gazeau, C. Ménager, C. Wilhelm and J.-C. Bacri, *et al.*, Magnetic targeting of magnetoliposomes to solid tumors with MR imaging monitoring in mice: feasibility 1, *Radiology*, 2006, **239**(2), 415–424.
31. M. Gonzales and K. M. Krishnan, Synthesis of magnetoliposomes with monodisperse iron oxide nanocrystal cores for hyperthermia, *J. Magn. Magn. Mater.*, 2005, **293**(1), 265–270.
32. Y. Ling, K. Wei, Y. Luo, X. Gao and S. Zhong, Dual docetaxel/superparamagnetic iron oxide loaded nanoparticles for both targeting magnetic resonance imaging and cancer therapy, *Biomaterials*, 2011, **32**(29), 7139–7150.
33. N. Schleich, P. Sibret, P. Danhier, B. Ucakar, S. Laurent and R. N. Muller, *et al.*, Dual anticancer drug/superparamagnetic iron oxide-loaded PLGA-based nanoparticles for cancer therapy and magnetic resonance imaging, *Int. J. Pharm.*, 2013, **447**(1), 94–101.

34. M. Ashjari, S. Khoee and A. R. Mahdavian, Controlling the morphology and surface property of magnetic/cisplatin-loaded nanocapsules *via* w/o/w double emulsion method, *Colloids Surf., A*, 2012, **408**, 87–96.
35. R. Thomas, I.-K. Park and Y. Y. Jeong, Magnetic iron oxide nanoparticles for multimodal imaging and therapy of cancer, *Int. J. Mol. Sci.*, 2013, **14**(8), 15910–15930.
36. J. K. Oh and J. M. Park, Iron oxide-based superparamagnetic polymeric nanomaterials: design, preparation, and biomedical application, *Prog. Polym. Sci.*, 2011, **36**(1), 168–189.
37. S. Laurent, L. V. Elst and R. N. Muller, Superparamagnetic Iron Oxide nanoparticles for MRI, *The Chemistry of Contrast Agents in Medical Magnetic Resonance Imaging*, John Wiley & Sons, Ltd, 2013, pp. 427–447.
38. D. L. Thorek, A. K. Chen, J. Czupryna and A. Tsourkas, Superparamagnetic iron oxide nanoparticle probes for molecular imaging, *Ann. Biomed. Eng.*, 2006, **34**(1), 23–38.
39. R. Qiao, C. Yang and M. Gao, Superparamagnetic iron oxide nanoparticles: from preparations to *in vivo* MRI applications, *J. Mater. Chem.*, 2009, **19**(35), 6274–6293.
40. S. Laurent, S. Dutz, U. O. Häfeli and M. Mahmoudi, Magnetic fluid hyperthermia: focus on superparamagnetic iron oxide nanoparticles, *Adv. Colloid Interface Sci.*, 2011, **166**(1), 8–23.
41. Y.-X. J. Wang, Current status of superparamagnetic iron oxide contrast agents for liver magnetic resonance imaging, *World J. Gastroenterol.*, 2015, **21**(47), 13400–13402.
42. S. Maurea, P. P. Mainenti, A. Tambasco, M. Imbriaco, C. Mollica and E. Laccetti, *et al.*, Diagnostic accuracy of MR imaging to identify and characterize focal liver lesions: comparison between gadolinium and superparamagnetic iron oxide contrast media, *Quant. Imaging Med. Surg.*, 2014, **4**(3), 181–189.
43. Y.-X. J. Wang, Superparamagnetic iron oxide based MRI contrast agents: current status of clinical application, *Quant. Imaging Med. Surg.*, 2011, **1**(1), 35.
44. C. Klein, E. Nagel, B. Schnackenburg, A. Bornstedt, S. Schalla and V. Hoffmann, *et al.*, The intravascular contrast agent clariscan (NC 100150 injection) for 3D MR coronary angiography in patients with coronary artery disease, *MAGMA*, 2000, **11**(1–2), 65–67.
45. I. Berry, S. Benderbous, J. P. Ranjeva, D. Gracia-Meavilla, C. Manelfe and D. Le Bihan, Contribution of Sinerem® used as blood-pool contrast agent: detection of cerebral blood volume changes during apnea in the rabbit, *Magn. Reson. Med.*, 1996, **36**(3), 415–419.
46. B. Bonnemain, Superparamagnetic agents in magnetic resonance imaging: physicochemical characteristics and clinical applications a review, *J. Drug Targeting*, 1998, **6**(3), 167–174.
47. Y.-w. Jun, Y.-M. Huh, J.-s. Choi, J.-H. Lee, H.-T. Song and S. Kim, *et al.*, Nanoscale size effect of magnetic nanocrystals and their utilization for cancer diagnosis *via* magnetic resonance imaging, *J. Am. Chem. Soc.*, 2005, **127**(16), 5732–5733.

48. A. Roch, Y. Gossuin, R. N. Muller and P. Gillis, Superparamagnetic colloid suspensions: water magnetic relaxation and clustering, *J. Magn. Magn. Mater.*, 2005, **293**(1), 532–539.
49. J.-H. Lee, Y.-M. Huh, Y.-w. Jun, J.-w. Seo, J.-t. Jang and H.-T. Song, *et al.*, Artificially engineered magnetic nanoparticles for ultra-sensitive molecular imaging, *Nat. Med.*, 2007, **13**(1), 95–99.
50. J. Jt, H. Nah, J. H. Lee, S. H. Moon, M. G. Kim and J. Cheon, Critical enhancements of MRI contrast and hyperthermic effects by dopant-controlled magnetic nanoparticles, *Angew. Chem., Int. Ed.*, 2009, **121**(7), 1260–1264.
51. L. Sandiford, A. Phinikaridou, A. Protti, L. K. Meszaros, X. Cui and Y. Yan, *et al.*, Bisphosphonate-anchored PEGylation and radiolabeling of superparamagnetic iron oxide: long-circulating nanoparticles for *in vivo* multimodal (T1 MRI-SPECT) imaging, *ACS Nano*, 2012, 7(1), 500–512.
52. U. I. Tromsdorf, O. T. Bruns, S. C. Salmen, U. Beisiegel and H. Weller, A highly effective, nontoxic T1 MR contrast agent based on ultrasmall PEGylated iron oxide nanoparticles, *Nano Lett.*, 2009, **9**(12), 4434–4440.
53. Z. Zhou, L. Wang, X. Chi, J. Bao, L. Yang and W. Zhao, *et al.*, Engineered iron-oxide-based nanoparticles as enhanced T1 contrast agents for efficient tumor imaging, *ACS Nano*, 2013, **7**(4), 3287–3296.
54. B. Gleich and J. Weizenecker, Tomographic imaging using the nonlinear response of magnetic particles, *Nature*, 2005, **435**(7046), 1214–1217.
55. M. H. Pablico-Lansigan, S. F. Situ and A. C. S. Samia, Magnetic particle imaging: advancements and perspectives for real-time *in vivo* monitoring and image-guided therapy, *Nanoscale*, 2013, **5**(10), 4040–4055.
56. R. M. Ferguson, K. R. Minard and K. M. Krishnan, Optimization of nanoparticle core size for magnetic particle imaging, *J. Magn. Magn. Mater.*, 2009, **321**(10), 1548–1551.
57. K. M. Krishnan, Biomedical nanomagnetics: a spin through possibilities in imaging, diagnostics, and therapy, *IEEE Trans. Magn.*, 2010, **46**(7), 2523–2558.
58. R. M. Ferguson, A. P. Khandhar and K. M. Krishnan, Tracer design for magnetic particle imaging, *J. Appl. Phys.*, 2012, **111**(7), 07B318.
59. J. Weizenecker, B. Gleich, J. Rahmer, H. Dahnke and J. Borgert, Three-dimensional real-time *in vivo* magnetic particle imaging, *Phys. Med. Biol.*, 2009, **54**(5), L1.
60. Y. Du, P. T. Lai, C. H. Leung and P. W. Pong, Design of superparamagnetic nanoparticles for magnetic particle imaging (MPI), *Int. J. Mol. Sci.*, 2013, **14**(9), 18682–18710.
61. M. Mehrmohammadi, M. Qu, L. L. Ma, D. K. Romanovicz, K. P. Johnston and K. V. Sokolov, *et al.*, Pulsed magneto-motive ultrasound imaging to detect intracellular accumulation of magnetic nanoparticles, *Nanotechnology*, 2011, **22**(41), 415105.
62. M. Mehrmohammadi, K. Yoon, M. Qu, K. Johnston and S. Emelianov, Enhanced pulsed magneto-motive ultrasound imaging using superparamagnetic nanoclusters, *Nanotechnology*, 2010, **22**(4), 045502.

63. M. Mehrmohammadi, T.-H. Shin, M. Qu, P. Kruizinga, R. L. Truby and J.-H. Lee, *et al.*, *In vivo* pulsed magneto-motive ultrasound imaging using high-performance magnetoactive contrast nanoagents, *Nanoscale*, 2013, **5**(22), 11179–11186.
64. T.-H. Shin, Y. Choi, S. Kim and J. Cheon, Recent advances in magnetic nanoparticle-based multi-modal imaging, *Chem. Soc. Rev.*, 2015, **44**(14), 4501–4516.
65. M. Mehrmohammadi, J. Oh, S. Mallidi and S. Y. Emelianov, Pulsed magneto-motive ultrasound imaging using ultrasmall magnetic nanoprobes, *Mol. Imaging*, 2011, **10**(2), 102.
66. M. Qu, M. Mehrmohammadi, R. Truby, I. Graf, K. Homan and S. Emelianov, Contrast-enhanced magneto-photo-acoustic imaging *in vivo* using dual-contrast nanoparticles, *Photoacoustics*, 2014, **2**(2), 55–62.
67. M. Qu, S. Mallidi, M. Mehrmohammadi, R. Truby, K. Homan and P. Joshi, *et al.*, Magneto-photo-acoustic imaging, *Biomed. Opt. Express*, 2011, **2**(2), 385–396.
68. D. J. Grootendorst, J. Jose, R. M. Fratila, M. Visscher, A. H. Velders and B. Ten Haken, *et al.*, Evaluation of superparamagnetic iron oxide nanoparticles (Endorem®) as a photoacoustic contrast agent for intra-operative nodal staging, *Contrast Media Mol. Imaging*, 2013, **8**(1), 83–91.
69. E. I. Galanzha, E. V. Shashkov, T. Kelly, J.-W. Kim, L. Yang and V. P. Zharov, *In vivo* magnetic enrichment and multiplex photoacoustic detection of circulating tumour cells, *Nat. Nanotechnol.*, 2009, **4**(12), 855–860.
70. O. L. Gobbo, K. Sjaastad, M. W. Radomski, Y. Volkov and A. Prina-Mello, Magnetic nanoparticles in cancer theranostics, *Theranostics*, 2015, **5**(11), 1249.
71. J. Tian, J. Bai, X. P. Yan, S. Bao, Y. Li and W. Liang, *et al.*, Multimodality molecular imaging, *IEEE Eng. Med. Biol. Mag.*, 2008, **27**(5), 48–57.
72. I. Brigger, C. Dubernet and P. Couvreur, Nanoparticles in cancer therapy and diagnosis, *Adv. Drug Delivery Rev.*, 2002, **54**(5), 631–651.
73. A. Szpak, S. Fiejdasz, W. Prendota, T. Strączek, C. Kapusta and J. Szmyd, *et al.*, T1–T2 Dual-modal MRI contrast agents based on superparamagnetic iron oxide nanoparticles with surface attached gadolinium complexes, *J. Nanopart. Res.*, 2014, **16**(11), 1–11.
74. J. T. W. Wang, L. Cabana, M. Bourgognon, H. Kafa, A. Protti and K. Venner, *et al.*, Magnetically decorated multiwalled carbon nanotubes as dual MRI and SPECT contrast agents, *Adv. Funct. Mater.*, 2014, **24**(13), 1880–1894.
75. H. Xu, L. Cheng, C. Wang, X. Ma, Y. Li and Z. Liu, Polymer encapsulated upconversion nanoparticle/iron oxide nanocomposites for multimodal imaging and magnetic targeted drug delivery, *Biomaterials*, 2011, **32**(35), 9364–9373.
76. H.-Y. Lee, Z. Li, K. Chen, A. R. Hsu, C. Xu and J. Xie, *et al.*, PET/MRI dual-modality tumor imaging using arginine-glycine-aspartic (RGD)–conjugated radiolabeled iron oxide nanoparticles, *J. Nucl. Med.*, 2008, **49**(8), 1371–1379.

77. A. Moore, Z. Medarova, A. Potthast and G. Dai, *In vivo* targeting of underglycosylated MUC-1 tumor antigen using a multimodal imaging probe, *Cancer Res.*, 2004, **64**(5), 1821–1827.
78. J. Xie, K. Chen, J. Huang, S. Lee, J. Wang and J. Gao, *et al.*, PET/NIRF/MRI triple functional iron oxide nanoparticles, *Biomaterials*, 2010, **31**(11), 3016–3022.
79. J. Bai, J. T.-W. Wang, N. Rubio, A. Protti, H. Heidari and R. Elgogary, *et al.*, Triple-modal imaging of magnetically-targeted nanocapsules in solid tumours *in vivo*, *Theranostics*, 2016, **6**(3), 15.
80. R. M. Hoffman, The multiple uses of fluorescent proteins to visualize cancer *in vivo*, *Nat. Rev. Cancer*, 2005, **5**(10), 796–806.
81. N. C. Shaner, P. A. Steinbach and R. Y. Tsien, A guide to choosing fluorescent proteins, *Nat. Methods*, 2005, **2**(12), 905–909.
82. H. Tada, H. Higuchi, T. M. Wanatabe and N. Ohuchi, *In vivo* real-time tracking of single quantum dots conjugated with monoclonal anti-HER2 antibody in tumors of mice, *Cancer Res.*, 2007, **67**(3), 1138–1144.
83. X. Chen and S. Wong, *Cancer Theranostics*, Academic Press, 1st edn, 2014, p. 544.
84. J. Gao, K. Chen, Z. Miao, G. Ren, X. Chen and S. S. Gambhir, *et al.*, Affibody-based nanoprobes for HER2-expressing cell and tumor imaging, *Biomaterials*, 2011, **32**(8), 2141–2148.
85. R. T. Greenlee, T. Murray, S. Bolden and P. A. Wingo, Cancer statistics, 2000, *Ca-Cancer J. Clin.*, 2000, **50**(1), 7–33.
86. R. Kumar, I. Roy, T. Y. Ohulchanskky, L. A. Vathy, E. J. Bergey and M. Sajjad, *et al.*, *In vivo* biodistribution and clearance studies using multimodal organically modified silica nanoparticles, *ACS Nano*, 2010, **4**(2), 699–708.
87. E. E. Uzgiris, A. Sood, K. Bove, B. Grimmond, D. Lee and S. Lomnes, A multimodal contrast agent for preoperative MR Imaging and intraoperative tumor margin delineation, *Technol. Cancer Res. Treat.*, 2006, **5**(4), 301–309.
88. V. Saxena, M. Sadoqi and J. Shao, Enhanced photo-stability, thermal-stability and aqueous-stability of indocyanine green in polymeric nanoparticulate systems, *J. Photochem. Photobiol., B*, 2004, **74**(1), 29–38.
89. S. P. Foy, R. L. Manthe, S. T. Foy, S. Dimitrijevic, N. Krishnamurthy and V. Labhasetwar, Optical imaging and magnetic field targeting of magnetic nanoparticles in tumors, *ACS Nano*, 2010, **4**(9), 5217–5224.
90. J. V. Frangioni, *In vivo* near-infrared fluorescence imaging, *Curr. Opin. Chem. Biol.*, 2003, **7**(5), 626–634.
91. I. L. Medintz, H. T. Uyeda, E. R. Goldman and H. Mattoussi, Quantum dot bioconjugates for imaging, labelling and sensing, *Nat. Mater.*, 2005, **4**(6), 435–446.
92. C. Y. Ang, L. Giam, Z. M. Chan, A. W. Lin, H. Gu and E. Devlin, *et al.*, Facile synthesis of $Fe_2O_3$ nanocrystals without $Fe(CO)_5$ precursor and one-pot synthesis of highly fluorescent $Fe_2O_3$–CdSe nanocomposites, *Adv. Mater.*, 2009, **21**(8), 869–873.

93. G. Chen, H. Qiu, P. N. Prasad and X. Chen, Upconversion nanoparticles: design, nanochemistry, and applications in theranostics, *Chem. Rev.*, 2014, **114**(10), 5161–5214.
94. K. M. Tsoi, Q. Dai, B. A. Alman and W. C. Chan, Are quantum dots toxic? Exploring the discrepancy between cell culture and animal studies, *Acc. Chem. Res.*, 2012, **46**(3), 662–671.
95. F. Erogbogbo, K.-T. Yong, R. Hu, W.-C. Law, H. Ding and C.-W. Chang, *et al.*, Biocompatible magnetofluorescent probes: luminescent silicon quantum dots coupled with superparamagnetic iron (III) oxide, *ACS Nano*, 2010, **4**(9), 5131–5138.
96. D. Psimadas, P. Georgoulias, V. Valotassiou and G. Loudos, Molecular nanomedicine towards cancer: $^{111}$In-labeled nanoparticles, *J. Pharm. Sci.*, 2012, **101**(7), 2271–2280.
97. B. I. Kharisov, O. V. Kharissova and S. Berdonosov, Radioactive nanoparticles and their main applications: Recent advances, *Recent Pat. Nanotechnol.*, 2014, **8**(2), 79–96.
98. V. Torchilin and M. M. Amiji, *Handbook of Materials for Nanomedicine*, Pan Stanford Publishing, 2010.
99. P. W. Miller, N. J. Long, R. Vilar and A. D. Gee, Synthesis of 11C, 18F, 15O, and 13N radiolabels for positron emission tomography, *Angew. Chem., Int. Ed.*, 2008, **47**(47), 8998–9033.
100. M. Shokeen and C. J. Anderson, Molecular imaging of cancer with copper-64 radiopharmaceuticals and positron emission tomography (PET), *Acc. Chem. Res.*, 2009, **42**(7), 832–841.
101. T. J. Wadas, E. H. Wong, G. R. Weisman and C. J. Anderson, Coordinating radiometals of copper, gallium, indium, yttrium, and zirconium for PET and SPECT imaging of disease, *Chem. Rev.*, 2010, **110**(5), 2858–2902.
102. J. P. Holland, M. J. Williamson and J. S. Lewis, Unconventional nuclides for radiopharmaceuticals, *Mol. Imaging*, 2010, **9**(1), 1.
103. S. Liu and D. S. Edwards, 99mTc-labeled small peptides as diagnostic radiopharmaceuticals, *Chem. Rev.*, 1999, **99**(9), 2235–2268.
104. G. Sugiura, H. Kühn, M. Sauter, U. Haberkorn and W. Mier, Radiolabeling strategies for tumor-targeting proteinaceous drugs, *Molecules*, 2014, **19**(2), 2135–2165.
105. D. B. Weiner and W. V. Williams, *Chemical and Structural Approaches to Rational Drug Design*, CRC Press, 1994.
106. S. Z. Lever and J. R. Lever, Technetium-99m pharmaceuticals: preparation and quality control in nuclear medicine, *J. Nucl. Med.*, 2009, **50**(5), 831.
107. Y. Liu, G. Liu and D. J. Hnatowich, A brief review of chelators for radiolabeling oligomers, *Materials*, 2010, **3**(5), 3204–3217.
108. M. Ogawa, I. O. Umeda, M. Kosugi, A. Kawai, Y. Hamaya and M. Takashima, *et al.*, Development of $^{111}$In-labeled liposomes for vulnerable atherosclerotic plaque imaging, *J. Nucl. Med.*, 2014, **55**(1), 115–120.
109. W. Phillips, A. Rudolph, B. Coins and R. Klipper, Biodistribution studies of liposome encapsulated hemoglobin (LEH) studied with a newly developed 99m-technetium liposome label, *Biomater., Artif. Cells, Immobilization Biotechnol.*, 1992, **20**(2–4), 757–760.

110. M. O. Oyewumi, R. A. Yokel, M. Jay, T. Coakley and R. J. Mumper, Comparison of cell uptake, biodistribution and tumor retention of folate-coated and PEG-coated gadolinium nanoparticles in tumor-bearing mice, *J. Controlled Release*, 2004, **95**(3), 613–626.
111. M. J. Hamamura, S. Ha, W. W. Roeck, L. T. Muftuler, D. J. Wagenaar and D. Meier, *et al.*, Development of an MR-compatible SPECT system (MRSPECT) for simultaneous data acquisition, *Phys. Med. Biol.*, 2010, **55**(6), 1563.
112. H. Zolata, F. A. Davani and H. Afarideh, Synthesis, characterization and theranostic evaluation of Indium-111 labeled multifunctional superparamagnetic iron oxide nanoparticles, *Nucl. Med. Biol.*, 2015, **42**(2), 164–170.
113. R. Torres Martin de Rosales, R. Tavaré, A. Glaria, G. Varma, A. Protti and P. J. Blower, 99mTc-bisphosphonate-iron oxide nanoparticle conjugates for dual-modality biomedical imaging, *Bioconjugate Chem.*, 2011, **22**(3), 455–465.
114. R. Torres Martin de Rosales, R. Tavaré, R. L. Paul, M. Jauregui-Osoro, A. Protti and A. Glaria, *et al.*, Synthesis of 64CuII–bis (dithiocarbamatebisphosphonate) and its conjugation with superparamagnetic iron oxide nanoparticles: *in vivo* evaluation as dual-modality PET–MRI agent, *Angew. Chem., Int. Ed.*, 2011, **50**(24), 5509–5513.
115. A. Hanini, A. Schmitt, K. Kacem, F. Chau, S. Ammar and J. Gavard, Evaluation of iron oxide nanoparticle biocompatibility, *Int. J. Nanomed.*, 2011, **6**, 787–794.
116. P. Bourrinet, H. H. Bengele, B. Bonnemain, A. Dencausse, J.-M. Idee and P. M. Jacobs, *et al.*, Preclinical safety and pharmacokinetic profile of ferumoxtran-10, an ultrasmall superparamagnetic iron oxide magnetic resonance contrast agent, *Invest. Radiol.*, 2006, **41**(3), 313–324.
117. M. Babincova, P. Sourivong, D. Leszczynska and P. Babinec, Blood-specific whole-body electromagnetic hyperthermia, *Med. Hypotheses*, 2000, **55**(6), 459–460.
118. H. Markides, M. Rotherham and A. El Haj, Biocompatibility and toxicity of magnetic nanoparticles in regenerative medicine, *J. Nanomater.*, 2012, **2012**, 13.
119. W. Ra, D. Stark, B. Engelstad, B. Bacon, C. Compton and D. White, *et al.*, Superparamagnetic iron oxide: pharmacokinetics and toxicity, *Am. J. Roentgenol.*, 1989, **152**(1), 167–173.
120. C. C. Berry, S. Wells, S. Charles, G. Aitchison and A. S. Curtis, Cell response to dextran-derivatised iron oxide nanoparticles post internalisation, *Biomaterials*, 2004, **25**(23), 5405–5413.
121. C. C. Berry, Progress in functionalization of magnetic nanoparticles for applications in biomedicine, *J. Phys. D: Appl. Phys.*, 2009, **42**(22), 224003.
122. S. H. Hu, T. Y. Hsieh, C. S. Chiang, P. J. Chen, Y. Y. Chen and T. L. Chiu, *et al.*, Surfactant-free, lipo-polymersomes stabilized by iron oxide nanoparticles/polymer interlayer for synergistically targeted and magnetically guided gene delivery, *Adv. Healthcare Mater.*, 2014, **3**(2), 273–282.

123. N. Schleich, C. Po, D. Jacobs, B. Ucakar, B. Gallez and F. Danhier, *et al.*, Comparison of active, passive and magnetic targeting to tumors of multifunctional paclitaxel/SPIO-loaded nanoparticles for tumor imaging and therapy, *J. Controlled Release*, 2014, **194**, 82–91.
124. W.-M. Li, C.-S. Chiang, W.-C. Huang, C.-W. Su, M.-Y. Chiang and J.-Y. Chen, *et al.*, Amifostine-conjugated pH-sensitive calcium phosphate-covered magnetic-amphiphilic gelatin nanoparticles for controlled intracellular dual drug release for dual-targeting in HER-2-overexpressing breast cancer, *J. Controlled Release*, 2015, **220**, 107–118.
125. A. N. Nacev, *Magnetic Drug Targeting: Developing the Basics*, University of Maryland, Maryland, 2013.
126. J. Bai, J. T. Wang, K. C. Mei, W. T. Al-Jamal and K. T. Al-Jamal, Real-time monitoring of magnetic drug targeting using fibered confocal fluorescence microscopy, *J. Controlled Release*, 2016, **244**(Pt B), 240–246.
127. K. T. Al-Jamal, J. Bai, J. T. Wang, A. Protti, P. Southern and L. Bogart, *et al.*, Magnetic drug targeting: preclinical *in vivo* studies, mathematical modeling, and extrapolation to humans, *Nano Lett.*, 2016, **16**(9), 5652–5660.
128. K. C. Mei, J. Bai, S. Lorrio, J. T. Wang and K. T. Al-Jamal, Investigating the effect of tumor vascularization on magnetic targeting *in vivo* using retrospective design of experiment, *Biomaterials*, 2016, **106**, 276–285.
129. B. Chertok, A. E. David and V. C. Yang, Polyethyleneimine-modified iron oxide nanoparticles for brain tumor drug delivery using magnetic targeting and intra-carotid administration, *Biomaterials*, 2010, **31**(24), 6317–6324.
130. A. J. Cole, A. E. David, J. Wang, C. J. Galbán, H. L. Hill and V. C. Yang, Polyethylene glycol modified, cross-linked starch-coated iron oxide nanoparticles for enhanced magnetic tumor targeting, *Biomaterials*, 2011, **32**(8), 2183–2193.
131. L. Zhang, F. Yu, A. J. Cole, B. Chertok, A. E. David and J. Wang, *et al.*, Gum arabic-coated magnetic nanoparticles for potential application in simultaneous magnetic targeting and tumor imaging, *AAPS J.*, 2009, **11**(4), 693–699.
132. S. Maassen, E. Fattal, R. Müller and P. Couvreur, Cell cultures for the assessment of toxicity and uptake of polymeric particulate drug carriers, *STP Pharma Sci.*, 1993, **3**(1), 11–22.
133. A. Nacev, C. Beni, O. Bruno and B. Shapiro, Magnetic nanoparticle transport within flowing blood and into surrounding tissue, *Nanomedicine (Lond.)*, 2010, **5**(9), 1459–1466.

CHAPTER 4

# Photodynamic Therapy

CORMAC HALLY[a], BEATRIZ RODRÍGUEZ-AMIGO[a], ROGER BRESOLÍ-OBACH[a], ORIOL PLANAS[a], JAUME NOS[a], ESTER BOIX-GARRIGA[a,b], RUBÉN RUIZ-GONZÁLEZ[a] AND SANTI NONELL*[a]

[a]Institut Químic de Sarrià, Universitat Ramon Llull, Via Augusta 390, 08017 Barcelona, Spain; [b]School of Pharmaceutical Sciences, University of Geneva, University of Lausanne, Geneva, Switzerland
*E-mail: santi.nonell@iqs.url.edu

## 4.1 Fundamentals of Photodynamic Therapy

### 4.1.1 The Paradigm

Photodynamic therapy (PDT) is a particular type of phototherapy based on the combination of three factors: light, oxygen and a drug referred to as photosensitiser (PS). None of them is toxic *per se*, however their combination leads to the production of reactive oxygen species (ROS) capable of inflicting severe oxidative damage to neighbouring cells and tissues.

PDT is endowed with several advantages as compared to other therapeutic modalities. First and foremost, PDT shows higher, though still not perfect, selectivity due to its dual mode of action: selective localization of the PS in diseased lesions together with the capacity to deliver light in a spatially confined and focused manner.[1,2] Second, PDT is a local treatment, which greatly reduces any side effects. Indeed, PDT has been reported to show reduced toxicity, minor side effects and in some cases enhance the quality of life and

Drug Discovery Series No. 63
Theranostics and Image Guided Drug Delivery
Edited by Maya Thanou

Published by the Royal Society of Chemistry, www.rsc.org

lengthen survival.[3] Finally, PDT has a multi-target mode of action, derived from the broad reactivity of ROS against a wide range of cell components, which decreases the onset of resistances and, thus, enables repeated treatments if required.

Light is required to promote the PS to a higher-energy electronically-excited state, from which ROS are generated. Thus, the quality of light, particularly its wavelength, and the amount of light reaching the PS are crucial aspects.[4] Thus, for efficient light distribution to a deeply-seated solid tumour, red and near-infrared light (NIR) are ideal whereas for superficial microbial infections shorter wavelengths can be preferred.[5,6] Even though the therapeutic window in the red is expected to allow several mm penetration, a homogenous light distribution and propagation through tissues is more complex than in the surface,[7,8] which has been reported to end up in differences within treatments.[9,10] In addition, differences in light penetration can also be subjected to parameters such as skin pigmentation.[11] Thus, light dosimetry is crucial for inducing the appropriate level of ROS.[8,12]

In turn, the PS must meet a number of demands essential for its photoinduced activity: it must absorb light with high efficiency, be able to generate ROS with high *quantum yield* (number of ROS molecules formed per absorbed light photon) and be resistant to photoinduced self-damage. Commonly to any drug, it must also meet the requirements of solubility in biological media, biocompatibility, selective localization at the treatment site, lack of side effects and rapid clearance from the body.

The third component, oxygen, is ubiquitous in tissues; however, it must be borne in mind that it will be consumed during PDT therefore the light-delivery regime must be adjusted to avoid depleting its level significantly. This is particularly important for the treatment of solid tumours that are notoriously hypoxic.[13–15]

### 4.1.2 Applications

Cancer is the number-one application of PDT and accounts for its recognition and expansion, with the approval of 15 PSs for clinical use and several others undergoing clinical trials, which anticipates further impact of this approach. Targeted cancer and pre-cancer lesions range from superficial (skin, head, melanoma, actinic keratosis, basal/squamous cell carcinoma, or Bowen's disease) to those in internal cavities/organs such as lungs, stomach, colon, bladder or breast.[2,16]

Despite the fact that the photodynamic effect was originally discovered in paramecia,[17] the application of PDT to fight microbial infections was relegated to a second stage when antibiotics seemed to be the definitive solution for bacterial infections. The emerging threat of antibiotic resistance, which PDT is largely devoid of, together with a better understanding of the photodynamic action and the development of cationic PSs has led to a rebirth of the PDT approach to fight microbial infections. There are only three FDA-approved photoantimicrobial agents so far, namely Methylene Blue (MB),

Toluidine Blue O and Indocyanine Green (ICG), with indications for superficial and mouth-related infections and dentistry,[16,18,19] though many other PSs are in the pipeline for approval. Combined approaches using PDT both as antimicrobial and disinfectant are also under scrutiny.[20] In the same line, ablative anticancer and antimicrobial therapies may be combined into a single treatment.[21]

The scope of PDT expands also into other fields such as ophthalmology, with age-related macular degeneration as its most successful indication,[22,23] dermatology and cosmetics,[24,25] cardiovascular diseases, for correcting vascular function under diseased conditions,[26–29] rheumatoid arthritis,[30,31] blood sterilization (extracorporeal PDT/photopheresis),[32,33] cell proliferation and regeneration (*i.e.*, wound healing),[6,34–36] and lymphatic ablation.[37]

### 4.1.3 Mechanisms of Action

From a photochemical point of view, light absorption by the PS is the first step in a line of events that end up with the production of ROS. When the PS absorbs light of the appropriate wavelength this energy pumps the molecule to its electronic singlet-excited state ($^1PS^*$). It then either returns to its ground state giving the energy back through heat/fluorescence or undergoes intersystem crossing (ISC), in which a spin change occurs leading to the longer-lived PS's triplet excited state ($^3PS^*$). At this stage, two markedly different forward pathways can take place (Figure 4.1). In Type I processes, the $^3PS^*$ undergoes an electron transfer reaction with a suitable acceptor leading to the formation of radicals, which, in turn, may further react with other electron acceptors such as oxygen to generate ROS such as superoxide radical anion ($O_2^{\bullet -}$), hydrogen peroxide ($H_2O_2$) or the hydroxyl radical ($^{\bullet}OH$). In a second scenario, known as Type II processes, the PS transfers

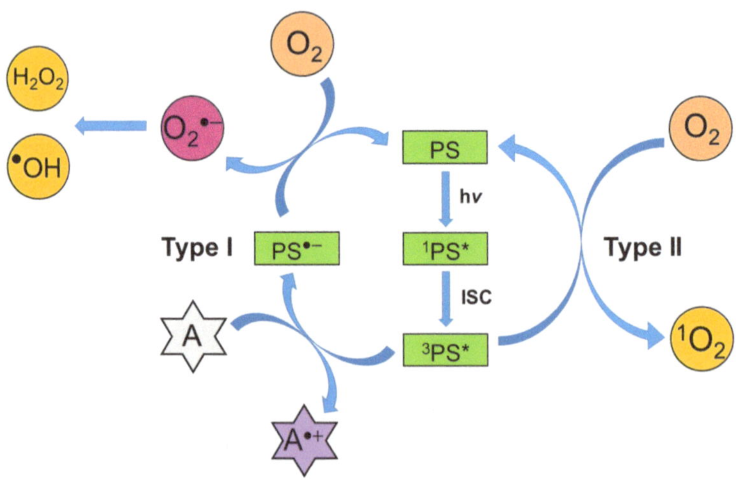


**Figure 4.1** Photophysical cycle of the PS and mechanisms of ROS generation.

its energy to molecular oxygen leading to the formation of singlet oxygen (hereafter $^1O_2$), a neutral, non-radical ROS. In both cases the ground-state PS is recovered, ready to accept another photon to start the photosensitisation cycle again.[38]

ROS formed either *via* Type I or Type II processes have in common their intrinsic high oxidising nature, which can damage most biomolecules (*i.e.* lipids, DNA, proteins or carbohydrates).[39] However, differences in lifetime, diffusion and reactivity will define the final outcome in the biomolecule and also fingerprint the nature of the oxidising reagent responsible for the damage. Due to ROS reactivity, the degradation of the PS is a concern in both mechanisms. Redox reactions (Type I) imply a second electron reaction to restore the PS, which makes photobleaching more likely. Thus, the Type II mechanism ($^1O_2$ production) is preferred.[40]

Both Type I and Type II processes can occur simultaneously, the ratio of which depends on the type of PS used, and the concentration of the electron-transfer substrates and oxygen.[41] A field of recurrent interest is unravelling the mechanism of action of the PS, which is an important feature in assessing the potential of novel photoactive compounds.[42–44] The photodynamic cycle ends whenever any of the three participants is no longer present or their combination is blocked (*i.e.* in the presence of any ROS scavenger/quenchers).

Type I mechanism is more complex than type II since it involves a broader variety of acceptors. This fact can be harnessed by rendering final reactive species other than those in which oxygen is the final acceptor, as for instance, nitrogen reactive species such as nitric oxide derivatives.[45,46] Other interesting oxygen-independent pathways have been reported, including the combination of azide with an excited PS, whose proposed mechanism is appended as a Type I process.[40,47–49] The use of iodide and bromide are also shown to be effective in order to enhance PDT inactivation of bacteria, which mechanism seems to be oxygen independent.[50,51]

### 4.1.4 Biological Actions

Different biological events have been identified which can contribute to the observed success in PDT.

In a first case scenario, the PDT-generated ROS can destroy cells directly. In bacteria, several attempts have endeavoured to assess the specific *loci* responsible for the final damage. When an external PS is delivered, damage occurs from the outside, starting from the outer and/or inner membrane, which can be enough to result in bacterial cell death. In such small cells, however, it is hard to ascertain where the PS is acting.[52,53] Fluorescent proteins as genetically-encoded PSs provide further insight in the mechanism responsible for direct cell killing. In this case, the PS is expressed within the cell and so the damage occurs internally. Some studies report that inner membrane damage is enough to exert cell death.[54,55] In eukaryotic cells on the other hand, ROS damage can result in a wider range of cell-death pathways including necrosis, apoptosis, autophagy or mitotic catastrophe.[56,57] The biological consequences and final cell-death outcome will be dictated by the type of ROS generated, its quantity and, as a critical factor, the subcellular site of generation.[39]

Studies in tumour systems have shown that direct PDT cell photokilling is not sufficient to explain the success achieved when applied to real tumours in mice.[58,59] ROS may also damage the tumour-associated vasculature, a capital necessity for the growth of the tumour mass and progression of cancer to other parts of the body.[60,61] Evidences of vascular-targeted PDT alterations include thrombus formation, blood vessel occlusion and vascular function shutdown; these events result in disruption/blocking of the channels which supply oxygen and nutrients to the tumour, leading to tumour eradication.[60,62]

Finally, studies in animal models have demonstrated that PDT can induce an immune response in the host.[63,64] This fact, *per se*, is not surprising since other approaches such as surgery, chemotherapy and ionizing radiation also activate an immune response. However, whereas the previous methods develop immunosuppression, PDT has shown immunostimulation, which makes it an advantage for cancer treatment. Thus, after PDT, a concomitant inflammatory response occurs inducing the release of immunomodulators, such as cytokines and growth factors. This activity response of the host is reported to have implications in further destruction of remaining tumour cells and preventing recurrence and metastasis. Given this fact, approaches to further enhance the immune response in combination with PDT have been tried or proposed.[65] Examples include macrophage-targeted PDT[65,66] or the use of immunostimulant adjuvants such as β-D-glucan schizophyllan and microbial vaccines or imiquimod (for skin diseases) among others.[67–70]

On the other hand, immunosuppressors have also been proposed in combination with PDT after treatment for macular degeneration to prevent regeneration of vasculature, exemplifying the immune effect also in non-tumoural diseases.[23]

## 4.2 Theranostic Features of PDT Drugs

### 4.2.1 Fluorescent Properties of PDT Drugs

Theranostics is based on the combination of a diagnosic imaging agent with a therapeutic drug on the same multifunctional platform.[71] This is indeed the case of PDT drugs that produce cytotoxic ROS upon light absorption and whose fluorescence is also used for molecular imaging, locating many diseases and photodiagnosis. Therefore, PDT is ideally suited for theranostic purposes, since obtaining a fluorescence image of the PS in the target tissue can give rise to image-guided dosimetry and aid surgical resections as well as monitoring of the treatment response.[72] In fact, PDT is nowadays being routinely used for the detection and diagnosis of bladder cancer and fluorescence-guided resection in surgical operations.[73] Photophysical and photochemical properties of the approved PSs for cancer treatments are collected in Table 4.1.

### 4.2.2 Selectivity of PDT Drugs

An essential element of theranostics is the selectivity of the agent. Developing PSs with ever better selectivity for the diseased tissue is crucial for effectiveness of the treatment and the avoidance of undesired side effects.[95] The preferential accumulation of the PSs in malignant tissues may be either intrinsic or induced by other targeting moieties that convey the PS to its active site.

#### *4.2.2.1 Photosensitisers with Intrinsic Selectivity*

Although most of the clinically approved PSs show some inherent selectivity for neoplastic tissues, this is far from ideal since the PS is being largely retained in other tissues such as the skin.[2] Thus, there are only a few examples of intrinsic selectivity related to the nature of the PS.[73,96,97]

5-Aminolevulinic acid (ALA) is a precursor of the PS protoporphyrin IX (PpIX). It is intracellularly metabolised in the mitochondria as part of the haem biosynthetic pathway (Figure 4.2).[98] This prodrug penetrates non-selectively into all cells leading to an accumulation of PpIX when the rate of conversion of ALA to PpIX is greater than the conversion of PpIX to haem.[74,83] The mechanism by which PpIX selectively accumulates in neoplastic regions is not completely understood; however it seems clear that multiple factors are responsible for this effect. Several authors reported a higher activity of enzymes present in neoplastic cells that facilitate the conversion of ALA to PpIX, increasing the rate of PpIX accumulation by 20 fold.[99] Furthermore, a lower concentration of iron in carcinoma cells and a down-regulation of ferrochelatase activity also induce PpIX accumulation.[100] All in all, there is a greater uptake of ALA by neoplastic cells, inducing larger PpIX concentration in comparison with healthy ones.

The esterified derivatives of ALA, such as 5-Aminolevulinate or hexaminolevulinate, increase the lipophilicity and facilitate the crossing through cell membranes, leading to a homogeneous distribution in tissues.[101] Owing to this inherent selectivity, ALA and its family derivatives have been used worldwide for PDT and fluorescence imaging. However, focus is now centred in further improving selectivity and penetrability of PSs towards tumorous cells.[96]

#### *4.2.2.2 Passive Targeting in PDT*

The need to increase selectivity of PDT drugs, skin photosensitivity,[102] as well as the hydrophobicity of many PSs (which limits *in vivo* administration and provokes the loss of their photophysical properties in aqueous medium) led to the development of a wide variety of drug-delivery strategies.[103]

**Table 4.1** Clinically approved PSs with their excitation wavelength, maximum emission wavelength, fluorescent quantum yield, $^1O_2$ quantum yield and its main subcellular location.[a]

| Class | Photosensitiser (INN Name – trade name) | $\lambda_{exc}$ nm$^{-1}$ | $\lambda_{max\ em}$ nm$^{-1}$ | $\Phi_F$ | $\Phi_\Delta$[b] | Subcellular localization | Ref. |
|---|---|---|---|---|---|---|---|
| **Hematoporphyrin (1st generation PS)** | Porfimer sodium Hematoporphyrin (HpD) - Photofrin | 630 | 690 | NA | 0.89 | Golgi apparatus, plasma membrane | 74–79 |
| **Protoporphyrin prodrugs** | δ-Aminolevulinic acid HCl (ALA) - Levulan | 635 | 635 | 0.05 | 0.56 | Mitochondria, cytosol, cytosolic membranes, cell membranes | 74,77,78,80–82 |
| | 5-Aminolevulinate (MAL) – Metvix | 635 | 635 | 0.05 | 0.56 | Mitochondria, cytosol, cytosolic membranes, cell membranes | 74,75,77,78 |
| | Hexamino-levulinate HCl (HAL) - Cysview or Hexvix | 375–450 | 635 | 0.05 | 0.56 | Mitochondria, cytosol, cytosolic membranes, cell membranes | 77,78 |
| **Chlorins** | Meta-tetra (hydroxyphenyl) chlorin or temoporfin (*m*-THPC) - Foscan | 652 | 652 | 0.22 | 0.01/0.30[c] | ER, mitochondria | 74,75,82–86 |
| | Talaporfin sodium, *N*-aspartyl chlorin e6, or LS11 (NPe6) - Laserphyrin | 664 | 670 | 0.003 | 0.77 | Lysosome, endosome | 74,75,80 |
| | Chlorin e6 (Ce6-PVP) - Photolon | 660–665 | 675 | 0.18 | 0.64 | Lysosomes | 83,87,88 |
| | Chlorin e6, chlorin p6 and purpurins - Photoditazin | 660–662 | ~700 | 0.05 | 0.6 | Lysosomes | 83,89,90 |
| | Chlorin e6, chlorin p6 and purpurins - Radachlorin | 662 | 668 | 0.04 | 0.75 | Lysosomes | 83,91 |
| **Porphycenes** | 9-Acetoxy-2,7,12,17-tetrakis-(β-methoxyethyl) - ATMPn | 638 | ~650 | 0.27 | 0.38 | Mitochondria, lysosomes, ER, plasma membrane | 82,83,85,92 |
| **Phthalocyanines** | Mixture of chloroaluminium sulphonic phthalocyanines (AlPcS4) - Photosens | 675 | 689 | 0.559 | 0.38 | Mitochondria | 80,82,83,85,93 |
| **Psoralens** | 8-Methoxypsoralen, methoxsalen (8-MOP) - Uvadex | 303 | 505 | 0.0028 | 0.004[d] | Nuclei | 82,94 |

[a] $\lambda_{exc}$: clinical excitation wavelength; $\lambda_{max\ em}$: maximum emission wavelength.
[b] Most are given in phosphate buffer pH ≈ 7.4 or in a mixture of phosphate buffer with 1% Triton X-100.
[c] EtOH.

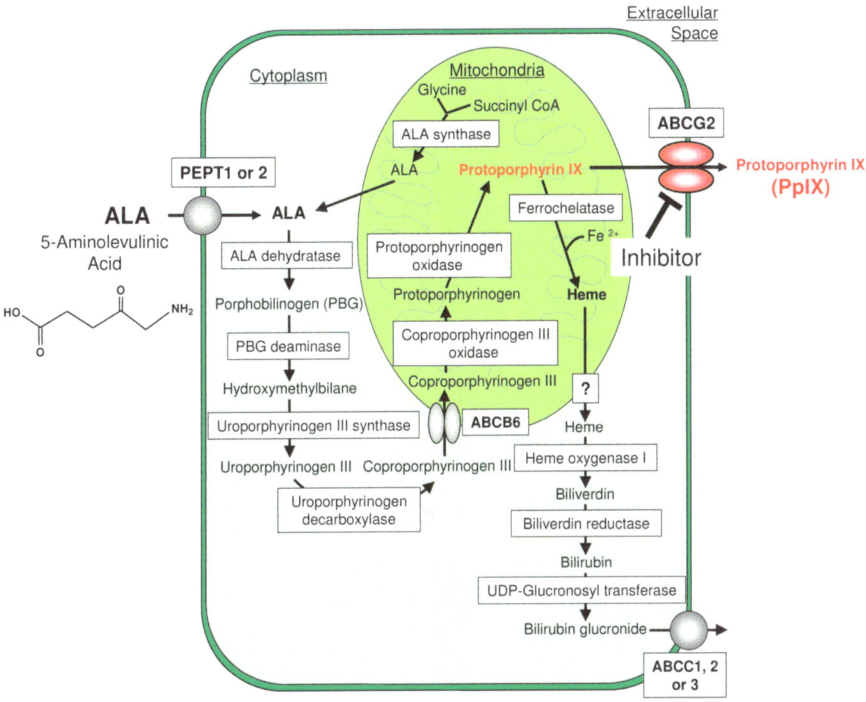


**Figure 4.2** Scheme of the haem biosynthesis in mammalian cells from exogenous administration of ALA, with the enzymes (represented in rectangles) involved. Reproduced with permission from T. Ishikawa *et al.*, *Pharmaceutics*, 2011, **3**, 615, DOI: 10.3390/pharmaceutics3030615,[98] © 2011 by the authors. Published under the terms of the CC BY 3.0 license, https://creativecommons.org/licenses/by/3.0/.

Nanoparticles (NPs) began being scrutinised as vehicles for PDT in order to solve the aforementioned disadvantages. NPs supply an anchor or a core *milieu* where the PS can be well accommodated, rendering a water compatible and stable system.[104] NPs also have the exquisite ability of being retained in tumour tissues thanks to the enhanced and permeability retention effect (EPR),[105] phenomenon known as passive targeting and which clearly potentiates localization of NPs at the site of action, increasing treatment tolerability.

A vast array of nanodelivery systems have been studied for PDT, including liposomes,[106–108] organic polymeric NPs, micelles or dendrimers such as those of poly-(acrylamide)[109–111] or poly-(lactic-*co*-glycolic) acid,[112–114] nanovehicles made of inorganic materials such as silica[115,116] or gold,[117–119] or even systems which are PSs *per se*, such as fullerenes.[120] Remarkable advances have even led to some clinically approved formulations such as Foslip® or Visudyne®.[121,122]

Despite a general improvement of PDT when NPs are used, a higher selectivity is still pursued for cell targeting, since healthy peritumoural cells frequently experience the same phototoxic effects due to unfortunate PS uptake. It is becoming more evident that the PS needs to be activated once internalised inside the target cells in order to achieve a more selective treatment, either by internal stimuli such as pH or temperature,[123] by an external stimulus such as light[124] or through recovery of the photophysical activity by specific enzymes at the target site[125] (Figure 4.3).

The theranostic ability of the PS thereby becomes more specific, since fluorescence is only switched on when the PS reaches its target, which facilitates diseased tissue recognition and its rejection and/or treatment *in situ*. Such an example is the system of porphyrin-lipoprotein nanovesicles (PLP) developed by Cui *et al.*[126] The loaded porphyrin presents more than 95% of fluorescence quenching inside the PLP vesicles, but once PLP are internalised by cells, fluorescence is restored. PLP show excellent passive targeting and

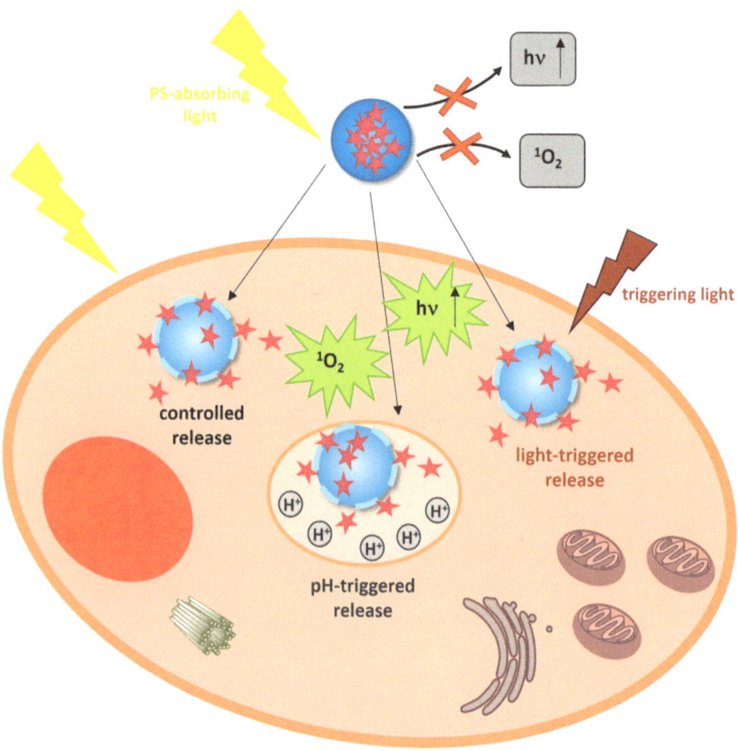


**Figure 4.3** Some common mechanisms employed in passive targeted NPs to trigger the precise release of the PS once inside the target cells. In controlled release, the NP is likely disrupted in contact with the cellular membrane or inside some cellular organelle, or the PS affinity for a cell organelle is superior and its outwards diffusion is facilitated. pH- and light-triggered NPs contain a specific material sensitive to these stimuli, the alteration of which produces a certain sort of cleavage resulting in PS delivery. Regardless of the mechanism, it is desirable that the PS be self-quenched inside the NP to observe a "switch on" of its photophysical properties only inside the required cell.

theranostic ability *in vivo* in xenograft and orthotopic glioma mice models as well as in primary prostate tumour and metastatic ovarian cancer bearing mice, which guides tumour detection and PDT.

Another strategy to improve phototoxic effects is to exploit surface plasmon resonance[127] of gold (Au) NPs to obtain more potent PSs. This has been the case of Au nanospheres and Au nanoaggregates loaded with MB described by Simon *et al.*,[128] or the case of Au NPs coated with silica shells of varying thickness, loaded as well with the same PS.[119] MB loses its photoactivity when administered in solution, since cell enzymatic activity metabolises it to a non-photoactive form. Au NPs, apart from avoiding its degradation, enhance its photoactivity by the plasmon resonance effect, which altogether increases MB phototoxicity. Another approach of exploiting the optical properties of Au NPs is a lipid-porphyrin vesicle, or porphysome, coated with Au NPs reported by Farhadi *et al.*,[129] in which the ability of Au NPs to produce Surface Enhanced Raman Scattering (SERS) is coupled with the high $^1O_2$ production of a Pd(II)-chlorin, which is, however, not fluorescent. SERS imaging thus serves to track the otherwise quiescent NPs *in vivo* as well as to probe real-time $^1O_2$ dose since SERS intensity decreases with an increase in $^1O_2$ production.

In order to enhance the detection of the PS-loaded nanovehicles, there is the tendency to develop or potentiate those PSs that emit in the red/NIR window, *i.e.* 650–900 nm, since at this spectral window there is minimal autofluorescence background from cells. In an elegant and innovative study, Sheng *et al.*[130] reported a nanovehicle made of the self-assembly of Human Serum Albumin (HSA) molecules encapsulating ICG. These biodegradable NPs showed an enhanced uptake and phototoxic effect *in vitro* and in various *in vivo* models. NIR fluorescence enabled clear detection of the tumour and its margins, which improved PDT application *in vivo* and enhanced the theranostics of HSA-ICG compared to ICG administered in solution.

Another current strategy to ameliorate the final cytotoxic effect of PDT, is the combination of PDT with a chemotherapeutic agent or another anticancer drug. Lin *et al.* reported on a new type of nanovehicle consisting in polyethylene glycol (PEG) chains chemically linked through lysine amino acids to pyropheophorbide a (Ppa) and cholic acid, which self-assembles in aqueous medium forming NPs additionally loaded with an HSP90 inhibitor. Fluorescence of Ppa is strongly quenched in the NPs, but restored once internalised by prostate cancer cells. *In vivo*, the NPs accumulate preferentially in the tumour, and activatable Ppa fluorescence enables an effective tumour localization and PDT. Moreover, the combination of PDT with the chemotherapeutic drug synergistically enhances tumour eradication and mice survivial.[131]

### 4.2.2.3 Active Targeting: Photosensitiser-ligand Conjugates

Third generation PSs, or the association of PSs with supramolecular structures, aim to increase selectivity of PDT, requiring smaller amounts of drug and, most importantly, reducing side effects of the treatment. Selectivity is achieved by taking advantage of physiological differences between target and healthy cells, such as overexpressed membrane proteins and glycans or accelerated metabolism. The linking of the PS must be performed so that the targeting entity, *i.e.* antibodies, is not altered. It is of utter importance that they do not lose their selectivity in order to ensure an efficient binding with its target.

Some of the most employed targeting elements to enhance PDT selectivity are folate molecules, peptides, lectins, and antibodies (or its fragments). 40% of human tumours present overexpressed folate receptors while being silent in healthy cells.[132,133] Peptides such as F3 (which has cell penetrant properties) and GE11 (that binds to the cancer overexpressed membrane epidermal growth factor receptor, EGFR) are used as targeting entities.[134–137] Additionally, 90% of human carcinomas present an overexpression of the Oncofetal Thomsen-Friedenreich Dissacharide (T antigen), to which Jacalin lectin selectively binds.[138,139]

### 4.2.2.4 *Active Targeting: Photoimmunoconjugates*

Another strategy developed to increase PDT selectivity is the use of PS-labelled antibodies, or its fragments, as targeting agents.[140,141] Therefore, only those cells overexpressing the corresponding antigen should be affected when light is applied.

Monoclonal antibodies (mAb) are large modular proteins, formed by four subunits, with two identical heavy chains, and two also identical light chains. Figure 4.4 presents the structure of a typical immunoglobulin (Ig).

The variable region of the antibody performs binding with the antigen while the constant region enables host-cell recognition and immune system activation.[142]

There are different types of antibody fragments depending on the size of the targeting entity. Fragment Antigen-Binding (Fab) are the largest and most studied antibody fragment. Single Chain Variable Fragments (scFv) are recombinant molecules which gather the variable part of the antibody. The light and heavy variable chains are linked by means of a flexible linker, which

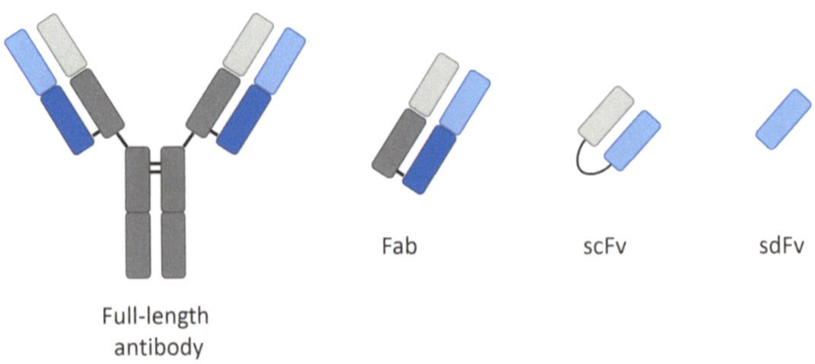


**Figure 4.4** Antibody and fragments structure and nomenclature.

allow the two peptides to interact and assemble through non-covalent interactions. Single Domain Variable Fragments (sdFv) are the smallest fragments in which a single domain (a protein lacking tertiary structure) of the variable part of the antibody preserves specificity for the antigen.

The use of antibody fragments has gained popularity due to the greater mobility and penetrability into tissues of these smaller molecules and to the fact that shorter protein chains can be produced faster and easily by prokaryotic cells; antibody fragments require no further posttranslational modifications, such as glycosylation.[141] Another characteristic of these targeting entities is the partial or total absence of the constant fragment, which reduces the immunological recognition of the host.[142]

On the other hand, antibody fragments have less affinity to their antigen than regular antibodies and are less stable due to the lack of disulphide bridges between domains and monomers. Pharmacokinetics and biodistribution of fragments are also different when compared with full length antibodies due to the size difference. Cellular sublocalisation is also another important factor to take into consideration when applying PDT.[143]

This promising therapy has some drawbacks which must be solved before it becomes routinely applied clinically. Most PSs are hydrophobic molecules and, therefore, it may be difficult to obtain the conjugate without destabilizing the Ig. The number of PSs per antibody must be optimised in order to achieve maximum selectivity and cell death. Also, the drug must absorb at the wavelength of the light applied, have high extinction coefficient values, whilst considering that more energetic lights are less tissue penetrant than red and infrared radiation.

Mitsunaga *et al.* achieved tumour size reduction by using trastuzumab (mAb directed against human epidermal growth factor) as a targeting entity for PDT using IRDye 700DX.[144] This same PS was used by Gao *et al.* in which they used targeted PDT in order to activate the immune system to treat cancer. Fluorescence was used as a control tool for cancer treatment.[145] Planas *et al.* reported the discovery of a PS which shifts its absorption spectra once it bonds with an amine (to a lysine amino acid in a mAb). This shift enables selective excitation of the molecules which are covalently bound to the targeting entity, increasing selectivity.[146]

Photoimmunotherapy can benefit greatly from the fluorescent properties of PSs, being able to add the diagnostic component to the therapy. The visual delimitation of the target enables visualisation of the affected area and to easily assess and control the progress of the treatment.

### *4.2.2.5 Active-targeting: Multifunctional Nanoparticles*

As for the case of targeted PSs, a ligand moiety can also be anchored to nanovehicles to target a specific cell marker and thus achieve enhanced selectivity.[147] Although some studies about active-targeted NPs for PDT exist, literature in this domain is not extensive, possibly due to unexpected low targeting achieved for certain NP-ligand systems or some differing results obtained for certain ligands, such as folic acid. For instance, folate-targeted liposomes have been reported to produce just a moderate increase in the phototoxic effect (~2 fold),[107,132] generally attributed to the competition between receptor-mediated and non-specific uptake of targeted NPs. Additionally, Selvestrel *et al.*[148] reported no uptake differences between non-targeted or folate-decorated silica NPs loaded with *m*-tetrahydroxyphenylchlorin (*m*THPC). On the contrary, Tian *et al.*[149] reported a satisfying specificity of folate-targeted PLGA-DSPE micelles when incubated in HeLa cells (positive folate overexpression) compared to the lack of phototoxicity found for HaCaT cells (folate negative). As suggested by Selvestrel *et al.*, the PEG shell could shield certain small targeting ligands, especially if they are slightly hydrophobic, as it is the case of folate. This phenomenon has also been described by other authors[150] and would correlate with the satisfactory targeting obtained by the folate-targeted free-PEG NPs from Tian *et al.*[149]

In their study, Selvestrel *et al.*[148] also investigated in detail the active targeting of the peptide RGD and the mAb cetuximab in PEGylated silica NPs loaded with *m*THPC. In both cases, despite a higher internalization of the targeted NPs when compared to the non-targeted ones, no superior phototoxic effect was observed, pointing out that subcellular localization may also play an important role in the correlation between uptake and PDT efficacy. On the other hand, when cetuximab was adsorbed on the surface of liposomes and performed as a dual targeting moiety-chemotherapeutic drug, the treatment was rendered up to six-fold more effective than merely PDT for targeted cancer cells.[151] Therefore, the combination of targeting moieties which are additionally anticancer agents seems promising and opens a new venue of research to explore.

In another study, Obaid *et al.*[152] reported that PEGylated Au NPs conjugated with a phthalocyanine (Pc) derivative and targeted with the lectin Jacalin or with the mAb HER-2 were at least five-fold more phototoxic than non-targeted NPs in two target cell lines, HT-29 and SK-BR-3 cells. In this case, the targeting efficiency of both ligands was similar, which was attributed to their lysosome subcellular localization in both cases.

Related to the theranostic properties of PDT targeted NPs, the PLGA-DSPE-folate micelles of Tian *et al.* were loaded with the activatable PS selenium-rubyrin.[149] Its photophysical activity is quenched at basic or neutral pH due to photon-induced electron transfer, but fluorescence and $^1O_2$ production are restored at acidic pH, which provokes activation of the PS only when internalised in lysosomes through folate-receptor endocytosis. Additionally, Jin *et al.* demonstrated that fluorescence imaging of folate-targeted activatable porphysomes intravenously injected in KB xenograft bearing mice could guide the treatment protocol and indicate the time window for best PDT, which was 24 h after injection.[153] Effective tumour reduction was only achieved with the folate-targeted porphysomes, whereas non-targeted counterparts were not internalised and therefore did not show any phototoxicity.

In conclusion, the reports published seem to indicate that a selective internalization of targeted NPs *versus* the non-targeted counterparts and exclusively to the desired cancer cells is dependent on multiple factors, such as the presence or absence of PEG, type of NP and ligand or how the PS is loaded (covalent conjugation or physical entrapment). The promising results described must encourage scientists in continuing and broadening the research of targeted NPs for PDT and its theranostic applications.

### 4.2.3 Subcellular Localization

Photosensitiser subcellular localization has long been recognised as a key parameter not only for the efficiency of the photodynamic treatment but also for the type of cell death mechanism triggered.[154] $^{1}O_2$ has a very short lifetime in living cells,[155,156] and therefore the damage caused remains mainly confined in the vicinity of its photosensitisation. Mitochondria have largely been considered the preferred organelle destination for a PS since they are prone to induce apoptosis. Further investigations, though, suggest that other organelles such as lysosomes, Golgi apparatus or endoplasmic reticulum may also be important in prompting apoptosis.[157,158]

Intracellular distribution of PSs can be observed thanks to their fluorescence without requiring any additional attachment of labels in order to track them. Cellular fluorescence microscopy images, compared with those of other specific-organelle-fluorescent markers, are the most common experiments performed to elucidate the organelle-PS location (co-localisation),[154] and have been largely applied not only to study free PS internalization,[159–162] but also the cellular *loci* of PSs loaded in NPs.[114,148,152,163] Steady-state fluorescence microscopy is, however, limited when related to the elucidation of more complex or dynamic events, for instance PS release from a NP in a living cell. Fluorescence Lifetime Imaging Microscopy (FLIM),[164] in which a time-resolved image of the emitting species is obtained, can provide valuable information on these dynamic processes on the basis that PS fluorescence lifetime is very sensitive to the surrounding environment of the PS.[165] FLIM has permitted researchers to observe that free *m*THPC (Foscan) and its liposomal formulation (Foslip®) have similar fluorescence lifetimes (and organelle localization) once inside the cell. This fact implicates that internalization in both cases should be through similar pathways, therefore rejecting an endocytic entry of Foslip® and rather suggesting a disintegration of the liposome in contact with the plasma membrane thereby releasing free *m*THPC into the cell.[166] On the contrary, HSA NPs loaded with Ppa were shown to be first internalised into the cell to subsequently release Ppa in a time gap of 24 h.[165] FLIM has also served to monitor real-time NIR light-triggered release of Ce6 from silica-coated Au nanorods in living cells (Figure 4.5),[167] and two-photon FLIM has been employed to follow the cellular uptake of Photofrin® in real time.[168]

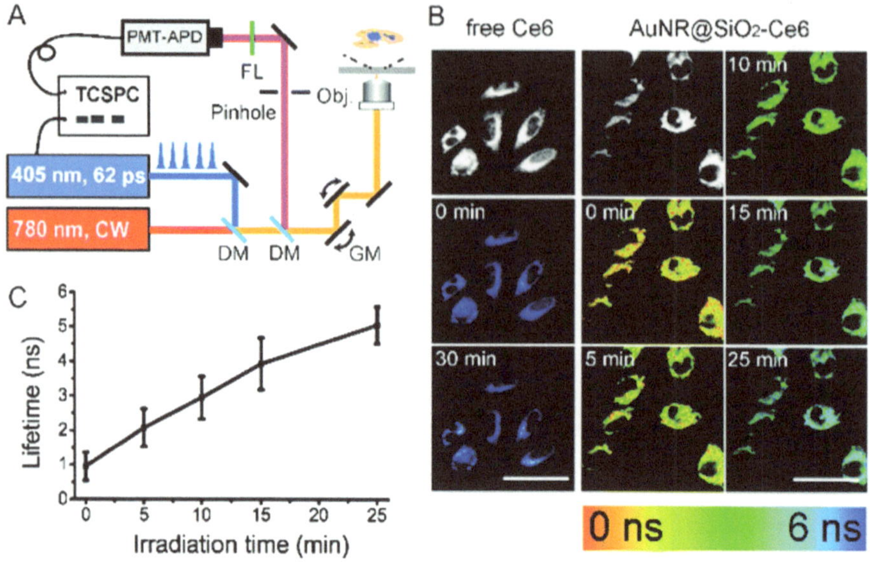


**Figure 4.5** (A) Setup schematics of the fluorescence lifetime and FLIM experiments. (B) Fluorescence images of KB cells incubated with free Ce6 (first column) and silica-coated Au nanorods loaded with Ce6 (AuNR@$SiO_2$-Ce6) (second and third columns). Fluorescence intensity images are shown in grey, and FLIM images of these cells under different irradiation times of 2 mW 780 nm laser are shown in false-colour with the colour map indicating the fluorescence lifetime. (C) The time-dependent mean fluorescence lifetime of all pixels in the fluorescence lifetime images of the Au NR@$SiO_2$–Ce6 incubated KB cells after irradiation. DM: dichroic mirror, GM: galvo mirror, FL: filter. Scale bar: 100 μm. Reproduced from ref. 167 with permission from The Royal Society of Chemistry.

This technique has also been valuable to report the existence of Förster Resonance Energy Transfer (FRET) between quantum dot-PS complexes *in vitro*.[169] Altogether, the differences encountered depending on the PS and the NP system highlights the importance in elucidating for each PS and each particular NP its specific release and internalization pattern, and these examples support FLIM being a valuable tool to study the real-time release of PSs from NPs in living cells.

## 4.3 PDT Drugs Combined with Additional Imaging Agents

Fluorescence imaging is a valuable and useful technique for diagnosis. Despite this, fluorescence may not always be a valid option when needing deep-tissue imaging. To address this problem, strategies which do not depend on PS fluorescence are being explored, such as NPs or the use of other contrast agents detected by different imaging techniques.

### 4.3.1 Nanoparticles as Contrast Agents in PDT

Compared with conventional contrast agents, NPs can be easily surface-decorated with cancer target molecules, which enhance their cell-targeting capabilities as well as medical diagnosis for a wide range of *in vivo* imaging techniques. Some of these are near-infrared fluorescence imaging (NIRF), magnetic resonance imaging (MRI), positron emission tomography (PET), computed tomography (CT), ultrasound (US), and photoacoustic imaging (PAI) and different PSs which make them suitable for PDT or photothermal therapy (PTT).

Some of the most commonly used NPs as contrast agents are quantum dots (QDs), which are crystalline semiconductors with diameters smaller than 10 nm. The most common composition consists of a CdSe core with a coating shell of ZnS. Other alloys such as CdTe, CdS or InP can also be used. Their large molar extinction coefficients, broad band absorption spectra, size-tuneable narrow emission spectra (CdSe/ZnS QDs with 2 nm diameter long result in blue emission whereas 7 nm diameter emit red light),[170] high fluorescence quantum yields ($\Phi_F$) and high photostability make them an excellent choice for NIRF. Despite this, the potential toxicity of heavy metal ions precludes their use in clinical bioimaging. However, QDs have been attached with tumour targeting molecules such as radiolabelled prostate-specific membrane antigen (PSMA) antibody[171] and arginine-lysine-aspartic acid (RGD) peptides.[172]

Gold nanospheres and nanoshells have been used for drug delivery and PTT in clinical trials.[173] The plasmonic effect of Au NPs has a big potential for imaging and cancer therapy. Among Au NPs, gold nanostars (Au NS), with their multiple arms, have been more relevant thanks to their enhanced plasmonic properties in the NIR which are suitable for biomedical applications, as also in *in vivo* lymphatic system mapping with photoacoustic tomography (PAT).[174] Au NS surface functionalisation with $Gd^{3+}$ cations,[175] by means of 1,4,7,10-tetraazacyclododecane-1,4,7,10-tetraacetic acid (DOTA) chelator, have been proven to be excellent MRI contrast agents. Shi *et al.* have synthesised chemically deposited Au NPs onto fullerene ($C_{60}$) in which PEG-5000 was covalently bound and doxorubicin adsorbed, giving rise to a radio frequency controlled drug delivery system.[176]

PEG-glycated superparamagnetic MnO NPs with PpIX has been reported as a promising strategy.[177] They can be used as a contrast enhancing agent in MRI due to their magnetic properties, as well as fluorescent PpIX that allows simultaneous tumour detection and PDT.

Porphysomes are obtained by self-assembly of porphyrin lipids into liposomes. Porphysomes convert absorbed photons into heat as efficiently as gold nanorods[178] due to the high porphyrin packing density. This property makes them promising candidates for PTT and PAI because they give rise to strong photoacoustic signals. The aqueous core can be loaded with different molecules, such as doxorubicin or carboxyfluorescein.[178] In addition, metal ions can be chelated into the porphyrin ring for PET (*i.e.* $^{69}Cu$ radioisotope)[179] or MRI (Mn atom).[180]

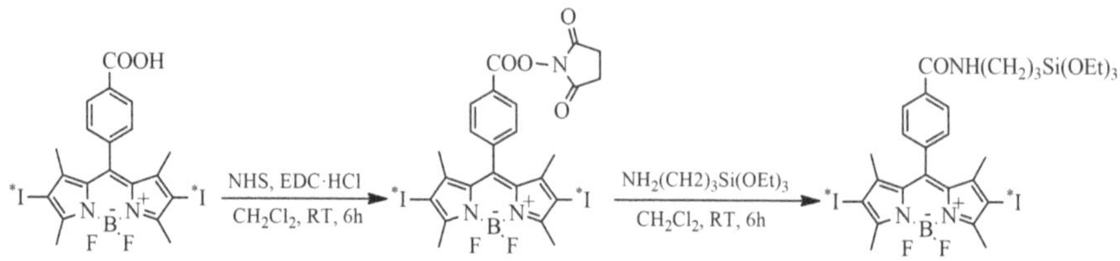


**Scheme 4.1** Synthesis of silane radiolabelled 2,6-diiode substituted dipyrromethene boron difluoride BODIPY (Iodine 124 or 125 radioactive isotopes [*I]).

Silica NPs can be very useful when developing new contrast agents due to their versatile surface functionalisation. This is the case of 2,6-diiode substituted dipyrromethene boron difluoride (BODIPY) doped silica NPs.[181] The incorporation of two heavy iodine atoms increases the population of the excited triplet states, which greatly enhances the efficiency of $^1O_2$ production. Moreover, if the iodine atoms are replaced by its radioactive isotopes ($^{124}I$ or $^{125}I$), the NPs become good PET contrast agents (Scheme 4.1).[181]

The combination of chemotherapy, PTT, and MRI in one single nanocarrier has been achieved with a novel magnetic Gd silica NP.[182] These are composed by a Gd core (as MRI contrast agent) covered with a mesoporous silica layer with adsorbed doxorubicin (as a chemical drug), ICG (as a PTT agent) and a positively charged coating which improves cellular uptake and prevents drug release. Drug release takes place under acidic conditions, but also when temperature increases due to photothermal effects after laser irradiation.[182]

### 4.3.2 Photosensitiser Conjugates with Contrast Agents

The combination of a PS with a contrast agent has become a prominent strategy in cancer treatment because cancer cells can be visually exposed as well as selectively destroyed without affecting the surrounding healthy tissues. Consequently, it is crucial to properly select the most suitable PS.

MRI-PS combination is the most frequently-used contrast technique, in which $Gd^{3+}$ complexes are bound at the meso-position of a porphyrin ring (Scheme 4.2).[183] Several $Gd^{3+}$ and lanthanide complexes have been attached to different kinds of porphyrins and reported as attractive MRI[184,185] and NIRF[186] contrast agents for PDT.

Pcs are very well-known PSs, but they are also valuable as photoacoustic agents, which is a huge advantage for deep tissue and tomographic imaging. Among them, tetrasulfonated PcS4 accumulates on tumours and presents strong photoacoustic signals. Good tomographic contrast imaging can be achieved between different organs, such as kidneys, intestines and spleen, within 1 h of injection.[187] This confers an excellent applicability as

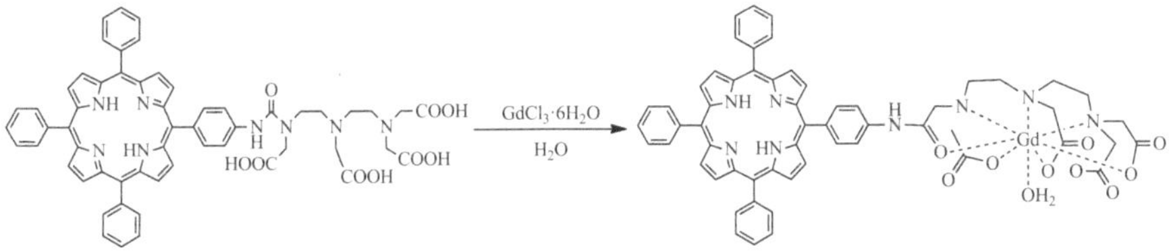


**Scheme 4.2** $Gd^{3+}$ chelation with diethylenetriamine pentaacetic acid bonded to the porphyrin meso position.

a bi-functional agent for PDT as well as PAI or Multispectral Optoacoustic Tomography (MSOT).[187]

Other PSs, such as PpIX, bis[4-(*N*,*N*-dibenzylamino)-2,6-dihydroxyphenyl] squaraine, chlorin e6 and MB, have been studied as potential photoacoustic agents because of their low $\Phi_F$, which raised the expectation that they could perform a relevant thermal relaxation from the excited state to the ground state.[188] However, none has as strong a photoacoustic signal as ZnPc.[188]

# 4.4 Theranostic Applications of PDT

When performing a surgical procedure with the goal of completely removing a tumour, it is critical to finely differentiate the tumour margins. Moreover, it is also of great interest to have tools to quantify the *in situ* production of ROS, thereby yielding tools to clinicians to assess the end of the treatment.

## 4.4.1 Tumour Delimitation

Since its discovery and approval in the 1990's, induction of endogenous PpIX synthesis using ALA and its derivatives has become a methodology of choice to perform PDT and photodynamic diagnosis (PDD) of malignant tumours.[121] As discussed above, ALA triggers the biosynthesis of PpIX which shows fluorescence emission in the red spectral range and photosensitises the production of ROS, mainly $^1O_2$, upon blue and red light exposure. Notwithstanding, the key to the successful use of ALA and its derivatives for phototheranostic purposes lies in the preferential accumulation of PpIX in tumours and malignant cells following ALA administration.[189]

Tumour diagnosis using ALA or its derivatives are usually performed by administration of the drug either topically (in superficial cancers) or as an oral dose several hours prior to the surgical procedure. After PpIX expression, the area of dissection is irradiated with purple/blue light (400–410 nm) yielding fluorescent visibility in the red region and thus allowing the surgeon to better resect the malignant tissue while preserving the healthy one (see Figure 4.6).[147,190]

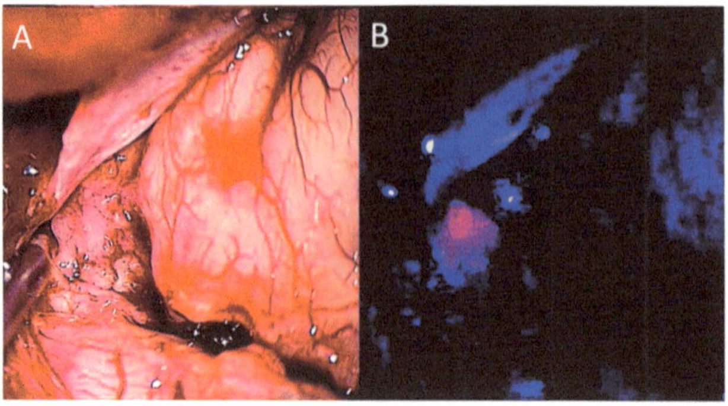


**Figure 4.6** (A) Appearance of glioma tissue treated with 5-ALA under white light in the microscopic view, (B) whilst under violet-blue light excitation PpIX it presents strong fluorescence emission. Reproduced from Widhalm *et al.*, *PLoS ONE*, 2013, **8**(10), e76988, DOI: 10.1371/journal.pone.0076988.[191] Published under the terms of the CC BY 4.0 license, https://creativecommons.org/licenses/by/4.0/.

Given the first approval of ALA as topical PS to remove actinic keratosis, it comes as no surprise that dermatology is the medical speciality where ALA and its derivatives have revolutionised the way of detecting and treating dermatologic disorders,[192,193] including squamous cell carcinomas (SCCs)[194] and basal cell carcinomas (BCCs).[195] However, other medical specialities have successfully implemented the usage of ALA for the detection and treatment of cancers. Phase III clinical trials of ALA for the resection of malignant glioma using fluorescence surgery revealed that ALA enables more complete resection of tumours, leading to improved progression-free survival in patients.[196] Lapini *et al.* reported clinical evidence of the outperforming capacity of PDD using Hexvix® in detecting bladder cancers compared to the standard white-light cystoscopy.[197]

Although phototheranostic treatments for cancer based on biosynthesis boosting of endogenous PpIX show several appealing perspectives, their application is still limited due to some unresolved issues. The expression of PpIX in cancerous cells is not always selective, as PpIX can also accumulate in inflammatory tissues yielding false positives[198,199] and specific cancers do not show sufficient contrast due to low PpIX expression compared to normal healthy tissues. In these cases, fluorescence of PpIX is masked by the autofluorescence of other endogenous porphyrins, as well as collagen or flavin.[200]

MB is another approved PS that has been used as a red/NIR fluorescence contrast agent in image guided fluorescence surgery. MB is a small dye with absorption in the red and low fluorescence emission ($\Phi_F \approx 0.04$ in EtOH)[201] at the edge of the red/NIR region. In fact, the low emission intensity of MB precluded for several years its use as a fluorescence contrast agent. In the field of tumour identification, MB has been used in the detection of solid tumours, such as insulinoma,[202] solitary fibrous tumour on pancreas,[203] parathyroid adenomas,[204] breast cancer[205] and paraganglioma.[206] However MB is not an ideal NIR contrast agent because of its low signal intensity, nonspecific uptake, and short retention times.[207]

### 4.4.2 Fluorescence Image Guided Surgery and PDT

The first-line treatment for the vast majority of solid tumours involves surgery followed by several sessions of chemotherapy and/or radiotherapy. In a typical surgical procedure, the surgeon, aided with CT and MRI images recorded previously, removes all the malignant tissue and sends it to the pathologist for analysis. The standard pathology report normally takes several days to be obtained, while the patient is recovering from the operation. If the pathologic report indicates the presence of positive tumour margins, then the patient is often asked to go through another surgical procedure, which is neither convenient nor practical. In contrast, a more desirable strategy would be to provide the surgeon with tools to intra-operatively discern the malignant tissue and, especially, eliminate the possible tumour margins left after resection. The combination of PDT with fluorescence image-guided surgery (FIGS) procedures would allow solving of the previous inconveniences.

Bogaards *et al.* analysed the technical feasibility of the combination of FIGS with PDT to treat VX2 carcinomas grown in rabbit brain. The authors used ALA as PDT drug, which was administered *i.v.* 4 h prior to the procedure. Their results showed that the technique was viable, yet no improvement between the outcome of rabbits undergoing FIGS and those receiving the combined treatment was observed, suggesting that further adjustment of the doses and procedure was required.[208] Zilidis *et al.* successfully applied FIGS and repetitive PDT in brain metastatic malignant melanomas in 6 patients, all of them remaining free of brain-disease until death.[209] In this case, a tandem of two PS was used: Photofrin® was administered intravenously 48 h before surgery and ALA administered orally 24 h before surgery.

Evidence of higher performance of PDT of residual margins after incomplete surgical resection in head and neck cancer models has been provided by Moore *et al.*[210] The authors used panitumumab-IRDye700DX to guide the resection of SCC implemented in the head and neck of mice. According to their results even when 10% or 50% of the tumour was left in the wound bed, long-term growth inhibition of residual disease was observed, suggesting that PDT is capable of eliminating malignant cells left in the surgical procedure. Interestingly, greater suppression of tumour recurrence was achieved in the 90% resection group, indicating that the efficacy of the combined treatment is superior when decreasing amounts of residual tumour.

The combined resection of head and neck cancer in mice followed by photodynamic treatment of the wound bed has also been reported by Muhanna *et al.* using multimodal porphyrin lipoprotein-mimicking NPs labelled with $^{64}Cu$.[211] The presence of the radioactive $^{64}Cu$ allowed the authors to combine fluorescence imaging and PET to guide the tumour resection and completed it with PDT.

Also, Mitsunaga *et al.* developed a cancer cell-selective PIT[144] using PC IR700 dye[212] in a mAb against human EGFR (HER2 cells). They were able to induce phototoxicity only where the mAb had bound, as well as assessing the efficacy of the treatment by monitoring fluorescence of the PS.

## 4.4.3 Dosimetry

A real-time assessment of the amount of ROS generated during PDT is essential to remove the malignant entity, whilst minimising the harm done to healthy tissues.[8,213,214] Explicit conventional dosimetry only takes into consideration the amount of light irradiated and the local concentration of PS and $O_2$. This is a simple approximation to dosimetry, but in many cases is insufficient. Dosimetry can be greatly improved by taking advantage of the theranostic properties of the PS (Figure 4.7).[215] For example, fluorescence can be used to obtain quality information on the amount of $^1O_2$ generated.

### 4.4.3.1 *Photobleaching-based PDT Dosimetry*

Implicit dosimetry monitors PDT by measuring the extent of photodegradation of the PS or the formation of photo-byproducts. Fluorescence is usually the property chosen to assess dosimetry. Implicit methods deliver combined information on the three parameters which affect PDT.

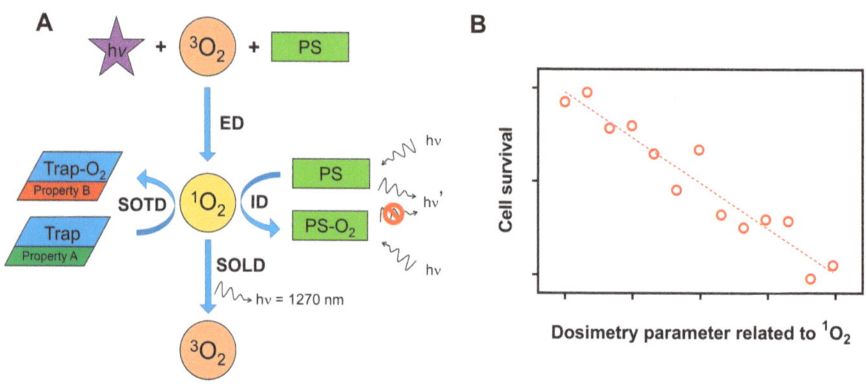


**Figure 4.7** (A) Scheme of the four types of PDT dosimetry. ED: Explicit dosimetry, where the concentration of $O_2$, PS and the irradiated light are used to calculate the $^1O_2$ dose. ID: Implicit dosimetry, where the PS photobleaching is measured in order to calculate the $^1O_2$ dose. SOLD: Singlet oxygen luminescence dosimetry, where the $^1O_2$ dose is calculated directly from $^1O_2$ phosphorescence at 1270 nm. SOTD: Singlet oxygen trapping dosimetry where the $^1O_2$ dose is calculated from the evolution of a $^1O_2$ trapping moiety. (B) Cellular survival response in front of a dosimetry parameter, which should be related to the amount of $^1O_2$ generated.

PS photobleaching has been correlated by means of biophysical methods with the amount of $^1O_2$ generated (eqn (4.1)).[216,217] These models have enabled the definition of "*PDT dose*" as: the total cumulative $^1O_2$ produced during the treatment time ($t$).

$$^1O_2\text{Dose} = \frac{1}{\tau_\Delta}\int_0^t {}^1O_2 \mathrm{d}t = -\frac{1}{\tau_\Delta k_{os}}\int_{[PS]_0}^{[PS]_t}\frac{\mathrm{d}[PS]}{[PS]} = \frac{1}{\tau_\Delta k_{os}}\ln\left(\frac{[PS]_0}{[PS](t)}\right) \quad (4.1)$$

Eqn (4.1) assumes that PS molecules react with $^1O_2$ ($k_{os}$) (valid affirmation at high PS concentrations). Several studies have confirmed this correlation with 5-ALA or PpIX as PSs, both *in vitro* and *in vivo*.[218,219] Other works with ALA and PpIX confirm these findings while others correlate "*fluorescence loss integrated over the time interval of the photobleaching*" with cellular death or tissue damage.[218,219] Correlation has later on been used as a predictive tool, to the extent of reaching "*Phase I*" clinical trials. For example, Valentine *et al.* proved that all thier patients presenting superficial BCC which responded positively to PDT with ALA presented with more than 90% photobleaching activity.[220,221]

Despite its promising applications, implicit dosimetry methods present some drawbacks: (i) Cell and tissue autofluorescence can partially mask photobleaching fluorescence;[222] (ii) certain PSs and physiological conditions, such as hypoxia (<5 $O_2$ μM), present negligible $^1O_2$-mediated photobleaching, non-$^1O_2$-mediated photobleaching being the dominant phenomenon. In these cases, implicit measurements no longer deliver a reliable dose metric;[223] (iii) formation of photo-byproducts with different photophysical properties which could hinder PDT efficiency.[224]

### 4.4.3.2 *Singlet Oxygen Luminescence Dosimetry*

A second approach to quantify $^1O_2$ generation is by taking advantage of $^1O_2$ phosphorescence in the NIR ($\lambda_{em}$ = 1270 nm).[225] Singlet oxygen luminescence dosimetry (SOLD) can be used to directly monitor the extent and evolution of PDT (eqn (4.2)).[226]

$$^1O_2\text{Dose} = \frac{1}{\tau_\Delta}\int_0^t \left[{}^1O_2\right]\mathrm{d}t \quad \alpha \quad \frac{1}{\tau_\Delta}\int_0^t L_{1270}\mathrm{d}t \quad (4.2)$$

SOLD only began to be a feasible option in $^1O_2$ quantification 15 years ago due to the difficulty in measuring its emission in biologic media. Liquid nitrogen cooled germanium detectors were only capable of detecting intracellular $^1O_2$ luminescence in very favourable conditions ($D_2O$ and high PS concentrations) with poorly reproducible results.[227–229] The newly appeared "NIR photomultiplier tube" (*e.g.* R5509, Hamamatsu) increased in more than one order of magnitude the sensitivity of the detector of the time.[226]

Hirano *et al.* reported the detection of $^1O_2$ phosphorescence generated using Photofrin® from murine tumours cells suspended in non-deuterated.[230] In the same year, Niedre *et al.* proved correlation between the integrated intensity of $^1O_2$ phosphorescence and leukemia cell mortality (OCI-AML5) *in vitro*.[231,232] They also demonstrated that the correlation is not affected by the dose of light nor the concentration of PS and $O_2$.[232] These preliminary studies have given rise to additional investigations using different cell lines (*e.g.*, Jurkat cells, human colon carcinoma (HT29 and SW480) and pancreas cancer (AsPC-1 and MiaPaCa-2) together with different PSs (ALA, PpIX, Ce6 or pheophorbide a)).[223,233,234]

As a first approach of *in vivo* SOLD, Niedre *et al.* found correlation between the number of *"counts"* and the photodynamic response of the skin of *SKH1-HR hairless mice*.[235] Human oedema and erythema disorders also present the same correlation.[236] Additionally, the presence of $^1O_2$ quenchers[237] or triplet oxygen[238] can be assessed and diagnosed by means of time resolved phosphorescence, a technique which delivers information of the microenvironment of the PS. Other *in vivo* cancer studies have discovered that larger amounts of SOLD detected $^1O_2$ correlate with a greater reduction in a cancer tumour, such as in L-gliosarcosoma.[234,239] Additionally, instrumental setups have improved immensely in the past years to increase signal to noise ratio, such as including the use of more sensitive detectors and incorporation of optical fibre.

Photobleaching and SOLD dosimetry techniques are able to monitor and even predict cell viability in the majority of studied conditions. SOLD has had difficulties entering clinical studies due to its cost, technological complexity and low signal to noise ratio. On the other hand, photobleaching presents a more practical, cheap and simpler monitoring system, which has enabled this methodology to already be used in clinical studies.

### *4.4.3.3 Singlet Oxygen Chemical Trapping Dosimetry*

An alternative to the two previously explained dosimetry techniques would be the use of chemical compounds which selectively react with $^1O_2$ and also have a *"singular property"* which is measured. To this end, two major strategies have been pursued: the self-reporting PSs and the use of $^1O_2$ detection probes.

Self-reporting PSs are supramolecular entities with three embedded functionalities: a PS responsible for $^1O_2$ generation, a scissile linker that breaks when $^1O_2$ is formed, and a fluorophore ($^1O_2$ reporter) whose fluorescence is quenched by the PS. Once $^1O_2$ is generated, it cleaves the linker and releases the fluorophore, which then recovers its inherent fluorescence.[240,241] Thus, fluorescence evolution is used to monitor $^1O_2$ production. A variation of this strategy is to use a reporter whose fluorescence increases upon reacting with $^1O_2$.[242]

Molecular probes are another option in order to detect intracellular $^1O_2$. A popular approach is to use $^1O_2$ chemiluminescent probes, such as fluorescenyl cypridina luciferin analogues (FCLA).[243] When FCLA probes react with $^1O_2$, an excited photoproduct is generated that decays emitting light. *In vitro* and *in vivo* PDT studies have reported good correlations between cell viability and cumulative chemiluminescence.[243–245] Likewise, $^1O_2$ fluorescent probes are based on the fluorescent change of the probe after reacting with $^1O_2$. Fluorescence can either decrease (*e.g.* anthracene fluorescence bleaching) or increase (based on molecule dyads in which the interaction between a fluorophore and a quencher changes once $^1O_2$ is generated).[44] An example of an *in vivo* "turn-on" fluorescent probe is a dyad in which anthracene interacts with a silicon-containing rhodamine, which emits fluorescent radiation in the NIR region, the region in which tissues present a spectral window.[246] Other groups have used luminescent probes based on rare earth complexes in order to monitor intracellular $^1O_2$ in PDT conditions.[247,248]

The main advantage of $^1O_2$ fluorescent probes is that their $\Phi_F$ are several orders of magnitude larger than $^1O_2$ phosphorescence quantum yield. Moreover, fluorescent photons emitted in the visible region of the spectrum can be detected using simpler and more sensitive detectors than NIR photons.[44] On the other hand, the most concerning drawback of the probe is its complexation with proteins. For example, the commercially available probe SOSG (Singlet Oxygen Sensor Green)[249] forms complexes with globular proteins, such as bovine serum albumin.[250] This causes SOSG to be cell-impermeant in regular growth media (typically supplemented with foetal calf serum), whilst in protein-free media SOSG is cell-permeant.[250,251] Several strategies have been displayed to overcome these drawbacks.

A first approach is based on the use of genetically-encoded proteins as $^1O_2$ probes. These proteins do not need the addition of any external cofactor and thus provides unprecedented spatiotemporal control. Some examples have already been reported such as IFP 1.4 and *holo*UnaG, which contain biliverdin and bilirubin as fluorescent chromophores respectively.[252,253] Both chromophores react selectively with $^1O_2$ generating photo-oxidation products which are not fluorescent, and therefore a decrease of fluorescence is detected. The main advantage lies in the fact that cells by themselves can biosynthesise the $^1O_2$ probe and can be used as a $^1O_2$ dosimeter.

Nanotechnology has become a promising path in order to protect the probe against protein complexation and facilitate cellular internalisation. Lately, anthracene dipropionic acid (ADPA) conjugated with silica NPs have shown to prevent probe-protein interaction, increasing $^1O_2$ detectability inside HeLa cells.[254] This nanoprobe has been used for *S. aureus* photoinactivation in which a correlation was observed between fluorescent change of ADPA and bacterial cellular death.

## 4.5 Outlook and Concluding Remarks

The intrinsic fluorescence of many PSs and the use of light for drug activation make PDT an ideal theranostic tool as a single molecular entity provides diagnosis and therapy. PDT is an active field of research, both preclinical and clinical, aiming at increasing the efficiency and safety of treatments, and ultimately, the number of clinical indications. Recent discoveries in fields as diverse as molecular engineering, nanotechnology, photonics and plasmonics offer new opportunities to further expand its scope and increase its value. Of special interest is the development of novel dosimetry approaches that will allow a better spatiotemporal control of light and, especially, $^1O_2$ delivery. Likewise, the ability to use PSs to visualize tumour margins during resection followed by photodynamic elimination of residual cancer cells in the tumour bed may have a strong impact on patient prognosis.

## Abbreviations

| | |
|---|---|
| $^1PS^*$ | Singlet Excited State Photosensitiser |
| $^1O_2$ | Singlet Oxygen |
| $^3PS^*$ | Triplet Excited State Photosensitiser |
| ADPA | Anthracene Dipropionic Acid |
| ALA | 5-Aminolevulinic Acid |
| BCC | Basal Cell Carcinoma |
| CT | Computed Tomography |
| EGFR | Epidermal Growth Factor Receptor |
| EPR | Enhanced Permeabilty and Retention Effect |
| FCLA | Fluorescenyl Cypridina Luciferin Analogues |
| IGFS | Fluorescence Image Guided Surgery |
| FLIM | Fluorescence Lifetime Imaging Microscopy |
| HSA | Human Serum Albumin |
| ICG | Indocyanine Green |
| Ig | Immunoglobulin |
| ISC | Intersystem Crossing |
| mAb | Monoclonal Antibody |
| MB | Methylene Blue |
| MRI | Magnetic Resonance Imaging |
| MSOT | Multispectral Optoacoustic Tomography |
| *m*THPC | *m*-Tetrahydroxyphenylchlorin |
| NIR | Near Infrared |
| NIRF | Near Infrared Fluorescence |
| NP | Nanoparticle |
| PAI | Photoacoustic Imaging |
| PAT | Photoacoustic Tomography |
| PDD | Photodynamic Diagnosis |
| PDT | Photodynamic Therapy |
| PEG | Polyethyleneglycol |
| PET | Positron Emission Tomography |
| PLP | Porphyrin Lipoprotein |
| Ppa | Pyropheophorbide a |
| PpIX | Protoporphyrin IX |

| | |
|---|---|
| PS | Photosensitiser |
| QD | Quantum Dot |
| ROS | Reactive Oxygen Species |
| SCC | Squamous Cell Carcinomas |
| SERS | Surface Enhanced Raman Scattering |
| SOLD | Singlet Oxygen Luminescence Dosimetry |
| SOSG | Singlet Oxygen Sensor Green |
| SOTD | Singlet Oxygen Trapping Dosimetry |
| US | Ultrasound |
| $\Phi_F$ | Fluorescent Quantum Yield |
| $\Phi_\Delta$ | Singlet Oxygen Quantum Yield |

## Acknowledgements

Part of the research described in this chapter has been supported by the Spanish Ministerio de Economía y Competitividad through grant no. CTQ2013-48767-C3-1-R. C. H. thanks IQS and Eurofragance for a fellowship. B. R.-A. thanks FERRER and O. P. and R. B.-O. thanks the SUR del DEC de la Generalitat de Catalunya (grants No. 2015 FI_B1 00063 and 2015 FI_B 00315) for their respective predoctoral fellowships. S. N. is very grateful to his former students and collaborators for their contribution to some of the work described herein. Finally we would like to thank all the researchers who have contributed to this field and whose names are listed in the references.

## References

1. C. A. Robertson, D. H. Evans and H. Abrahamse, *J. Photochem. Photobiol., B*, 2009, **96**, 1.
2. P. Agostinis, K. Berg, K. A. Cengel, T. H. Foster, A. W. Girotti, S. O. Gollnick, S. M. Hahn, M. R. Hamblin, A. Juzeniene, D. Kessel, M. Korbelik, J. Moan, P. Mroz, D. Nowis, J. Piette, B. C. Wilson and J. Golab, *Ca-Cancer J. Clin.*, 2011, **61**, 250.
3. D. E. J. G. J. Dolmans, D. Fukumura and R. K. Jain, *Nat. Rev. Cancer*, 2003, **3**, 380.
4. B. C. Wilson and M. S. Patterson, *Phys. Med. Biol.*, 2008, **53**, 61.
5. M. Ochsner, *J. Photochem. Photobiol., B.*, 1997, **39**, 1.
6. S. C. Gupta, D. Hevia, S. Patchva, B. Park, W. Koh and B. B. Aggarwal, *Antioxid. Redox Signaling*, 2012, **16**, 1295.
7. A. E. Karsten, A. Singh and M. W. Braun, *Lasers Med. Sci.*, 2012, **27**, 79.
8. B. W. Pogue, J. T. Elliott, S. C. Kanick, S. C. Davis, K. S. Samkoe, E. V. Maytin, S. P. Pereira and T. Hasan, *Phys. Med. Biol.*, 2016, **61**, R57.
9. H.-B. Ris, *Lung Cancer*, 2005, **49**(1), S65.
10. C. M. Moore, A. R. Azzouzi, E. Barret, A. Villers, G. H. Muir, N. J. Barber, S. Bott, J. Trachtenberg, N. Arumainayagam, B. Gaillac, C. Allen, A. Schertz and M. Emberton, *BJU Int.*, 2015, **116**, 888.

11. F. H. Mustafa and M. S. Jaafar, *Indian J. Phys.*, 2013, **87**, 203.
12. T. C. Zhu, M. M. Kim, X. Liang, J. C. Finlay and T. M. Busch, *Photonics Lasers Med.*, 2015, **4**, 59.
13. T. H. Foster, R. S. Murant, R. G. Bryant, R. S. Knox, S. L. Gibson and R. Hilf, *Radiat. Res.*, 1991, **126**, 296.
14. R. B. Veenhuizen and F. A. Stewart, *Radiother. Oncol.*, 1995, **37**, 131.
15. T. M. Sitnik, J. A. Hampton and B. W. Henderson, *Br. J. Cancer*, 1998, **77**, 1386.
16. B. Rodríguez-Amigo, O. Planas, R. Bresolí-Obach, J. Torra, R. Ruiz-González and S. Nonell, *Photosensitisers for Photodynamic Therapy: State of the Art and Perspectives*, The Royal Society of Chemistry, London, 2016.
17. J. D. Spikes, *Photochem. Photobiol.*, 1997, **65S**, 142S.
18. G. B. Kharkwal, S. K. Sharma, Y. Y. Huang, T. Dai and M. R. Hamblin, *Lasers Surg. Med.*, 2011, **43**, 755.
19. H. Gursoy, C. Ozcakir-Tomruk, J. Tanalp and S. Yilmaz, *Clin. Oral Investig.*, 2013, **17**, 1113.
20. C. Vassena, S. Fenu, F. Giuliani, L. Fantetti, G. Roncucci, G. Simonutti, C. L. Roman, R. De Francesco and L. Drago, *Int. J. Antimicrob. Agents*, 2014, **44**, 47.
21. H. Qiu, Y. Mao, Y. Gu, J. Zhu, Y. Wang, J. Zeng, N. Huang, Q. Liu and Y. Yang, *J. Photochem. Photobiol., B*, 2014, **130**, 305.
22. F. Ziemssen and H. Heimann, *Expert Opin. Drug Metab. Toxicol.*, 2012, **8**, 1023.
23. A. Kawczyk-Krupka, A. M. Bugaj, M. Potempa, K. Wasilewska, W. Latos and A. Sieroń, *Photodiagn. Photodyn. Ther.*, 2015, **12**, 161.
24. P. Babilas, S. Schreml, M. Landthaler and R. M. Szeimies, *Photodermatol., Photoimmunol. Photomed.*, 2010, **26**, 118.
25. M. T. Wan and J. Y. Lin, *Clin., Cosmet. Invest. Dermatol.*, 2014, **7**, 145.
26. E. Rosàs, P. Santomá, M. Duran-Frigola, B. Hernandez, M. C. Llinàs, R. Ruiz-González, S. Nonell, D. Sánchez-García, E. R. Edelman and M. Balcells, *Langmuir*, 2013, **29**, 9734.
27. T. M. Chou, K. W. Woodburn, W. F. Cheong, S. A. Lacy, K. Sudhir, D. C. Adelman and D. Wahr, *Catheter. Cardiovasc. Interv.*, 2002, **57**, 387–394.
28. M. O. Senge and M. W. Radomski, *Photodiagn. Photodyn. Ther.*, 2013, **10**, 1–16.
29. J. E. Muller, *J. Am. Coll. Cardiol.*, 2008, **52**, 1033.
30. D. Gabriel, N. Lange, V. Chobaz-Peclat, M. F. Zuluaga, R. Gurny, H. van den Bergh and N. Busso, *J. Controlled Release*, 2012, **163**, 178.
31. D. Gabriel, N. Busso, A. So, H. van den Bergh, R. Gurny and N. Lange, *J. Controlled Release*, 2009, **138**, 225.
32. R. S. Boumédine and D. C. Roy, *Cytotherapy*, 2005, **7**, 134.
33. J. Seghatchian and G. de Sousa, *Transfus. Apher. Sci.*, 2006, **35**, 189.
34. R. Spitler and M. W. Berns, *J. Biomed. Opt.*, 2014, **19**, 038001.
35. F. F. Sperandio, A. Simões, L. Corrêa, A. C. C. Aranha, F. S. Giudice, M. R. Hamblin and S. C. O. M. Sousa, *J. Biophotonics*, 2015, **8**, 795.

36. P. A. Swanson, A. Kumar, S. Samarin, M. Vijay-Kumar, K. Kundu, N. Murthy, J. Hansen, A. Nusrat and A. S. Neish, *Proc. Natl. Acad. Sci. U. S. A.*, 2011, **108**, 8803.
37. W. W. Kilarski, A. Muchowicz, M. Wachowska, R. Mężyk-Kopeć, J. Golab, M. A. Swartz and P. Nowak-Sliwinska, *Angiogenesis*, 2014, **17**, 347.
38. C. S. Foote, *Photochem. Photobiol.*, 1991, **54**, 659.
39. I. O. L. Bacellar, T. M. Tsubone, C. Pavani and M. S. Baptista, *Int. J. Mol. Sci.*, 2015, **16**, 20523.
40. L. Huang, Y. Xuan, Y. Koide, T. Zhiyentayev, M. Tanaka and M. R. Hamblin, *Lasers Surg. Med.*, 2012, **44**, 490.
41. A. P. Castano, P. Mroz and M. R. Hamblin, *Nat. Rev. Cancer*, 2006, **6**, 535.
42. N. S. Gsponer, M. L. Agazzi, M. B. Spesia and E. N. Durantini, *Methods*, 2016, **109**, 167.
43. M. Garcia-Diaz, Y.-Y. Huang and M. R. Hamblin, *Methods*, 2016.
44. R. Ruiz-González and A. L. Zanocco, in *Singlet Oxygen: Applications in Biosciences and Nanosciences*, ed. S. Nonell and C. Flors, The Royal Society of Chemistry, 2016, vol. 2, 30, pp. 103–120.
45. B. Halliwell, K. Zhao and M. Whiteman, *Free Radical Res.*, 1999, **31**, 651.
46. B. Halliwell, *Plant Physiol.*, 2006, **141**, 312.
47. L. Huang, T. G. St. Denis, Y. Xuan, Y.-Y. Huang, M. Tanaka, A. Zadlo, T. Sarna and M. R. Hamblin, *Free Radical Biol. Med.*, 2012, **53**, 2062.
48. R. Yin, M. Wang, Y. Y. Huang, G. Landi, D. Vecchio, L. Y. Chiang and M. R. Hamblin, *Free Radical Biol. Med.*, 2015, **79**, 14.
49. K. R. Kasimova, M. Sadasivam, G. Landi, T. Sarna and M. R. Hamblin, *Photochem. Photobiol. Sci.*, 2014, **13**, 1541.
50. Y. Zhang, T. Dai, M. Wang, D. Vecchio, L. Y. Chiang and M. R. Hamblin, *Nanomedicine*, 2015, **10**, 603.
51. D. Vecchio, A. Gupta, L. Huang, G. Landi, P. Avci, A. Rodas and M. R. Hamblin, *Antimicrob. Agents Chemother.*, 2015, **59**, 5203.
52. M. Salmon-Divon, Y. Nitzan and Z. Malik, *Photochem. Photobiol. Sci.*, 2004, **3**, 423.
53. M. B. Spesia, D. A. Caminos, P. Pons and E. N. Durantini, *Photodiagn. Photodyn. Ther.*, 2009, **6**, 52.
54. R. Ruiz-González, J. H. White, M. Agut, S. Nonell and C. Flors, *Photochem. Photobiol. Sci.*, 2012, **11**, 1411.
55. R. Ruiz-González, J. H. White, A. L. Cortajarena, M. Agut, S. Nonell and C. Flors, *Progress in Biomedical Optics and Imaging–Proceedings of SPIE*, Barcelona, Spain, 2013, vol. 8596, pp. 1–7.
56. J. O. Yoo and K. S. Ha, *Int. Rev. Cell Mol. Biol.*, 2012, **295**, 139.
57. P. Mroz, A. Yaroslavsky, G. B. Kharkwal and M. R. Hamblin, *Cancers*, 2011, **3**, 2516.
58. B. W. Henderson, S. M. Waldow, T. S. Mang, W. R. Potter, P. B. Malone and T. J. Dougherty, *Cancer Res.*, 1985, **45**, 572.
59. W. S. Chan, N. Brasseur, C. La Madeleine and J. E. Van Lier, *Anticancer Res.*, 1996, **16**, 1887.
60. B. Chen, B. W. Pogue, J. M. Luna, R. L. Hardman, P. J. Hoopes and T. Hasan, *Clin. Cancer Res.*, 2006, **12**, 917.

61. A. Weiss, H. van Den Bergh, A. W. Griffioen and P. Nowak-Sliwinska, *Biochim. Biophys. Acta, Rev. Cancer*, 2012, **1826**, 53.
62. R. Bhuvaneswari, Y. Y. Gan, K. C. Soo and M. Olivo, *Cell. Mol. Life Sci.*, 2009, **66**, 2275.
63. N. Yamamoto, S. Homma, T. W. Sery, L. A. Donoso and J. K. Hoober, *Eur. J. Cancer*, 1991, **27**, 467.
64. M. Korbelik, G. Krosl, J. Krosl and G. J. Dougherty, *Cancer Res.*, 1996, **56**, 5647.
65. M. Korbelik and M. R. Hamblin, *Photochem. Photobiol. Sci.*, 2015, **14**, 1403.
66. M. Korbelik, *Cancer Lett.*, 2009, **281**, 232.
67. G. Krosl and M. Korbelik, *Cancer Lett.*, 1994, **84**, 43.
68. M. Uehara, K. Sano, Z. L. Wang, J. Sekine, H. Ikeda and T. Inokuchi, *Cancer Immunol. Immunother.*, 2000, **49**, 401.
69. M. Korbelik, J. Sun and J. J. Posakony, *Photochem. Photobiol.*, 2001, **73**, 403.
70. M. Korbelik, J. Banáth, J. Sun, D. Canals, Y. A. Hannun and D. Separovic, *Int. Immunopharmacol.*, 2014, **20**, 359.
71. S. S. Kelkar and T. M. Reineke, *Bioconjugate Chem.*, 2011, **22**, 1879.
72. P. Huang, J. Lin, X. Wang, Z. Wang, C. Zhang, M. He, K. Wang, F. Chen, Z. Li, G. Shen, D. Cui and X. Chen, *Adv. Mater.*, 2012, **24**, 5104.
73. P. Rai, S. Mallidi, X. Zheng, R. Rahmanzadeh, Y. Mir, S. Elrington, A. Khurshid and T. Hasan, *Adv. Drug Delivery Rev.*, 2010, **62**, 1094.
74. A. E. O'Connor, W. M. Gallagher and A. T. Byrne, *Photochem. Photobiol.*, 2009, **85**, 1053.
75. D. E. J. G. J. Dolmans, D. Fukumura and R. K. Jain, *Nat. Rev. Cancer*, 2003, **3**, 380.
76. A. Kamkaew, S. H. Lim, H. B. Lee, L. V. Kiew, L. Y. Chug and K. Burgess, *Chem. Soc. Rev.*, 2013, **42**, 77.
77. S. Singh, A. Aggarwal, N. V. S. D. K. Bhupathiraju, G. Arianna, K. Tiwari and C. M. Drain, *Chem. Rev.*, 2015, **115**, 10261.
78. J. M. Fernandez, M. D. Bilgin and L. I. Grossweiner, *J. Photochem. Photobiol., B.*, 1997, **37**, 131.
79. D. H. Crean, C. Liebow, R. B. Penetrante and T. S. Mang, *Cancer*, 1993, **72**, 3068.
80. A. B. Ormond and H. S. Freeman, *Materials*, 2013, **6**, 817.
81. B. Liu, T. J. Farrell and M. S. Patterson, *Phys. Med. Biol.*, 2010, **55**, 5913.
82. R. W. Redmond and J. N. Gamlin, *Photochem. Photobiol.*, 1999, **70**, 391.
83. M. Wachowska, A. Muchowicz, M. Firczuk, M. Gabrysiak, M. Winiarska, M. Wańczyk, K. Bojarczuk and J. Golab, *Molecules*, 2011, **16**, 4140.
84. M. O. Senge and J. C. Brandt, *Photochem. Photobiol.*, 2011, **87**, 1240.
85. S. L. Gundy, W. Van der Putten, A. Shearer, A. G. Ryder and M. Ball, *Diagn. Opt. Spectrosc. Biomed.*, 2001, **4432**, 299.
86. C. Hadjur, N. Lange, J. Rebstein, P. Monnier, H. van den Bergh and G. Wagnières, *J. Photochem. Photobiol., B.*, 1998, **45**, 170.

87. Y. P. Istomin, M. A. Kaplan, S. V. Shliakhtsin, T. P. Lapzevich, D. A. Cerkovsky, L. N. Marchanka, A. S. Fedulov and T. V. Trukhachova, ed. D. H. Kessel, *Proc. SPIE*, 2009, **7380**, 73806V.
88. L. Martínez-Fernández, J. González-Vázquez, L. González and I. Corral, *J. Chem. Theory Comput.*, 2015, **11**, 406.
89. I. M. Belousova, M. V. Dobrun, L. V. Galebskaya, S. I. Gorelov, I. M. Kislyakov, S. E. Kolbasov, A. V. Kris'ko, T. K. Kris'ko, M. A. Malkov, T. D. Murav'eva and N. N. Petrishchev, *Laser Opt.*, 2010, **7822**, 78220W.
90. V. Y. Panevin, A. N. Sofronov, D. A. Firsov and A. L. Ter-Martirosyan, *St. Petersburg. Polytech. Univ. J. Phys. Math.*, 2015, **1**, 395.
91. V. A. Privalov, A. V. Lappa, O. V. Seliverstov, A. B. Faizrakhmanov, N. N. Yarovoy, E. V. Kochneva, M. V. Evnevich, A. S. Anikina, A. V. Reshetnicov, I. D. Zalevsky and Y. V. Kemov, *Proc. SPIE-Int. Soc. Opt. Eng.*, 2002, 178.
92. S. Fickweiler, C. Abels, S. Karrer, W. Bäumler, M. Landthaler, F. Hofstädter and R. Szeimies, *J. Photochem. Photobiol., B*, 1999, **48**, 27.
93. B. Brozek-Pluska, M. Orlikowski and H. Abramczyk, in *Handbook of Porphyrin Science*, ed. K. M. Kadish, K. M. Smith and R. Guilard, The University of South Florida, USA, 2012, vol. 39, pp. 317–331.
94. S. C. Shim, H. K. Kang, S. K. Park and E. J. Shin, *J. Photochem.*, 1987, **37**, 125.
95. L. B. Josefsen and R. W. Boyle, *Theranostics*, 2012, **2**, 916.
96. N. Shirasu, S. O. Nam and M. Kuroki, *Anticancer Res.*, 2013, 2823.
97. R. Weijer, M. Broekgaarden, M. Kos, R. van Vught, E. A. J. Rauws, E. Breukink, T. M. van Gulik, G. Storm and M. Heger, *J. Photochem. Photobiol., C*, 2015, **23**, 103.
98. T. Ishikawa, K. Takahashi, N. Ikeda, Y. Kajimoto, Y. Hagiya, S. Ogura, S. Miyatake and T. Kuroiwa, *Pharmaceutics*, 2011, **3**, 615.
99. S. Collaud, A. Juzeniene, J. Moan and N. Lange, *Curr. Med. Chem.: Anti-Cancer Agents*, 2004, **4**, 301.
100. L. Bourré, F. Giuntini, I. M. Eggleston, M. Wilson and A. J. MacRobert, *Br. J. Cancer*, 2009, **100**, 723.
101. M. Kim, H. Y. Jung and H. J. Park, *Int. J. Mol. Sci.*, 2015, **16**, 23259.
102. L. B. Josefsen and R. W. Boyle, *Met.-Based Drugs*, 2007, **2008**, 1.
103. G. Valduga, A. Pifferi, P. Taroni, G. Valentini, E. Reddi, R. Cubeddu and G. Jori, *J. Photochem. Photobiol. B*, 1999, **49**, 198.
104. Y. N. Konan, R. Gurny and E. Allémann, *J. Photochem. Photobiol. B*, 2002, **66**, 89.
105. V. Torchilin, *Adv. Drug Delivery Rev.*, 2011, **63**, 131.
106. F. Postigo, M. Mora, M. A. de Madariaga, S. Nonell and M. L. Sagrista, *Int. J. Pharm.*, 2004, **278**, 239.
107. M. García-Díaz, S. Nonell, A. Villanueva, J. C. Stockert, M. Cañete, A. Casadó, M. Mora and M. L. Sagristá, *Biochim. Biophys. Acta*, 2011, **1808**, 1063.
108. H. Toledano, R. Edrei and S. Kimel, *J. Photochem. Photobiol. B*, 1998, **42**, 20.

109. M. Kuruppuarachchi, H. Savoie, A. Lowry, C. Alonso and R. W. Boyle, *Mol. Pharmaceutics*, 2011, **8**, 920.
110. M. Qin, H. J. Hah, G. Kim, G. Nie, Y.-E. K. Lee and R. Kopelman, *Photochem. Photobiol. Sci.*, 2011, **10**, 832.
111. D. Gao, H. Xu, M. A. Philbert and R. Kopelman, *Angew. Chem., Int. Ed.*, 2007, **46**, 2224.
112. A. Vargas, N. Lange, T. Arvinte, R. Cerny, R. Gurny and F. Delie, *J. Drug Targeting*, 2009, **17**, 599.
113. M. Zeisser-Labouèbe, N. Lange, R. Gurny and F. Delie, *Int. J. Pharm.*, 2006, **326**, 174.
114. E. Boix-Garriga, P. Acedo, A. Casadó, A. Villanueva, J. C. Stockert, M. Cañete, M. Mora, M. L. Sagristá and S. Nonell, *Nanotechnology*, 2015, **26**, 365104.
115. P. Couleaud, V. Morosini, C. Frochot, S. Richeter, L. Raehm and J. O. Durand, *Nanoscale*, 2010, **2**, 1083.
116. O. Planas, R. Bresolí-Obach, J. Nos, T. Gallavardin, R. Ruiz-González, M. Agut and S. Nonell, *Molecules*, 2015, **20**, 6284.
117. M. K. Khaing Oo, Y. Yang, Y. Hu, M. Gomez, H. Du and H. Wang, *ACS Nano*, 2012, **6**, 1939.
118. D. C. Hone, P. I. Walker, R. Evans-Gowing, S. FitzGerald, A. Beeby, I. Chambrier, M. J. Cook and D. A. Russell, *Langmuir*, 2002, **18**, 2985.
119. O. Planas, N. Macia, M. Agut, S. Nonell and B. Heyne, *J. Am. Chem. Soc.*, 2016, **138**, 2762.
120. G. P. Tegos, T. N. Demidova, D. Arcila-Lopez, H. Lee, T. Wharton, H. Gali and M. R. Hamblin, *Chem. Biol.*, 2005, **12**, 1127.
121. T. Kiesslich, J. Berlanda, K. Plaetzer, B. Krammer and F. Berr, *Photochem. Photobiol. Sci.*, 2007, **6**, 619.
122. J. G. Christie and U. B. Kompella, *Drug Discovery Today*, 2008, **13**, 124.
123. X.-S. Li, M.-R. Ke, W. Huang, C.-H. Ye and J.-D. Huang, *Chem.–Eur. J.*, 2015, **21**, 3310.
124. C. S. Jin and G. Zheng, *Lasers Surg. Med.*, 2011, **43**, 734.
125. J. F. Lovell, T. W. B. Liu, J. Chen and G. Zheng, *Chem. Rev.*, 2010, **110**, 2839.
126. L. Cui, Q. Lin, C. S. Jin, W. Jiang, H. Huang, L. Ding, N. Muhanna, J. C. Irish, F. Wang, J. Chen and G. Zheng, *ACS Nano*, 2015, **9**, 4484.
127. E. Hutter and J. H. Fendler, *Adv. Mater.*, 2004, **16**, 1685.
128. T. Simon, M. Potara, A.-M. Gabudean, E. Licarete, M. Banciu and S. Astilean, *ACS Appl. Mater. Interfaces*, 2015, **7**, 16191.
129. A. Farhadi, Á. Roxin, B. C. Wilson and G. Zheng, *Theranostics*, 2015, **5**, 469.
130. Z. Sheng, D. Hu, M. Zheng, P. Zhao, H. Liu, D. Gao, P. Gong, G. Gao, P. Zhang, Y. Ma and L. Cai, *ACS Nano*, 2014, **8**, 12310.
131. T. Lin, W. Guo, Q. Long, A. Ma, Q. Liu, H. Zhang, Y. Huang, S. Chandrasekaran, C. Pan, K. S. Lam and Y. Li, *Theranostics*, 2016, **6**, 1324.
132. F. Moret, D. Scheglmann and E. Reddi, *Photochem. Photobiol. Sci.*, 2013, **12**, 823.

133. P. X. Li, J. H. Mu, H. L. Xiao and D. H. Li, *Oncol. Rep.*, 2015, **33**, 125.
134. X. Feng, D. Jiang, T. Kang, J. Yao, Y. Jing, T. Jiang, J. Feng, Q. Zhu, Q. Song, N. Dong, X. Gao and J. Chen, *ACS Appl. Mater. Interfaces*, 2016, **8**, 17817.
135. A. M. Master, Y. Qi, N. L. Oleinick and A. Sen Gupta, *Nanomed. Nanotechnol. Biol. Med.*, 2012, **8**, 655.
136. G. R. Reddy, M. S. Bhojani, P. McConville, J. Moody, B. A. Moffat, D. E. Hall, G. Kim, Y. E. L. Koo, M. J. Woolliscroft, J. V. Sugai, T. D. Johnson, M. A. Philbert, R. Kopelman, A. Rehemtulla and B. D. Ross, *Clin. Cancer Res.*, 2006, **12**, 6677.
137. T. Nakajima, K. Sano, M. Mitsunaga, P. L. Choyke and H. Kobayashi, *Cancer Res.*, 2012, **72**, 4622.
138. G. Obaid, I. Chambrier, M. J. Cook and D. A. Russell, *Angew. Chem., Int. Ed.*, 2012, **51**, 6158.
139. J.-K. Rhee, M. Baksh, C. Nycholat, J. C. Paulson, H. Kitagishi and M. G. Finn, *Biomacromolecules*, 2012, **13**, 2333.
140. P. M. R. Pereira, B. Korsak, B. Sarmento, R. J. Schneider, R. Fernandes and J. P. C. Tomé, *Org. Biomol. Chem.*, 2015, **13**, 2518.
141. A. L. Nelson, *mAbs*, 2010, **2**, 77.
142. V. L. Stefanelli and T. H. Barker, *J. Mater. Chem. B*, 2015, **3**, 1177.
143. G. A. M. S. van Dongen, G. W. M. Visser and M. Vrouenraets, *Adv. Drug Delivery Rev.*, 2004, **56**, 31.
144. M. Mitsunaga, M. Ogawa, N. Kosaka, L. T. Rosenblum, P. L. Choyke and H. Kobayashi, *Nat. Med.*, 2011, **17**, 1685.
145. L. Gao, C. Zhang, D. Gao, H. Liu, X. Yu, J. Lai, F. Wang, J. Lin and Z. Liu, *Theranostics*, 2016, **6**, 627.
146. O. Planas, T. Gallavardin and S. Nonell, *Chem. Commun.*, 2015, **51**, 5586.
147. G. Obaid, M. Broekgaarden, A.-L. Bulin, H.-C. Huang, J. Kuriakose, J. Liu and T. Hasan, *Nanoscale*, 2016, **8**, 12471.
148. F. Selvestrel, F. Moret, D. Segat, J. H. Woodhams, G. Fracasso, I. M. R. Echevarria, L. Baù, F. Rastrelli, C. Compagnin, E. Reddi, C. Fedeli, E. Papini, R. Tavano, A. MacKenzie, M. Bovis, E. Yaghini, A. J. MacRobert, S. Zanini, A. Boscaini, M. Colombatti and F. Mancin, *Nanoscale*, 2013, **5**, 6106.
149. J. Tian, L. Ding, H. J. Xu, Z. Shen, H. Ju, L. Jia, L. Bao and J. S. Yu, *J. Am. Chem. Soc.*, 2013, **135**, 18850.
150. P. M. Valencia, M. H. Hanewich-Hollatz, W. Gao, F. Karim, R. Langer, R. Karnik and O. C. Farokhzad, *Biomaterials*, 2011, **32**, 6226.
151. Y. Mir, S. A. Elrington and T. Hasan, *Nanomedicine*, 2013, **9**, 1114.
152. G. Obaid, I. Chambrier, M. J. Cook and D. A. Russell, *Photochem. Photobiol. Sci.*, 2015, **14**, 737.
153. C. S. Jin, L. Cui, F. Wang, J. Chen and G. Zheng, *Adv. Healthcare Mater.*, 2014.
154. A. P. Castano, T. N. Demidova and M. R. Hamblin, *Photodiagn. Photodyn. Ther.*, 2004, **1**, 279.
155. A. Jiménez-Banzo, M. L. Sagristà, M. Mora and S. Nonell, *Free Radical Biol. Med.*, 2008, **44**, 1926.

156. X. Ragàs, M. Agut and S. Nonell, *Free Radical Biol. Med.*, 2010, **49**, 770.
157. L. Liu, Z. Zhang and D. Xing, *Free Radical Biol. Med.*, 2011, **51**, 53.
158. I. Moserova and J. Kralova, *PLoS One*, 2012, **7**, e32972.
159. J. C. Stockert, M. Cañete, A. Juarranz, A. Villanueva, R. W. Horobin, J. I. Borrell, J. Teixido and S. Nonell, *Curr. Med. Chem.*, 2007, **14**, 997.
160. R. Ruiz-González, P. Acedo, D. Sánchez-García, S. Nonell, M. Cañete, J. C. Stockert and A. Villanueva, *Eur. J. Med. Chem.*, 2013, **63**, 401.
161. A. Srivatsan, P. Pera, P. Joshi, Y. Wang, J. R. Missert, E. C. Tracy, W. A. Tabaczynski, R. Yao, M. Sajjad, H. Baumann and R. K. Pandey, *Bioorg. Med. Chem.*, 2015, **23**, 3603.
162. R. Ezzeddine, A. Al-Banaw, A. Tovmasyan, J. D. Craik, I. Batinic-Haberle and L. T. Benov, *J. Biol. Chem.*, 2013, **288**, 36579.
163. J. Soriano, M. García-Díaz, M. Mora, M. L. Sagristà, S. Nonell, A. Villanueva, J. C. Stockert and M. Canete, *Biochim. Biophys. Acta*, 2013, **1830**, 4611.
164. K. Suhling, P. M. W. French and D. Phillips, *Photochem. Photobiol. Sci.*, 2005, **4**, 13.
165. K. Chen, A. Preuß, S. Hackbarth, M. Wacker, K. Langer and B. Röder, *J. Photochem. Photobiol., B*, 2009, **96**, 66.
166. H.-P. Lassalle, M. Wagner, L. Bezdetnaya, F. Guillemin and H. Schneckenburger, *J. Photochem. Photobiol., B*, 2008, **92**, 47.
167. Y. Xu, R. He, D. Lin, M. Ji and J. Chen, *Nanoscale*, 2015, **7**, 2433.
168. S.-C. A. Yeh, K. R. Diamond, M. S. Patterson, Z. Nie, J. E. Hayward and Q. Fang, *Theranostics*, 2012, **2**, 817.
169. E. Yaghini, F. Giuntini, I. M. Eggleston, K. Suhling, A. M. Seifalian and A. J. MacRobert, *Small*, 2014, **10**, 782.
170. W. W. Yu, L. Qu, W. Guo and X. Peng, *Chem. Mater.*, 2003, **15**, 2854.
171. X. Gao, Y. Cui, R. M. Levenson, L. W. K. Chung and S. Nie, *Nat. Biotechnol.*, 2004, **22**, 969.
172. W. Cai, D. Shin, K. Chen, O. Gheysens, Q. Cao, S. X. Wang, S. S. Gambhir and X. Chen, *Nano Lett.*, 2006, **6**, 669.
173. S. C. Gad, K. L. Sharp, C. Montgomery, J. D. Payne and G. P. Goodrich, *Int. J. Toxicol.*, 2012, **31**, 584.
174. C. Kim, H.-M. Song, X. Cai, J. Yao, A. Wei and L. V. Wang, *J. Mater. Chem.*, 2011, **21**, 2841.
175. Y. Liu, Z. Chang, H. Yuan, A. M. Fales and T. Vo-Dinh, *Nanoscale*, 2013, **5**, 12126.
176. J. Shi, Z. Chen, L. Wang, B. Wang, L. Xu, L. Hou and Z. Zhang, *Acta Biomater.*, 2016, **29**, 282.
177. T. D. Schladt, K. Schneider, M. I. Shukoor, F. Natalio, H. Bauer, M. N. Tahir, S. Weber, L. M. Schreiber, H. C. Schröder, W. E. G. Müller and W. Tremel, *J. Mater. Chem.*, 2010, **20**, 8297.
178. J. F. Lovell, C. S. Jin, E. Huynh, H. Jin, C. Kim, J. L. Rubinstein, W. C. W. Chan, W. Cao, L. V. Wang and G. Zheng, *Nat. Mater.*, 2011, **10**, 324.
179. T. W. Liu, T. D. MacDonald, J. Shi, B. C. Wilson and G. Zheng, *Angew. Chem., Int. Ed.*, 2012, **51**, 13128.

180. T. D. Macdonald, T. W. Liu and G. Zheng, *Angew. Chem., Int. Ed.*, 2014, **53**, 6956.
181. Z. Wang, X. Hong, S. Zong, C. Tang, Y. Cui and Q. Zheng, *Sci. Rep.*, 2015, **5**, 12602.
182. M. Cao, P. Wang, Y. Kou, J. Wang, J. Liu, Y. Li, J. Li, L. Wang and C. Chen, *ACS Appl. Mater. Interfaces*, 2015, **7**, 25014.
183. M. N. Umar, K. Khan, M. N. Anjum and M. Yaseen, *J. Struct. Chem.*, 2014, **55**, 910.
184. O. Galindev, M. Dalantai, W. S. Ahn and Y. K. Shim, *J. Porphyrins Phthalocyanines*, 2009, **13**, 823.
185. A. Ort, J. Schmitt and B. Ventura, *Inorg. Chem.*, 2016, **55**, 4545.
186. T. Zhang, R. Lan, C. Chan, G. Law, W.-K. Wong and K. Wong, *Proc. Natl. Acad. Sci. U. S. A.*, 2014, **111**, E5492.
187. A. B. E. Attia, G. Balasundaram, W. Driessen, V. Ntziachristos and M. Olivo, *Biomed. Opt. Express*, 2015, **6**, 591.
188. C. J. H. Ho, G. Balasundaram, W. Driessen, R. McLaren, C. L. Wong, U. S. Dinish, A. B. E. Attia, V. Ntziachristos and M. Olivo, *Sci. Rep.*, 2014, **4**, 5342.
189. S. R. Millon, J. H. Ostrander, S. Yazdanfar, J. Q. Brown, J. E. Bender, A. Rajeha and N. Ramanujam, *J. Biomed. Opt.*, 2010, **15**, 018002.
190. J. T. Elliott, A. V. Dsouza, S. C. Davis, J. D. Olson, K. D. Paulsen, D. W. Roberts and B. W. Pogue, *Biomed. Opt. Express*, 2015, **6**, 3765.
191. G. Widhalm, B. Kiesel, A. Woehrer, T. Traub-Weidinger, M. Preusser, C. Marosi, D. Prayer, J. A. Hainfellner, E. Knosp and S. Wolfsberger, *PLoS One*, 2013, **8**(10), e76988.
192. C. A. Morton, *J. Dermatol. Treat.*, 2002, **13**(suppl. 1), S25.
193. J. Lin and M. T. Wan, *Clin., Cosmet. Invest. Dermatol.*, 2014, **30**, 145.
194. L. Shi, X. Wang, Q. Tu, H. Wang, H. Zhang, P. Wang, L. Zhang, Z. Huang, F. Zhao and H. Luan, *Int. J. Nanomed.*, 2015, 347.
195. E. Christensen, C. Mørk and E. Skogvoll, *Br. J. Dermatol.*, 2012, **166**, 1342.
196. W. Stummer, U. Pichlmeier, T. Meinel, O. D. Wiestler, F. Zanella and H. J. Reulen, *Lancet Oncol.*, 2006, **7**, 392.
197. A. Lapini, A. Minervini, A. Masala, L. Schips, A. Pycha, L. Cindolo, R. Giannella, T. Martini, G. Vittori, D. Zani, F. Bellomo and S. C. Cunico, *Surg. Endosc.*, 2012, **26**, 3634.
198. A. Leunig, M. Mehlmann, C. Betz, H. Stepp, S. Arbogast, G. Grevers and R. Baumgartner, *J. Photochem. Photobiol., B*, 2001, **60**, 44.
199. Z. Baozhong and Y. Y. He, *Expert Rev. Anticancer Ther.*, 2010, **10**, 1797.
200. T. Minamikawa, H. Matsuo, Y. Kato, Y. Harada, E. Otsuji, A. Yanagisawa, H. Tanaka and T. Takamatsu, *Sci. Rep.*, 2016, **6**, 25530.
201. J. Olmsted, *J. Phys. Chem.*, 1979, **83**, 2581.
202. J. H. Winer, H. S. Choi, S. L. Gibbs-Strauss, Y. Ashitate, Y. L. Colson and J. V. Frangioni, *Ann. Surg. Oncol.*, 2010, **17**, 1094.
203. J. R. van der Vorst, A. L. Vahrmeijer, M. Hutteman, T. Bosse, V. T. H. B. M. Smit, C. J. H. van de Velde, J. V. Frangioni and B. A. Bonsing, *World J. Gastrointest. Surg.*, 2012, **4**, 180.

204. J. R. van der Vorst, B. E. Schaafsma, F. P. R. Verbeek, R.-J. Swijnenburg, Q. R. J. G. Tummers, M. Hutteman, J. F. Hamming, J. Kievit, J. V. Frangioni, C. J. H. van de Velde and A. L. Vahrmeijer, *Head Neck*, 2014, **36**, 853.
205. Q. R. J. G. Tummers, F. P. R. Verbeek, B. E. Schaafsma, M. C. Boonstra, J. R. van der Vorst, G.-J. Liefers, C. J. H. van de Velde, J. V. Frangioni and A. L. Vahrmeijer, *Eur. J. Surg. Oncol.*, 2014, **40**, 850.
206. Q. R. J. G. Tummers, M. C. Boonstra, J. V. Frangioni, C. J. H. Van De Velde, A. L. Vahrmeijer and B. A. Bonsing, *Int. J. Surg. Case Rep.*, 2015, **6**, 150.
207. H. Wada, H. Hyun, C. Vargas, J. Gravier, G. Park, S. Gioux, J. V. Frangioni, M. Henary and H. S. Choi, *Theranostics*, 2015, **5**, 1.
208. A. Bogaards, A. Varma, K. Zhang, D. Zach, S. K. Bisland, E. H. Moriyama, L. Lilge, P. J. Muller and B. C. Wilson, *Photochem. Photobiol. Sci.*, 2005, **4**, 438.
209. G. Zilidis, F. Aziz, S. Telara and M. S. Eljamel, *Photodiagn. Photodyn. Ther.*, 2008, **5**, 264.
210. L. S. Moore, E. de Boer, J. M. Warram, M. D. Tucker, W. R. Carroll, M. L. Korb, M. S. Brandwein-Gensler, G. M. van Dam and E. L. Rosenthal, *Cancer Med.*, 2016.
211. N. Muhanna, L. Cui, H. Chan, L. Burgess, C. S. Jin, T. D. MacDonald, E. Huynh, F. Wang, J. Chen, J. C. Irish and G. Zheng, *Clin. Cancer Res.*, 2016, **22**, 961.
212. X. Peng, D. R. Draney, W. M. Volcheck, G. R. Bashford, D. T. Lamb, D. L. Grone, Y. Zhang and C. M. Johnson, *Proc. SPIE*, 2006, **6097**, 60970E.
213. B. Li, L. Lin, H. Lin and B. C. Wilson, *J. Biophotonics*, 2016, **9**, 1314.
214. M. A. Weston and M. S. Patterson, in *Singlet Oxygen: Applications in Biosciences and Nanosciences*, ed. S. Nonell and C. Flors, The Royal Society of Chemistry, 2016, vol. 2, 33, pp. 151–168.
215. B. C. Wilson, M. S. Patterson and L. Lilge, *Lasers Med. Sci.*, 1997, **12**, 182.
216. J. S. Dysart, M. S. Patterson, T. J. Farrell and G. Singh, *Photochem. Photobiol.*, 2002, **75**, 289.
217. J. S. Dysart, G. Singh and M. S. Patterson, *Photochem. Photobiol.*, 2005, **81**, 196.
218. D. J. Robinson, H. S. Bruijn, N. Veen, M. R. Stringer, S. B. Brown and W. M. Star, *Photochem. Photobiol.*, 1998, **67**, 140.
219. C. Sheng, P. Jack Hoopes, T. Hasan and B. W. Pogue, *Photochem. Photobiol.*, 2007, **83**, 738.
220. R. M. Valentine, S. H. Ibbotson, C. T. A. Brown, K. Wood and H. Moseley, *Photochem. Photobiol.*, 2011, **87**, 242.
221. R. M. Valentine, S. H. Ibbotson, K. Wood, C. T. A. Brown and H. Moseley, *Photochem. Photobiol. Sci.*, 2013, **12**, 203.
222. J. C. Finlay, S. Mitra, M. S. Patterson and T. H. Foster, *Phys. Med. Biol.*, 2004, **49**, 4837.
223. M. T. Jarvi, M. S. Patterson and B. C. Wilson, *Biophys. J.*, 2012, **102**, 661.

224. W. R. Potter, T. S. Mang and T. J. Dougherty, *Photochem. Photobiol.*, 1987, **46**, 97.
225. S. Nonell and C. Flors, in *Singlet Oxygen: Applications in Biosciences and Nanosciences*, ed. S. Nonell and C. Flors, The Royal Society of Chemistry, 2016, vol. 2, 25, pp. 7–26.
226. M. T. Jarvi, M. J. Niedre, M. S. Patterson and B. C. Wilson, *Photochem. Photobiol.*, 2006, **82**, 1198.
227. J. G. Parker, *IEEE Circuits Devices*, 1987, **3**, 10.
228. M. S. Patterson, S. J. Madsen and B. C. Wilson, *J. Photochem. Photobiol., B*, 1990, **5**, 69.
229. A. Baker and J. R. Kanofsky, *Arch. Biochem. Biophys.*, 1991, **286**, 70.
230. T. Hirano, E. Kohno and M. Nishiwaki, *J. Jpn. Soc. Laser Surg. Med.*, 2001, **22**, 99.
231. M. Niedre, M. S. Patterson and B. C. Wilson, *Photochem. Photobiol.*, 2002, **75**, 382.
232. M. J. Niedre, A. J. Secord, M. S. Patterson and B. C. Wilson, *Cancer Res.*, 2003, **63**, 7986.
233. B. Li, M. T. Jarvi, E. H. Moriyama and B. C. Wilson, *Proc. SPIE*, 2007, **6427**, 642707.
234. I.-W. Kim, J. M. Park, Y. J. Roh, J. H. Kim, M.-G. Choi and T. Hasan, *J. Photochem. Photobiol., B*, 2016, **159**, 14.
235. M. J. Niedre, C. S. Yu, M. S. Patterson and B. C. Wilson, *Br. J. Cancer*, 2005, **92**, 298.
236. S. Mallidi, S. Anbil, S. Lee, D. Manstein, S. Elrington, G. Kositratna, D. Schoenfeld, B. Pogue, S. J. Davis and T. Hasan, *J. Biomed. Opt.*, 2014, **19**, 028001.
237. J. C. Schlothauer, S. Hackbarth, L. Jäger, K. Drobniewski, H. Patel, S. M. Gorun and B. Röder, *J. Biomed. Opt.*, 2012, **17**, 115005.
238. M. T. Jarvi, M. J. Niedre, M. S. Patterson and B. C. Wilson, *Photochem. Photobiol.*, 2011, **87**, 223.
239. B. C. Wilson, M. S. Patterson, B. Li and M. T. Jarvi, *J. Innovative Opt. Health Sci.*, 2015, **08**, 1540006.
240. S. Erbas-Cakmak and E. U. Akkaya, *Org. Lett.*, 2014, **16**, 2946.
241. Y. Yuan, C.-J. Zhang, S. Xu and B. Liu, *Chem. Sci.*, 2016, **7**, 1862.
242. Y. You, E. J. Cho, H. Kwon, J. Hwang and S. E. Lee, *Chem. Commun.*, 2015, **52**, 780.
243. Y. Wei, J. Song and Q. Chen, *Photochem. Photobiol. Sci.*, 2011, **10**, 1066.
244. Y. Qin, D. Xing, S. Luo, J. Zhou, X. Zhong and Q. Chen, *Photochem. Photobiol.*, 2005, **81**, 1534.
245. Y. Wei, D. Xing, S. Luo, W. Xu and Q. Chen, *J. Biomed. Opt.*, 2008, **13**, 024023.
246. S. Kim, T. Tachikawa, M. Fujitsuka and T. Majima, *J. Am. Chem. Soc.*, 2014, **136**, 11707.
247. B. Song, G. Wang, M. Tan and J. Yuan, *J. Am. Chem. Soc.*, 2006, **128**, 13442.

248. Z. Dai, L. Tian, Y. Xiao, Z. Ye, R. Zhang and J. Yuan, *J. Mater. Chem. B*, 2013, **1**, 924.
249. C. Flors, M. J. Fryer, J. Waring, B. Reeder, U. Bechtold, P. M. Mullineaux, S. Nonell, M. T. Wilson and N. R. Baker, *J. Exp. Bot.*, 2006, **57**, 1725.
250. A. Gollmer, J. Arnbjerg, F. H. Blaikie, B. W. Pedersen, T. Breitenbach, K. Daasbjerg, M. Glasius and P. R. Ogilby, *Photochem. Photobiol.*, 2011, **87**, 671.
251. *Singlet Oxygen Sensor Green Product Information*, https://www.thermofisher.com/order/catalog/product/S36002, visited on May 10, 2016.
252. T.-L. To, M. J. Fadul and X. Shu, *Nat. Commun.*, 2014, **5**, 1.
253. F. M. Pimenta, J. K. Jensen, M. Etzerodt and P. R. Ogilby, *Photochem. Photobiol. Sci.*, 2015, **14**, 665.
254. R. Bresolí-Obach, J. Nos, M. Mora, M. L. Sagristà, R. Ruiz-González and S. Nonell, *Methods*, 2016, DOI: 10.1016/j.ymeth.2016.06.007, in press.

CHAPTER 5

# Radiolabelling Liposomal Nanomedicines for PET Imaging

RAFAEL T. M. DE ROSALES

Division of Imaging Sciences & Biomedical Engineering, King's College London, St. Thomas' Hospital, SE1 7EH, London, UK
*E-mail: rafael.torres@kcl.ac.uk

## 5.1 Liposomal Nanomedicines

Nanomedicines have the potential to improve drug delivery to diseased tissues, mainly by allowing targeted delivery of poorly water-soluble drugs to specific cells or tissues.[1] Other advantages include the simultaneous co-delivery of several drugs and, by adding imaging reporters, visualisation of their biodistribution and site(s) of drug delivery. Liposomal drugs are the type of nanomedicines that have made a major clinical impact in recent years. However, they have only made a modest clinical impact to date, and have only found applications in a limited set of indications without major improvements over established chemotherapies.[2] A major development, however, has been shown by liposomal nanomedicines in preventing or diminishing the side effects of their encapsulated drugs. This is due to the fact that the encapsulated (and often highly toxic) drug is differentially released in target and non-target sites. Doxil/Caelyx is the best example we have to date. These are PEGylated liposomes containing doxorubicin that have shown

Drug Discovery Series No. 63
Theranostics and Image Guided Drug Delivery
Edited by Maya Thanou

Published by the Royal Society of Chemistry, www.rsc.org

equal or slightly superior efficacy to conventional chemotherapies in several cancers such as AIDS-related Kaposi's sarcoma, ovarian cancer, breast cancer and multiple myeloma. However, it has also shown significant avoidance or major reduction of the common cardiotoxicity compared to an equivalent amount of doxorubicin.[3] Hence, most nanomedicines with clinical approval today, such as Doxil, are used only as second-line treatments or as part of combination therapies with the main benefit of lesser side effects compared to the free chemotherapeutic drug.

## 5.2 The Importance of Imaging in the Development and Evaluation of Liposomal Nanomedicines

As well as the intrinsic difficulties of delivering drug nanocarriers to specific tissues *in vivo*,[4,5] a key factor that contributes to the relative failure of nanomedicines to show their potential is patient and disease heterogeneity.[6] This has been recently demonstrated in oncology patients using SPECT imaging and $^{111}$In labelled liposomes.[7] Thus, PEGylated liposomes labelled with this isotope and known to accumulate in solid tumours as a consequence of the enhanced permeation and retention (EPR) effect, did not accumulate in tumours at the same level in patients with different tumour types. This result was also observed in patients of the same tumour type, and highlights the need to use imaging-based stratification strategies to selectively treat, in the context of oncology, those patients with tumour tissues that will take up the drug in therapeutic quantities and hence are likely to benefit maximally from the nanomedicinal treatment. It is predicted that such a personalised/stratification medicine approach may allow nanomedicines, on their own or in combination with other therapies, to demonstrate their full clinical potential.[6,8–10]

There are several imaging techniques available in clinics that would allow such a treatment response prediction tool for nanomedicines, but only positron emission tomography (PET) allows quantitative analysis of nanomedicine biodistribution with exquisite sensitivity (signal:background ratios) and adequate spatio-temporal resolution at the whole-body level. Other clinical imaging techniques with potential to perform whole-body scans are computed tomography (CT) and magnetic resonance imaging (MRI). However they suffer from three major issues: (*i*) low contrast ratio, making it difficult to identify the areas of uptake; (*ii*) quantification is difficult/challenging and often inaccurate and, (*iii*) more importantly, low sensitivity (*i.e.* amount of contrast agent required to detect a signal). This is an important factor to consider, as we want to avoid any potential therapeutic effects from the imaging study with the nanomedicine. In this respect, nuclear imaging techniques (single-photon emission computed tomography (SPECT) and PET) are unique, as they only require sub-microgram quantities of radionuclides and tracers in order to provide the required signal for a whole-body human imaging study. This is in contrast to MRI and CT where gram quantities are required in order to obtain contrast. This >$10^6$-fold difference makes nuclear imaging techniques ideal for microdosing imaging studies (defined as 1% of the therapeutic dose) and assure us that the pre-therapy imaging study will not have any undesirable chemotherapeutic effect on patients.

In addition to using radiolabelled nanomedicines and PET as a treatment response prediction tool, they can be enormously useful in the development of nanomedicines and the preclinical and clinical levels. Other imaging techniques such as optical, MRI, CT and ultrasound (US) could be used and have been used in this respect, but only nuclear imaging techniques and in particular PET would allow the transition from preclinical to clinical without modification of the nanomedicine/liposome components and whilst still providing whole-body quantitative information.

## 5.3 Radiolabelling Liposomal Nanomedicines for Nuclear Imaging: SPECT *versus* PET

One key issue to consider when aiming to study the biodistribution of a liposomal nanomedicine is its biological half-life. Liposomal nanomedicines, and in particular stealth liposomes (*e.g.* DOXIL) show very long circulation times, with half-lives in humans in the 50–80 h range.[11] Hence it is important we choose isotopes with decay properties that match these biological half-lives. The most obvious candidate is $^{111}$In, a gamma-emitting isotope (gamma-ray energies: 171.3 and 245.4 keV and with a radioactive half-live of 67.2 h), ideal for tracking long-circulating biomolecules such as antibodies and hence, stealth liposomal nanomedicines. Another gamma-emitter, $^{67}$Ga, also has a long radioactive half-live (78.2 h), but the major disadvantage is that it is also an Auger electron-emitter (*i.e.* highly cytotoxic nuclear emission), hence its use with long circulating tracers can lead to high-radiation doses to patients and, potentially, radiotoxicity effects.[12] The other main SPECT isotope is $^{99m}$Tc, with excellent availability from a clinical-grade generator and emission properties (140 keV), but relatively short radioactive half-live (6 h), which limits the study of liposomal nanomedicines to a very short timeframe. Liposomal nanomedicines have been radiolabeled with radionuclides detectable by SPECT or planar imaging to study their biodistribution in preclinical and clinical studies since the 1980's.[8,9,13–22] This nuclear imaging technique, however, has lower sensitivity and temporal/spatial resolution than PET, which makes the latter the ideal technique to study drug biodistribution *in vivo*. For this reason, in this chapter we will focus on the methods developed to date to radiolabel liposomal nanomedicines with PET isotopes.

In addition to the above-mentioned advantages of clinical PET in terms of sensitivity, quantification and spatial/temporal resolution, another advantage is the recent development and availability of long-lived radionuclides properties that match the biological half-life of nanomedicines such as $^{89}$Zr ($t_{1/2}$ = 78.4 h),[23] $^{64}$Cu ($t_{1/2}$ = 12.7 h),[24] and more recently, $^{52}$Mn ($t_{1/2}$ = 134.2 h).[25]

To date, radiolabeling of liposomal drugs with PET isotopes has been explored to a much lesser extent than with gamma-emitting radionuclides. In the next sections, we will review the approaches attempted to date, which has been mostly by introduction of exogenous chelators in the intraliposomal space/phospholipid bilayer,[26–36] or by using biorthogonal conjugation strategies.[37]

## 5.3.1 Radiolabelling of Liposomal Nanomedicines with PET Isotopes

### *5.3.1.1 Using Chelators/Radionuclides Attached to the Phospholipid Bilayer*

The first reports of liposomes radiolabelled with PET isotopes involved the addition of the radionuclide to components of the phospholipid bilayer by modifying lipids with chelators or molecules that had been previously radiolabelled with short half live isotopes such as $^{18}$F ($t_{1/2}$ = 1.8 h). Marik *et al.*, from the Ferrara group synthesised and characterised a radiolabeled diglyceride, 3-[$^{18}$F]fluoro-1,2-dipalmitoylglycerol [[$^{18}$F]fluorodipalmitin ([$^{18}$F]FDP)] (Figure 5.1).[26] This lipid was incorporated into the lipid bilayer during the formation of $^{18}$F-labelled long-circulating PEGylated liposomes. *In vivo* studies with these PET liposomes, in comparison with control studies using [$^{18}$F] FDP, demonstrated the incorporation and stability of the $^{18}$F-lipid and the utility of this method to track liposomes using PET. A disadvantage of this method, however, is that it only allows imaging for a few hours, due to the short half-life of $^{18}$F, which limits the study of the long-term stability and biodistribution of the liposomes.

Seo *et al.* from the Ferrara group also reported on a similar method using a $^{64}$Cu-specific chelator (TETA = 1,4,8,11-tetraazacyclotetradecane - *N,N′,N″,N‴*-tetraacetic acid) to allow liposomal PET tracking for longer periods of time.[27] Thus, they synthesised 6-(*p*-(bromoacetamido)benzyl)-1,4,8,11-tetraazacyclotetradecane-*N,N′,N″,N‴*-tetraacetic acid (BAT) (Figure 5.2), that was then was conjugated with an artificial lipid to form a BAT–PEG-lipid for liposomal synthesis. Importantly, this strategy allows performing radiolabelling as the last step after the liposomes have been synthesised, purified and characterised. High radiolabelling yields of 95% can be obtained, as well as *in vitro* serum stabilities (stability of radionuclide-liposome interaction) of 88%. Significant differences in a 48 h *in vivo* study between the radiolabelled liposomes and free/unchelated $^{64}$Cu confirmed the potential of this method to label and track preformed liposomes using PET imaging. Seo *et al.* and Ferrara also studied the effect of the lipid acyl chain length in the desorption, and hence stability, of the radionuclide-lipid conjugate from the liposome.[29] Importantly, they found that the length of the hydrophobic chain in radiolabeled phospholipids indeed affected the stability of the radiolabel-liposome stability *in vitro* and *in vivo.*

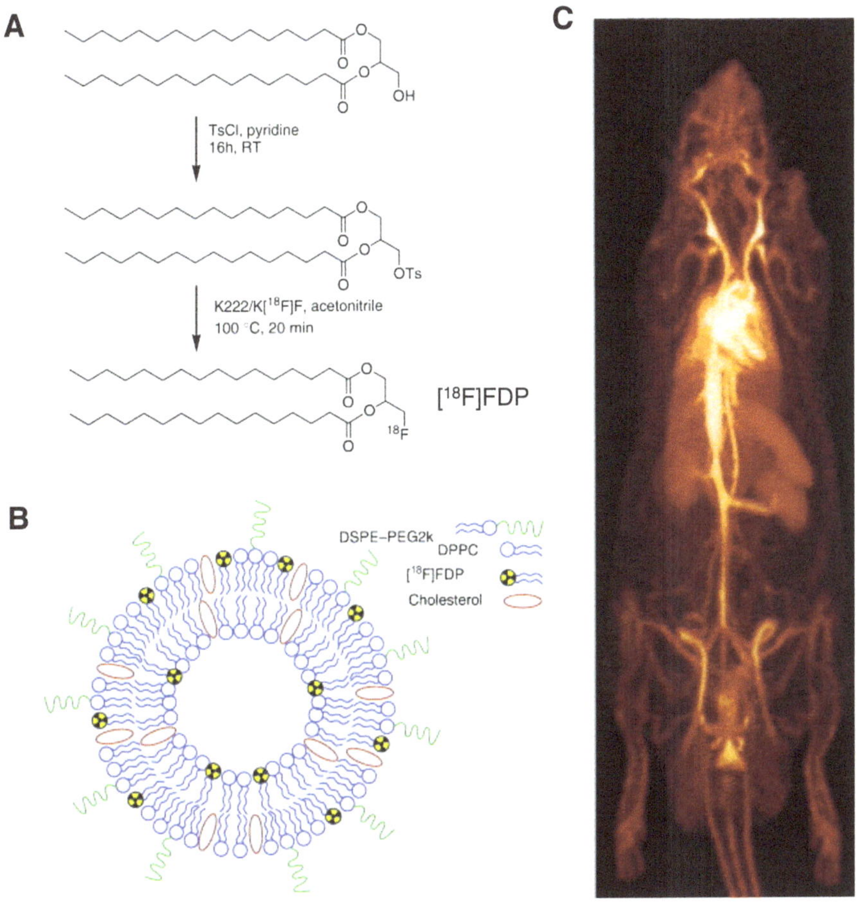


**Figure 5.1** (A) Synthesis of [$^{18}$F]FDP (3-[$^{18}$F]fluoro-1,2-dipalmitoylglycerol); (B) Schematic representation of the [$^{18}$F]FDP-radiolabeled liposome; (C) Maximum-intensity projection (MIP) PET image of a rat that was injected intravenously with the [$^{18}$F]FDP-radiolabeled liposome, showing the expected blood-pool image due to its long circulation time. Reprinted from *Nuclear Medicine and Biology*, **34**, J. Marik, M. S. Tartis, H. Zhang, J. Y. Fung, A. Kheirolomoom, J. L. Sutcliffe and K. W. Ferrara, Long-circulating liposomes radiolabelled with [$^{18}$F] fluorodipalmitin ([$^{18}$F] FDP), 165–171, Copyright 2007, with permission from Elsevier.

Li *et al.* from the Bao group used the same strategy (incorporation of radionuclides into the lipid bilayer) for generating multimodal liposomes amenable for detection using MRI, optical imaging and SPECT or PET by using 1,4,7,10-tetraazacyclododecane-1,4,7,10-tetraacetic acid (DOTA), a chelator that has been extensively used for radiolabelling with $^{64}$Cu.[30] The *in vitro* and *in vivo* studies reported using all these different modalities confirmed the utility of this method to study the biodistribution of

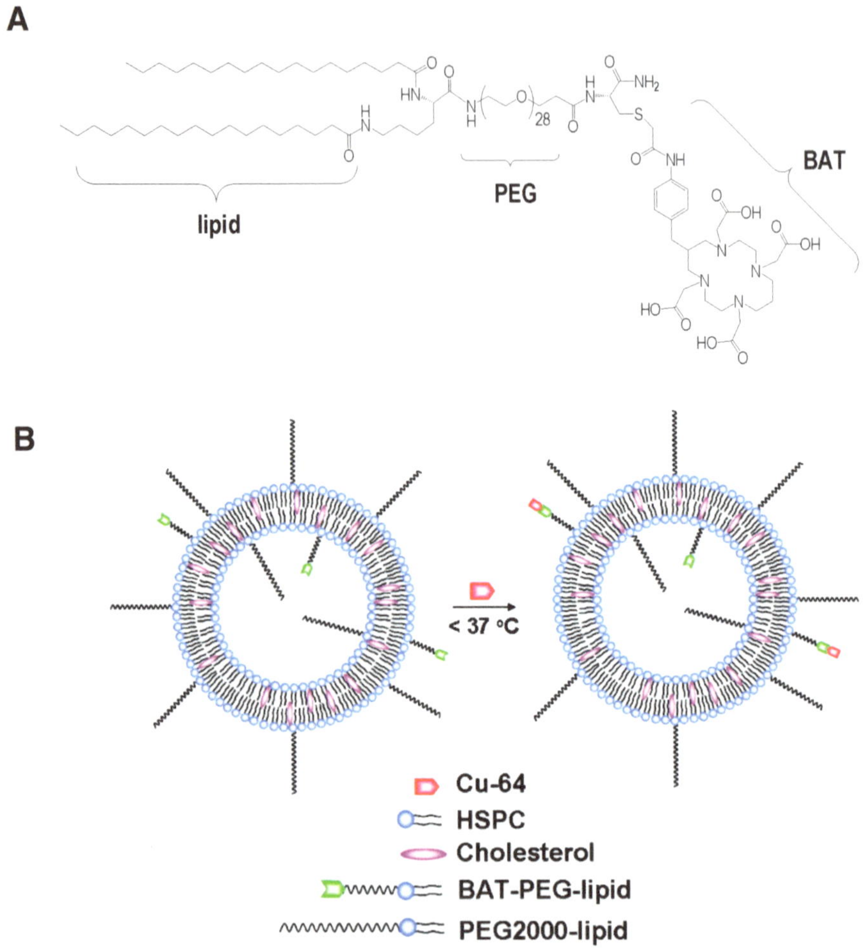


**Figure 5.2** (A) BAT-PEG-lipid; (B) Surface chelation model of $^{64}$Cu-labeled long circulating liposomes. Labeling was performed under mild conditions (<37 °C) to maximise liposome stability and loading. Figure adapted with permission from ref. 27. Copyright 2008 American Chemical Society.

liposomes using different imaging techniques and hence at different spatial and temporal scales. Kang *et al.* also used DOTA-conjugated lipids to track tetraiodothyroacetic acid (tetrac)-conjugated liposomes for PET imaging of tumour angiogenesis.[31] Tetrac was conjugated to the liposomes as it has been shown that this small molecule is a non-agonist thyroid hormone analogue that binds to the $\alpha_V\beta_3$ integrin receptor, overexpressed in many tumours undergoing angiogenesis.[38] The authors used PET imaging to report higher tumour uptakes (U87MG tumour-bearing mice) with tetrac-conjugated liposomes.

Despite the success using $^{64}Cu$ for tracking liposomes using PET, its half-life does not easily allow for imaging studies beyond 48 h, and as mentioned above many liposomal nanomedicines have circulation half-lives, in humans, in the 50–80 h range,[11] Hence, while the use of $^{64}Cu$ may be appropriate for preclinical animal studies, for human studies a PET isotope with a longer radioactive half-life is desirable. $^{89}Zr$ with its 78.4 h half-life has recently received increasing interest for monitoring long-circulating biomolecules in clinical PET studies, particularly antibodies.[23] For this reason, Seo *et al.* in the Ferrara group developed a radiolabelling method with this isotope that allowed them to image liposomes over 168 h (1 week).[32] This method relies on the lipid conjugation of deferrioxamine (DFO), a chelator that provides excellent stability to $^{89}Zr$ conjugates.[39] Excellent *in vitro* stabilities and the expected *in vivo* biodistribution for long circulating liposomes was observed for all the conjugates studied. Interestingly, the location of the $^{89}Zr$ label within the PEG polymers seems to have an effect on the clearance rate of intracellularly-trapped radioactivity, suggesting this is another important factor to take into account when using this labelling strategy. Perez-Medina *et al.* in the Reiner group also used DFO-conjugated to liposomes, but using a modular labelling strategy for achieving $^{89}Zr$ PET and near-infrared (NIR) fluorescence imaging instead.[33] Using this technique, the authors evaluated two different approaches, surface radionuclide chelation (using DFO conjugated to the liposomes), or a click-labelling strategy (using a DFO-conjugated clickable small molecule) (Figure 5.3). Their results show that the surface-chelation strategy provided more stable radiolabelled liposomes both *in vitro* and *in vivo*.

An interesting, chelator-free approach was recently explored by Abou *et al.* in the Lewis group by exploiting the structure and oxygen-rich nature of the phospholipid hydrophilic heads that could act as chelators and the oxophilicity of $^{89}Zr$ to yield radiolabelled liposomes.[40] This strategy has the major advantage that it could provide a broad method to radiolabel any liposomal nanomedicine. However, it suffers from modest stability in biological media, which is likely due to transchelation of the radiometal *in vivo*.[40]

### 5.3.1.2 Using Intraliposomal Chelators

Another strategy to radiolabel liposomes with PET isotopes is by introducing the radionuclide inside the intraliposomal space, rather than to the phospholipid bilayer. In order to do this, the radionuclide is commonly delivered in the intraliposomal space with the aid of an ionophore, a molecule that aids ions to cross lipid membranes. Ionophores are commonly used in the clinic for cell radiolabelling (*e.g.* white blood cell labelling with $^{111}In$).[41,42] Petersen *et al.* used this concept to radiolabelled DOTA-loaded liposomes with $^{64}Cu$ (Figure 5.4).[28] *In vivo* studies in tumour-bearing mice 24 h post injection of PEGylated liposomes radiolabelled using this method

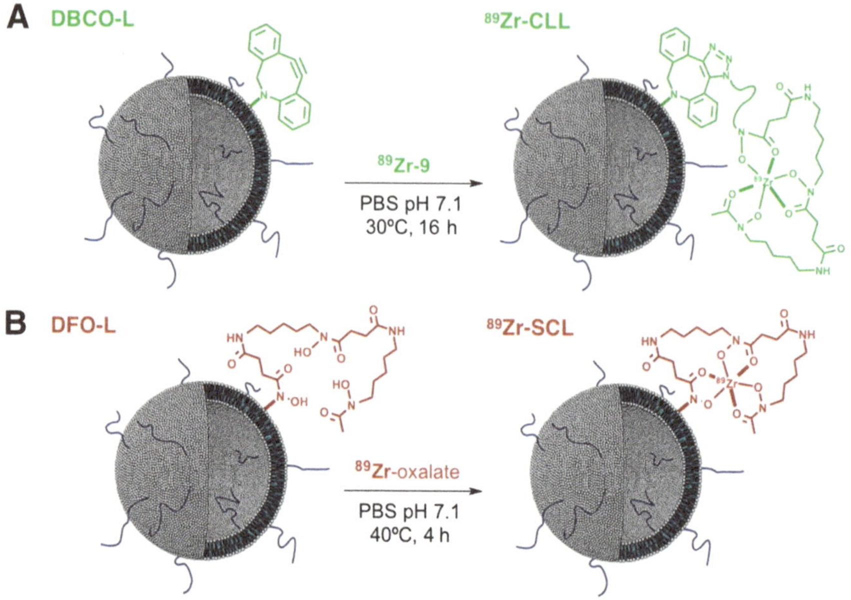


**Figure 5.3** Strategies used by Perez-Medina *et al.* for labeling liposomes with $^{89}Zr$ by click chemistry (**A**) and surface chelation (**B**) using DFO as the chelator. This research was originally published in JNM. C. Perez-Medina, D. Abdel-Atti, Y. Zhang, V. A. Longo, C. P. Irwin, T. Binderup, J. Ruiz-Cabello, Z. A. Fayad, J. S. Lewis, W. J. M. Mulder and T. Reiner. A Modular Labeling Strategy for *In Vivo* PET and Near-Infrared Fluorescence Imaging of Nanoparticle Tumor Targeting. *J Nucl Med.* 2014; **55**: 1706–1711. © by the Society of Nuclear Medicine and Molecular Imaging, Inc.

demonstrated the expected biodistribution pattern of liposomes, with high uptake in the spleen (32.4 ± 5.4%ID/g), liver (6.0 ± 0.9%ID/g), and moderate tumour uptake (2.5 ± 0.7%ID/g) as a result of the EPR effect. The same group recently reported remote loading of $^{64}Cu$ into DOTA-loaded liposomes without the use of an ionophore by incubating $^{64}CuCl_2$ with the liposomes at 55 °C for 75 min.[34] In *vitro/vivo* studies in dogs and tumour-bearing mice support their conclusions and potentially provide a simple method to radiolabel thermally stable liposomes that have a high intraliposomal chelator concentration.

Lee *et al.* recently proposed a very interesting method that uses a gradient-loadable chelator, diacetyl 4,4′-bis(3-(*N*,*N* diethylamino)propyl) thiosemicarbazone (4-DEAP-ATSC), for incorporating $^{64}Cu$ into liposomal nanotherapeutics (Figure 5.5).[35] This strategy seems to be highly efficient (>90% loading efficiency) and takes advantage of the property that, once bound to the radiometal, the metal complex formed is highly lipophilic which makes it permeable and furthermore, prone to protonation at the tertiary amines. This occurs inside the liposome due to increased acidity and inhibits its membrane permeability, providing a driving mechanism for the

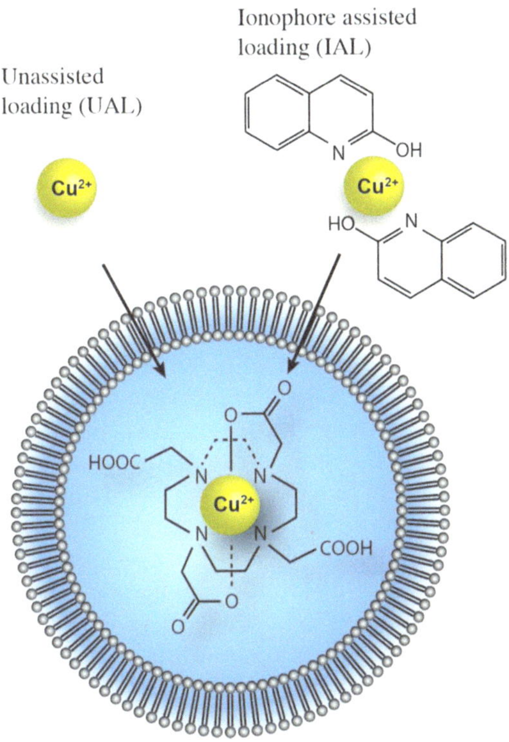


**Figure 5.4** Schematic representation of two remote loading methods for $^{64}$Cu (left, unassisted; right, using ionophores such as 2-hydroxyquinoline) developed by Petersen *et al.* that rely on the presence of intraliposomal DOTA chelators. Figure reproduced with permission from ref. 34. Copyright 2015 American Chemical Society.

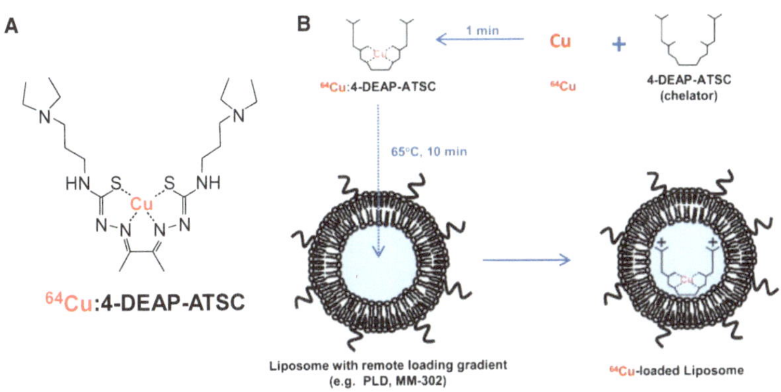


**Figure 5.5** (A) Structure of the $^{64}$Cu complex with diacetyl 4,4′-bis(3-(*N*,*N* diethylamino)propyl)thiosemicarbazone (4-DEAP-ATSC) and (B) protocol to radiolabel liposomes using this complex. Reprinted from *Nanomedicine: Nanotechnology, Biology and Medicine*, **11**, H. Lee, J. Zheng, D. Gaddy, K. D. Orcutt, S. Leonard, E. Geretti, J. Hesterman, C. Harwell, J. Hoppin, D. A. Jaffray, T. Wickham, B. S. Hendriks and D. Kirpotin, A gradient-loadable $^{64}$Cu-chelator for quantifying tumour deposition kinetics of nanoliposomal therapeutics by positron emission tomography, 155–165, Copyright 2017, with permission from Elsevier.

entrapment of the PET radionuclide and would be applicable to different liposomal nanomedicines, as the authors demonstrate.

#### 5.3.1.3 Exploiting the Metal-chelating Properties of the Encapsulated Drugs

Many drugs of interest in nanomedicine contain well-defined multidentate metal-binding motifs, also known chelating groups, within them (Figure 5.6A). This property, in combination with the high intraliposomal concentration of drug molecules, provides the ideal setting to radiolabel preformed liposomal drugs. Our group has recently exploited this idea to demonstrate the feasibility of this approach and its capabilities *in vitro* and *in vivo* with the PET radionuclides $^{89}$Zr, $^{64}$Cu and $^{52}$Mn.[43] The method involves the incubation of radiometal-ionophore complexes based on hydroxyquinoline ionophores (Figure 5.6B), with different PEGylated liposomal nanomedicines (containing metal-chelating drugs) for 30 min at 50 °C (Figure 5.6C). These conditions facilitate the permeation of the radiometal inside the liposome phospholipid bilayer, without changing their physicochemical properties of the liposomes tested, where it preferentially binds to the encapsulated drug molecules. *In vitro* stabilities of the radiometal-nanomedicine, as assessed by incubation in human serum at 37 °C for up to 72 h were excellent (>80% in all cases). Preclinical imaging studies using a metastatic mouse model of breast cancer demonstrated the expected *in vivo* stability and showed high uptake in the primary tumour (8.4 ± 1.6%ID/g whole tumour) and metastatic organs such as the lymph nodes (16.3 ± 7.1%ID/g) and lungs (12.4 ± 4.8%ID/g) due to the EPR (Figure 5.6D and E).

## 5.4 Challenges for Clinical Translation

Most of the radiolabeling strategies described above have demonstrated their value in the preclinical setting by allowing PET tracking of liposomal drugs *in vivo*, but their potential for clinical translation is less clear. This is because liposome modification by introduction of exogenous radiometal chelators or lipids can lead to regulatory barriers, changes in biodistribution, and the requirement of customised liposome manufacture. To decrease this barrier, an ideal method would work on preformed liposomal nanomedicines (without changes in the manufacture) that only contain approved components and using standard clinical reagents and methods. In addition, radiochemical yields and purities of >95% that are generally required for clinical use of radiopharmaceuticals should be achieved to avoid the requirement for purification. High specific activities (defined as the amount of radioactivity per amount of nanomedicinal carrier) should be achieved, allowing the radiolabeling of sub-therapeutic amounts of nanomedicine and hence microdosing PET studies.

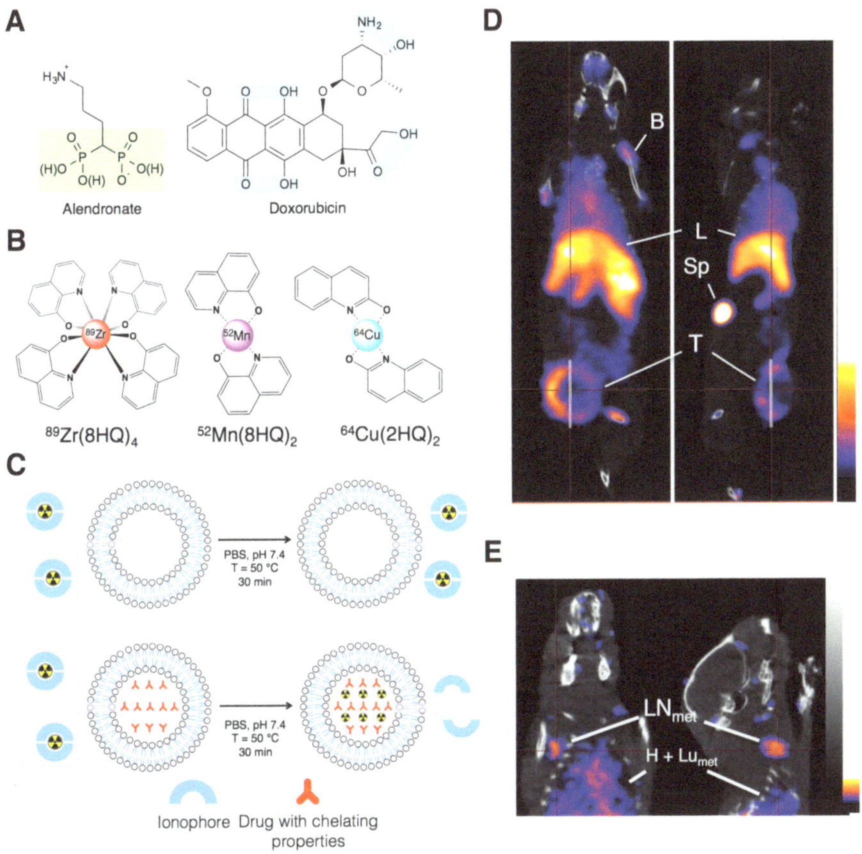


**Figure 5.6** (A) Chemical structure of drugs explored in this study with metal-binding motifs highlighted in color. (B) Radiometal–ionophore complexes used to radiolabel liposomal nanomedicines: 8HQ = 8-hydroquinoline; 2HQ = 2-hydroxyquinoline. (C) Schematic representation of the liposome radiolabeling method. Top: Incubation of empty liposomes, lacking encapsulated drug, do not radiolabel after incubating with the radiometal–ionophore complex in PBS or saline at 50 °C for 30 min. Bottom: Drug-loaded liposomes efficiently radiolabel under the same conditions. (D, E) *In vivo* PET-CT image of $^{89}$Zr-labelled PEGylated liposomal alendronate ($^{89}$Zr-PLA) in a metastatic breast cancer model; whole body image (D) and slice showing uptake in metastatic lymph node (E). T = tumour, L = liver, Sp = spleen, B = bone, Lnmet = metastatic lymph node, Lu = lungs, H = heart. Figure adapted with permission from S. Edmonds, A. Volpe, H. Shmeeda, A. C. Parente-Pereira, R. Radia, J. Baguna-Torres, I. Szanda, G. W. Severin, L. Livieratos, P. J. Blower, J. Maher, G. O. Fruhwirth, A. Gabizon and T. M. d. R. R, *ACS Nano*, 2016, **10**, 10294–10307 http://pubs.acs.org/doi/full/10.1021/acsnano.6b05935. These figures are available under the ACS AuthorChoice license. Further permissions related to the figures should be directed to the ACS.

## 5.5 Conclusion

The field of clinical nanomedicine would benefit by the introduction of image-guided patient selection strategies based on PET and its sensitivity, spatio-temporal resolution and quantification properties at the whole-body level. To this end, several techniques to radiolabel liposomal nanomedicines with $^{18}$F, and importantly $^{89}$Zr and $^{64}$Cu, have been recently developed and their value in the visualisation and quantification of liposomal biodistribution have been established. In order to translate these radiolabelling techniques to clinical use, however, requires careful consideration of several aspects, such as the impact of the chemicals used for labelling on the stability of the nanomedicine and the *in vivo* biodistribution/pharmacokinetics. The regulatory barriers due to the labelling process and the radiochemical purities and yields achievable are also key factors to take into account. The potential of using these methods for human PET microdosing studies is another factor that is often not measured and should be included in future studies. More importantly, radiolabelling methods with an aim to be used in the clinical setting should not be specific to one liposomal nanomedicine, but applicable to a wide number of nanomedicines to maximise the potential to be adopted by the nuclear imaging industry, radiopharmacies and clinical PET centres worldwide.

## References


1. O. C. Farokhzad and R. Langer, *ACS Nano*, 2009, **3**, 16–20.
2. R. Wang, P. S. Billone and W. M. Mullett, *J. Nanomater.*, 2013, **2013**, 1–12.
3. A. Gabizon, R. Catane, B. Uziely, B. Kaufman, T. Safra, R. Cohen, F. Martin, A. Huang and Y. Barenholz, *Cancer Res.*, 1994, **54**, 987–992.
4. R. K. Jain and T. Stylianopoulos, *Nat. Rev. Clin. Oncol.*, 2010, **7**, 653–664.
5. D. Peer, J. M. Karp, S. Hong, O. C. Farokhzad, R. Margalit and R. Langer, *Nat. Nanotechnol.*, 2007, **2**, 751–760.
6. U. Prabhakar, H. Maeda, R. K. Jain, E. M. Sevick-Muraca, W. Zamboni, O. C. Farokhzad, S. T. Barry, A. Gabizon, P. Grodzinski and D. C. Blakey, *Cancer Res.*, 2013, **73**, 2412–2417.
7. K. J. Harrington, S. Mohammadtaghi, P. S. Uster, D. Glass, A. M. Peters, R. G. Vile and J. S. W. Stewart, *Clin. Cancer Res.*, 2001, **7**, 243–254.
8. A. L. Petersen, A. E. Hansen, A. Gabizon and T. L. Andresen, *Adv. Drug Delivery Rev.*, 2012, **64**, 1417–1435.
9. S. Kunjachan, J. Ehling, G. Storm, F. Kiessling and T. Lammers, *Chem. Rev.*, 2015, **115**, 10907–10937.
10. T. Lammers, L. Y. Rizzo, G. Storm and F. Kiessling, *Clin. Cancer Res.*, 2012, **18**, 4889–4894.
11. A. Gabizon, H. Shmeeda and Y. Barenholz, *Clin. Pharmacokinet.*, 2003, **42**, 419–436.
12. A. R. Jonkhoff, P. C. Huijgens, R. T. Versteegh, E. B. van Dieren, G. J. Ossenkoppele, H. J. Martens and G. J. Teule, *Br. J. Cancer*, 1993, **67**, 693–700.

13. O. C. Boerman, P. Laverman, W. J. Oyen, F. H. Corstens and G. Storm, *Prog. Lipid Res.*, 2000, **39**, 461–475.
14. R. T. Proffitt, L. E. Williams, C. A. Presant, G. W. Tin, J. A. Uliana, R. C. Gamble and J. D. Baldeschwieler, *J. Nucl. Med.*, 1983, **24**, 45–51.
15. A. Gabizon, J. Hliberty, R. M. Straubinger, D. C. Price and D. Papahadjopoulos, *J. Liposome Res.*, 1988, **1**, 123–135.
16. H. Essien and K. J. Hwang, *Biochim. Biophys. Acta*, 1988, **944**, 329–336.
17. A. Gabizon, D. C. Price, J. Huberty, R. S. Bresalier and D. Papahadjopoulos, *Cancer Res.*, 1990, **50**, 6371–6378.
18. W. T. Phillips, A. S. Rudolph, B. Goins, J. H. Timmons, R. Klipper and R. Blumhardt, *Int. J. Rad. Appl. Instrum. B*, 1992, **19**, 539–547.
19. C. Tilcock, Q. F. Ahkong and D. Fisher, *Nucl. Med. Biol.*, 1994, **21**, 89–96.
20. A. Bao, B. Goins, R. Klipper, G. Negrete and W. T. Phillips, *J. Nucl. Med.*, 2003, **44**, 1992–1999.
21. P. Laverman, M. G. Carstens, O. C. Boerman, E. T. Dams, W. J. Oyen, N. van Rooijen, F. H. Corstens and G. Storm, *J. Pharmacol. Exp. Ther.*, 2001, **298**, 607–612.
22. M. Ogawa, I. O. Umeda, M. Kosugi, A. Kawai, Y. Hamaya, M. Takashima, H. Yin, T. Kudoh, M. Seno and Y. Magata, *J. Nucl. Med.*, 2013, **55**, 115–120.
23. M. A. Deri, B. M. Zeglis, L. C. Francesconi and J. S. Lewis, *Nucl. Med. Biol.*, 2013, **40**, 3–14.
24. M. Shokeen and C. J. Anderson, *Acc. Chem. Res.*, 2009, **42**, 832–841.
25. S. A. Graves, R. Hernandez, J. Fonslet, C. G. England, H. F. Valdovinos, P. A. Ellison, T. E. Barnhart, D. R. Elema, C. P. Theuer, W. Cai, R. J. Nickles and G. W. Severin, *Bioconjugate Chem.*, 2015, **26**, 2118–2124.
26. J. Marik, M. S. Tartis, H. Zhang, J. Y. Fung, A. Kheirolomoom, J. L. Sutcliffe and K. W. Ferrara, *Nucl. Med. Biol.*, 2007, **34**, 165–171.
27. J. W. Seo, H. Zhang, D. L. Kukis, C. F. Meares and K. W. Ferrara, *Bioconjugate Chem.*, 2008, **19**, 2577–2584.
28. A. L. Petersen, T. Binderup, P. Rasmussen, J. R. Henriksen, D. R. Elema, A. Kjær and T. L. Andresen, *Biomaterials*, 2011, **32**, 2334–2341.
29. J. W. Seo, S. Qin, L. M. Mahakian, K. D. Watson, A. Kheirolomoom and K. W. Ferrara, *J. Controlled Release*, 2011, **151**, 28–34.
30. S. Li, B. Goins, L. Zhang and A. Bao, *Bioconjugate Chem.*, 2012, **23**, 1322–1332.
31. C. M. Kang, H. J. Koo, S. Lee, K. C. Lee, Y. K. Oh and Y. S. Choe, *Nucl. Med. Biol.*, 2013, **40**, 1018–1024.
32. J. W. Seo, L. M. Mahakian, S. Tam, S. Qin, E. S. Ingham, C. F. Meares and K. W. Ferrara, *Nucl. Med. Biol.*, 2015, **42**, 155–163.
33. C. Perez-Medina, D. Abdel-Atti, Y. Zhang, V. A. Longo, C. P. Irwin, T. Binderup, J. Ruiz-Cabello, Z. A. Fayad, J. S. Lewis, W. J. M. Mulder and T. Reiner, *J. Nucl. Med.*, 2014, **55**, 1706–1711.
34. J. R. Henriksen, A. L. Petersen, A. E. Hansen, C. G. Frankær, P. Harris, D. R. Elema, A. T. Kristensen, A. Kjær and T. L. Andresen, *ACS Appl. Mater. Interfaces*, 2015, **7**, 22796–22806.

35. H. Lee, J. Zheng, D. Gaddy, K. D. Orcutt, S. Leonard, E. Geretti, J. Hesterman, C. Harwell, J. Hoppin, D. A. Jaffray, T. Wickham, B. S. Hendriks and D. Kirpotin, *Nanomedicine*, 2015, **11**, 155–165.
36. C. Perez-Medina, D. Abdel-Atti, J. Tang, Y. Zhao, Z. A. Fayad, J. S. Lewis, W. J. Mulder and T. Reiner, *Nat. Commun.*, 2016, 7, 11838.
37. F. Emmetiere, C. Irwin, N. T. Viola-Villegas, V. Longo, S. M. Cheal, P. Zanzonico, N. Pillarsetty, W. A. Weber, J. S. Lewis and T. Reiner, *Bioconjugate Chem.*, 2013, **24**, 1784–1789.
38. S. A. Mousa, J. J. Bergh, E. Dier, A. Rebbaa, L. J. O'Connor, M. Yalcin, A. Aljada, E. Dyskin, F. B. Davis, H. Y. Lin and P. J. Davis, *Angiogenesis*, 2008, **11**, 183–190.
39. L. R. Perk, M. J. Vosjan, G. W. Visser, M. Budde, P. Jurek, G. E. Kiefer and G. A. van Dongen, *Eur. J. Nucl. Med. Mol. Imaging*, 2010, **37**, 250–259.
40. D. S. Abou, D. L. J. Thorek, N. N. Ramos, M. W. H. Pinkse, H. T. Wolterbeek, S. D. Carlin, B. J. Beattie and J. S. Lewis, *Pharm. Res.*, 2012, **30**, 878–888.
41. J. G. McAfee and M. L. Thakur, *J. Nucl. Med.*, 1976, **17**, 480–487.
42. M. Roca, E. F. de Vries, F. Jamar, O. Israel and A. Signore, *Eur. J. Nucl. Med. Mol. Imaging*, 2010, **37**, 835–841.
43. S. Edmonds, A. Volpe, H. Shmeeda, A. C. Parente-Pereira, R. Radia, J. Baguna-Torres, I. Szanda, G. W. Severin, L. Livieratos, P. J. Blower, J. Maher, G. O. Fruhwirth, A. Gabizon and R. T. M. de Rosales, *ACS Nano*, 2016, **10**, 10294–10307.

CHAPTER 6

# *Liposomes for Hyperthermia Triggered Drug Release*

W. J. M. LOKERSE, B. KNEIDL, A. RYSIN, M. PETRINI AND L. H. LINDNER*

Department of Internal Medicine III, University Hospital of Munich, Ludwig-Maximilians University Munich, Marchioninistrasse 15, Munich, D-81377, Germany
*E-mail: Lars.Lindner@med.uni-muenchen.de

## 6.1 Introduction

Effective solid tumour treatment is challenging using standard treatment methods such as chemotherapy or radiotherapy. In both cases, the treatment usually comes together with pronounced side effects for the patients. For chemotherapy, research has paid attention to encapsulation of chemotherapeutic drugs into nanoparticles to reduce these side effects and prolong circulation time of the drug. The liposome is a type of nanoparticle ranging from approximately 50 to 500 nm in diameter that is capable of retaining drugs in the hydrophilic core or in the hydrophobic bilayer. These features have been exploited for establishment of liposomal chemotherapy that has made its introduction into the clinic. The best example of a clinically accepted liposomal chemotherapeutic is Doxil®/Caelyx®, a sterically stabilized liposome with Doxorubicin (DOX) as a payload. This liposome shows long circulation times in various species, including humans, which is caused by the sophisticated liposomal design.[1] Especially the surface modification

Drug Discovery Series No. 63
Theranostics and Image Guided Drug Delivery
Edited by Maya Thanou

Published by the Royal Society of Chemistry, www.rsc.org

of the liposome with polyethylene glycol 2000 ($PEG_{2000}$), which gives the particles a tremendous increase in blood half-life. Therapeutic effect of these particles would be induced by extravasation from the blood stream into the tumor interstitial space by gaps in the endothelial lining,[2] a process which is also known as the enhanced permeability and retention (EPR) effect.[3] However, a considerable amount of discussion rose on whether the drug would become bioavailable to the cancer cell and its nucleus when it is stably encapsulated in this liposomal carrier.[4,5] This resulted in an increase in research for liposomal formulations which could release the drug by an external trigger. Hyperthermia (HT) is one of these triggers that make use of the transition temperature of lipids for diffusion of drugs across a lipid bilayer. Yatvin and Weinstein were the first to describe this for liposomes in the 70s,[6,7] but with the introduction of $PEG_{2000}$ (pegylation) liposomal drug delivery systems, the establishment of pegylated thermosensitive liposomes (TSL) quickly followed.[8,9] One well-described pegylated TSL formulation is the low thermosensitive liposome (LTSL) which additionally contains lysolipids, resulting in membrane pore formation when exposed to mild HT (39–42 °C) for ultra-fast DOX release. These LTSLs were used to release DOX intravascularly, a methodology which significantly outperformed EPR-based drug delivery methods using Doxil.[10] Consecutively, the intravascular DOX release approach by LTSL quickly showed more success in murine,[11] rabbit[12] and canine tumor treatment.[13] These successes have caused TSL-based therapy to be one of the few stimuli-responsive drug delivery methodologies that have reached clinical trials.[14–17] This pioneering work gave a sharp rise in TSL related research, reaching a peak at this moment (Figure 6.1). Currently, there are many TSL formulations described with varying liposomal features, drugs encapsulated and testing procedures (Table 6.1). In this chapter,

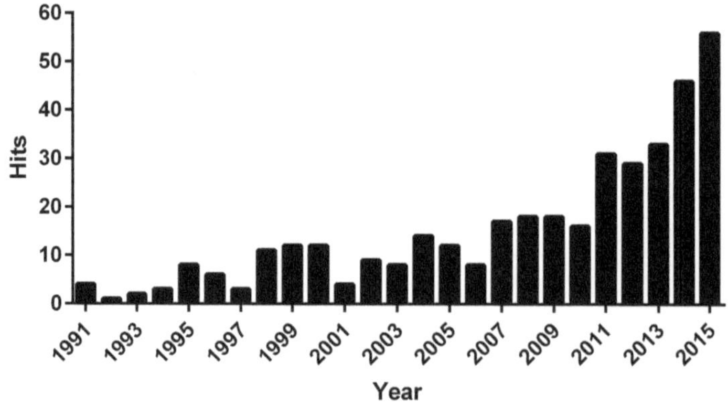


**Figure 6.1** Indicative figure of the increasing interest on drug delivery by thermosensitive liposomes (TSL) in recent years. Web Of Science (Thomson Reuters) hits on thermosensitive liposomes (*Y*-axis) per year (*X*-axis). Keywords: thermosensitive liposomes. Timespan: 1990 to 2015. Document type: article. Total number of hits: 366.

**Table 6.1** Chronological listing of novel developments in the TSL field based on base lipids of choice, encapsulated drugs, cholesterol use, surface modification, drug release improvement and additional active targeting moieties.

| Publication | Base lipids | Encapsulated compound | Cholesterol mol% | Surface modification | Release improvement | Targeting |
|---|---|---|---|---|---|---|
| Yatvin *et al.*[7] | DPPC : DSPC | Neomycin | - | - | - | - |
| Weinstein *et al.*[6] | DPPC : DSPC | Methotrexate | - | - | - | - |
| Sullivan *et al.*[27] | Pam2-PtdCho (DPPC) | Uridine | - | - | - | P3 |
| Iga *et al.*[28] | DPPC : DSPC | Cisplatin | - | - | - | - |
| Merlin *et al.*[29] | DPPC : DSPC | Doxorubicin | Yes 18–31% | - | - | - |
| Maruyama *et al.*[30] | DPPC : DSPC | Doxorubicin | - | $G_{M1}$ | - | - |
| Iga *et al.*[31] | DPPC : DSPC | Cisplatin | - | POE | - | - |
| Unezaki *et al.*[8] | DPPC : DSPC | Doxorubicin | - | PEG | - | - |
| Chelvi *et al.*[32] | Egg-PC | Melphalan | - | - | - | - |
| Gaber *et al.*[9] | DPPC : HSPC | Doxorubicin | Yes 16% | PEG | - | - |
| Merlin *et al.*[33] | DPPC : DSPC | Daunorubicin | Yes 18% | - | - | - |
| Sharma *et al.*[34] | Egg-PC | Taxol | Yes 12.5% | - | - | - |
| Hayashi *et al.*[35] | Egg-PC : DOPE | Calcein | - | NIPAM | NIPAM | - |
| Kono *et al.*[36] | Egg-PC : PA/ DOPE : DC-Chol | Calcein | Yes 48% | APr-co-NDDAM | APr-co-NDDAM | Cationic |
| Ishida *et al.*[37] | DPPC : DSPC | Doxorubicin | Yes 10% | PEG | - | - |
| Yuyama *et al.*[38] | DPPC : DSPC | TNF | - | PG1cUA | - | - |
| Needham *et al.*[39] | DPPC | Doxorubicin | - | PEG | MPPC | - |
| Gaber *et al.*[40] | DPPC : HSPC | Doxorubicin | Yes 16% | PEG | - | Folate |
| Tiwari *et al.*[41] | DPPC : DSPC | Plumbagin | - | - | - | - |

*(continued)*

**Table 6.1** (*continued*)

| Publication | Base lipids | Encapsulated compound | Cholesterol mol% | Surface modification | Release improvement | Targeting |
|---|---|---|---|---|---|---|
| Lindner *et al.*[42] | DPPC : DSPC | Carboxyfluorescein | - | $DPPG_2$ | - | - |
| Han *et al.*[43] | DPPC : HSPC | Doxorubicin | Yes 16–17% | PEG/PNIPAM | PEG/PNIPAM | - |
| Woo *et al.*[44] | DPPC | Cisplatin | - | PEG | MPPC | - |
| Paasonen *et al.*[45] | DPPC/DOPE : Egg-PC | Calcein | Yes 30% | HPMA | HPMA | - |
| Lindner *et al.*[46] | DPPC : DSPC | Miltefosine/ Carboxyfluorescein | - | $DPPG_2$ | Miltefosine | - |
| Puri *et al.*[47] | DPPC : Egg-PC | Calcein | - | PEG | - | HER2 |
| Zhang *et al.*[48] | DPPC : MPPC | Vinolrelbine bitartrate | | PEG | | |
| Winter *et al.*[49] | DPPC | Arsenic Trioxide | - | - | MPPC | - |
| Kono *et al.*[50] | Egg-PC | Doxorubicin | Yes 47% | PEG | Poly(EOEOVE) | - |
| Ta *et al.*[51] | DPPC : HSPC | Doxorubicin | Yes 16% | PEG | PAA | - |
| Wu *et al.*[52] | DPPC : DSPG/ HSPC : DSPG | Epirubicin | No/Yes 0–9% | PEG | MSPC/None | - |
| Tagami *et al.*[53] | DPPC | Doxorubicin | - | - | Brij78 | - |
| Smith *et al.*[54] | DPPC | Doxorubicin | - | PEG | - | HER2 |
| Zhang *et al.*[55] | DPPC | Albumin | - | PEG | MSPC | - |
| Hossann *et al.*[56] | DPPC : DSPC | Doxorubicin | - | $DPPG_2$ | - | - |
| Kulshrestha *et al.*[57] | DPPC : PG | Paclitaxel | - | - | - | - |
| Yudina *et al.*[58] | DPPC | TO-PRO-3 | - | PEG | MSPC | - |
| Park *et al.*[59] | DPPC | Doxorubicin | Yes 21% | PEG : ELP | - | - |
| May *et al.*[60] | DPPC | Gemcitabine/Oxaliplatin | - | - | Brij78 | - |

| | | | | | | |
|---|---|---|---|---|---|---|
| Al-Ahmady *et al.*[61] | DPPC : HSPC | Doxorubicin | Yes 16% | PEG | - | MUC-1 |
| Kim *et al.*[62] | DPPC : DSPC | Doxorubicin | Yes 15% | PEG : ELP | - | cRGD |
| Dou *et al.*[63] | DPPC : DPPG | Cisplatin | - | PEG | MSPC | - |
| Van Elk *et al.*[64] | DOPE : Egg-PC | Doxorubicin | Yes 2–10% | pHPMA1ac | - | - |
| Yang *et al.*[65] | DPPC | CPP-Dox | - | PEG | MSPC | NGR |
| Limmer *et al.*[66] | DPPC : DSPC | Gemcitabine | - | $DPPG_2$ | - | - |
| Shemesh *et al.*[67] | DPPC : Soy-PC | LPICG | Yes 17% | PEG | - | - |
| Al Sabbagh *et al.*[68] | DPPC | 5-FU | Yes 5–30% | PEG | - | - |
| Liang *et al.*[69] | DPPC | Doxorubicin | - | PEG | MSPC/CFL | - |
| Al-Ahmady *et al.*[70] | DPPC : DSPC | Doxorubicin | - | PEG | Lp-peptide | - |
| Rosca *et al.*[71] | DPPC : Eu.DOTA. DSA : XL750.DSA | Topotecan | - | PEG | MSPC | - |
| Wang *et al.*[72] | DPPC | Doxorubicin | - | PEG | MSPC | CREKA |
| Song *et al.*[73] | DPPC | Doxorubicin | Yes 35% | - | Brij78 | Pamidro-nate |
| Tagami *et al.*[74] | DPPC | Calcein | - | - | P188 | - |
| Peng *et al.*[75] | NA | Calcein/pDNA | NA | NA | NA | Cationic |
| Dicheva *et al.*[76] | DPPC : DSPC | Doxorubicin | - | PEG | - | Cationic |
| Lu *et al.*[77] | DPPC : DSPC | Idarubicin | - | PEG | - | - |

we will analyse each aspect of the TSL design and testing procedures to establish a fundamental understanding in creating an effective TSL formulation and which alterations on these particles have proven to be effective and which have not (Figure 6.2). Firstly, we will focus on how the choice of the base lipids influences thermosensitivity of the TSL bilayer (Section 6.2) and how the addition of cholesterol influences membrane stability (Section 6.3). Secondly, an in depth analysis is provided in surface modifications of TSL and how these influence particle stability, circulation time and drug release (Section 6.4). Thirdly, the addition of compounds to the TSL membrane to improve drug release (Section 6.5) and the comparative loading and release kinetics of a variety of drugs is described (Section 6.6). Finally, the effectivity of active targeting TSL to the tumor site (Section 6.7) and the influence of biological environment and testing procedures on TSL stability and drug release are addressed (Section 6.8).

## 6.2 Base Lipids of TSL

The first parameter that is important for TSL design is the lipid composition. The TSL is usually composed of a base lipid mixture that gives the thermoresponsive properties to a mild hyperthermic trigger (39–42 °C). The standard lipids that are usually chosen for a TSL formulation are 1,2-dipalmitoyl-*sn*-glycero-3-phosphocholine (DPPC), 1,2-distearoyl-*sn*-glycero-3-phosphocholine (DSPC), hydrogenated soy L-α-phosphatidylcholine (HSPC), Egg L-α-phosphatidylcholine (Egg-PC) and 1,2-dioleoyl-*sn*-glycero-3-phosphoethanolamine (DOPE). The advantage of using synthetic phospholipids (DPPC, DSPC, DOPE) over natural phospholipids (HSPC, Egg-PC) is the purity of the single components used, giving a high reproducibility between TSL batches. Making a liposome susceptible for drug release at mild hyperthermia (HT) depends on the molar ratio in which the base lipids are used. Release of drugs takes place when a lipid bilayer is exposed to a temperature close to its melting temperature ($T_m$). When a lipid bilayer is exposed to $\geq T_m$ temperatures, the carbon chains reorganize from an ordered gel state (nucleation) to a disordered liquid crystalline state (melting), making the bilayer more permeable. The length of the carbon chain, presence of double bonds, surface charge and lipid head group determines the $T_m$ of a specific lipid.[18] During liposome preparation, the lipids are frequently heated to temperatures $\geq T_m$ and are cooled down when preparation is complete. Nucleation of the lipids during the cool down occurs heterogeneously, as growing islets in a liquid membrane. Once the nucleation is complete, the solid islets coincide leaving so called grain boundaries in between.[19,20] During heating of the liposomes slightly below the $T_m$ (pre-transition), there will be lipid domains that undergo phase transition more effectively than adjacent domains. The grain boundaries between these domains are unstable areas which are particularly suitable for transmembrane transport of *e.g.* ions.[21] TSL formulations are usually composed of a mixture of phosphatidylcholines with 16:0 ($T_m$ = 41 °C) and 18:0 ($T_m$ = 55 °C) carbon chains. These lipids have been shown to equally distribute over the entire lipid bilayer[22] and when exposed to mild HT, the 16:0 carbon chains (C16) undergo a phase transition, while the 18:0 carbon chains (C18) remain unaltered. The grain boundaries between these C16

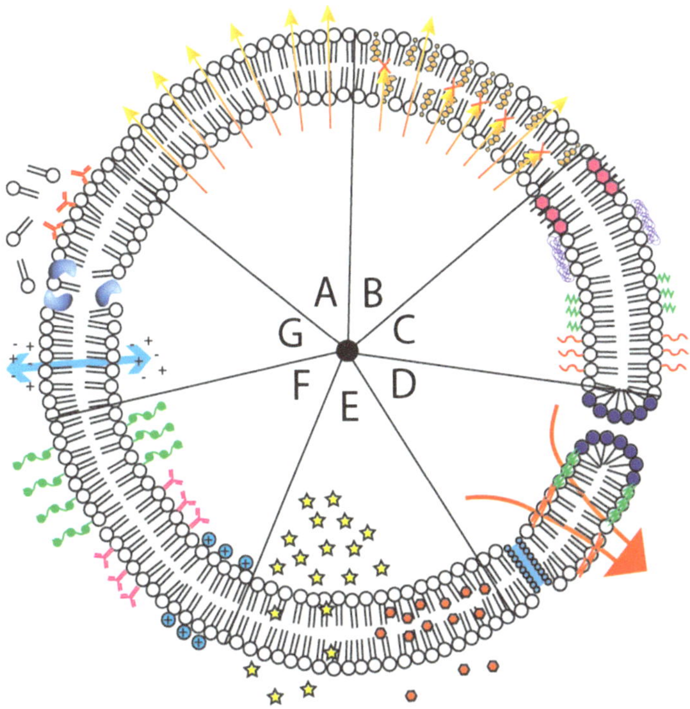


**Figure 6.2** Graphical summary of thermosensitive liposome (TSL) based drug delivery options described in the literature. The base lipids, usually lecithins, give the liposome its thermosensitive properties and because of solid gel to liquid disordered phase transition and through grain boundaries in the membrane, the encapsulated drug can be released around phase transition temperature (A; Section 6.2). Adding cholesterol to a TSL formulation can give the TSL membrane more rigidity to reduce premature drug leakage and improve particle circulation time in the blood stream (B; Section 6.3). Several surface modifications have been described for TSL to increase circulation time like gangliosides (pink) or poly-ethyleneglycol (PEG; purple). Other surface modifications like thermosensitive polymers (green) or peptides (red) have been introduced to improve drug release (C; Section 6.4). Upon hyperthermia (HT) exposure, these thermosensitive surface modification molecules undergo a hydrophilic-hydrophobic chemical transition, generating instability in the TSL membrane and enhancing drug release. Other release enhancing methods are based on pore formation into the TSL membrane at HT by using lysolipids, detergents (dark blue) or a leucine zipper peptide bundle (light blue) (D; Section 6.5). Compounds can be incorporated into the aqueous core (stars) or into the hydrophobic liposomal bilayer (hexagons). The loading and release of different compounds depends greatly on their chemical properties (E; Section 6.6). Several attempts have also been made to target TSLs to cancer cells by linking antibodies (pink) or targeting peptides (green) to the surface of the liposomes. Cationic lipids (light blue) were used to target overexpressed, negatively charged proteins on the surface of neovascular endothelial cells (F; Section 6.7). When TSLs are introduced in complex biological environments like serum or the blood stream, several factors like osmotic pressure (blue arrow), serum protein-membrane interactions (blue), IgG binding and lipid transfer can greatly influence TSL stability and circulation time (G; Section 6.8).

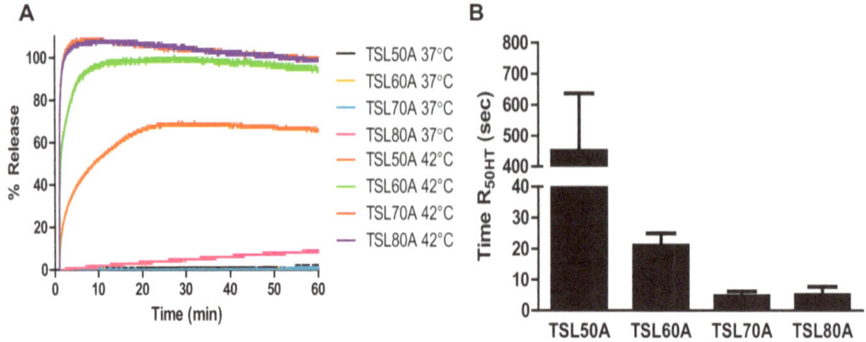


**Figure 6.3** Doxorubicin (DOX) release analysis of thermosensitive liposomes (TSLs) with alternating (X:(100-5-X):5) DPPC:DSPC:DSPE-PEG$_{2000}$ molar ratios (X:(100-5-X):5). The TSL formulation with 50 mol% DPPC (TSL50A) showed no leakage at 37 °C, but incomplete release at 42 °C. TSL80A gave fast triggered release at 42 °C, but premature DOX leakage at 37 °C. Therefore, TSLs with 70 mol% DPPC were concluded to be optimal in this case since there was fast release at 42 °C and no detectable leakage at 37 °C. Figure reprinted with permission from W. J. Lokerse, E. C. Kneepkens, T. L. ten Hagen, A. M. Eggermont, H. Grull and G. A. Koning, *Biomaterials*, 2016, **82**, 138–150.[25]

liquid-crystalline and C18 gel state carbon chains will improve drug release at mild HT (39–42 °C) as long as the mol% of the C18 lipid in the bilayer does not exceed threshold levels of ≥35 mol%.[23–25] Next to adapting TSL for release optimization properties, reducing drug leakage at physiological conditions is an important factor to consider. Increasing the C18:C16 ratio increases stability at physiological conditions but reduces drug release at mild HT, which shows that to obtain an optimal TSL, fine tuning of the base C18:C16 is recommended (Figure 6.3).[25,26]

## 6.3 Cholesterol

Cholesterol is a naturally occurring sterol that is found in eukaryotic cell membranes where it can stabilize membranes towards a liquid-ordered phase.[78] This stabilizing effect has also been shown in the liposomal field where cholesterol has been described to make membranes more permeable in gel phase, while in the liquid crystalline phase it decreases membrane permeability.[79] However, cholesterol-lipid interactions are more complex and the membrane permeability is very dependent on the membrane lipids used and the molar fraction of cholesterol.[80,81] When cholesterol is used up to 10 mol%, pre-transition of the bilayer can be abolished, while membrane permeability around $T_m$ remains largely unaltered.[82] However, at these quantities, multiple phase transitions can be found as rafts over the lipid membrane, which makes the membrane rigidity not homogenous, thereby making encapsulated compound leakage still possible.[83] Increasing mol% of cholesterol in a liposomal formulation will induce a more tight packing of the lipids in the liquid-ordered phase, which reduces serum protein induced disruption of the membrane, eventual drug leakage and particle clearance *in vivo*.[84,85] Incorporation of cholesterol in TSL was early described by Merlin and coworkers[29] where they attempted to create a more stable TSL formulation than the first described by Yatvin *et al.*[7] They showed that incorporation of up to 18 mol% of cholesterol could greatly stabilize a TSL composed of DPPC/DSPC and prevent unwanted drug release at physiological temperature. Further increasing cholesterol mol% to ≥30 mol% would be irrelevant as these membranes are thinner and have homogenous cholesterol distribution, thereby becoming liquid-ordered, resulting in a loss of thermal transition properties.[18,83,86] This basically led to the conclusion that the cholesterol/phospholipid ratio has to be near 2 : 9 mol : mol, to obtain a liposome with adequate stability and maintaining some membrane fluidity.[18,87] However, when PEG was introduced as surface modification (Section 6.4) to prolong circulation time of liposomes, it was also shown to assist in drug release from liposomes that have a mol% of cholesterol up to 30.[9] This led to a mindset in the 90s of designing TSL that should have an adequate stability and circulation time. Therefore, it was considered at the time that including 15–20 mol% of cholesterol in a TSL formulation was a necessity to obtain optimal stability for *in vivo* applications. When later the focus shifted more to ultrafast drug release,[5] cholesterol was only used at relatively low mol% or was totally absent from TSL formulations. Currently, cholesterol is only used in amounts above 20 mol% in TSL formulations when combined with release improving compounds[50,59,73] (Section 6.5).

In conclusion, cholesterol plays a key role in stabilizing lipid membranes. TSL formulations usually have some membrane instability and drug leakage at physiological temperature. Adding the right amount of cholesterol to the formulation can give the TSL a higher stability in circulation without greatly affecting drug release at HT. However, the TSL formulations more recently described make less use of cholesterol and focus more on release improvement for acquiring ultrafast drug release (Section 6.5).

## 6.4 Surface Modification

Surface modification implies the use of molecules that cover the liposome surface which reduces particle aggregation and interaction with serum components, thereby improving liposome stability and circulation time. When this proved to significantly increase circulation time and liposome stability for non-thermosensitive liposomes (NTSL),[88,89] surface modification of TSL quickly followed.[8,9] PEG has become a golden standard molecule of choice, but also other surface modifications have been used to increase circulation time. Furthermore, alternative surface modifications which improve drug release have also been adequately analyzed (Table 6.1). These release improving surface modifications will be discussed in this section in terms of circulation time and stability, followed by more in depth analysis on drug release in Section 6.5.

As stated above, PEG is the most commonly used surface modification in TSL research. The principle of so called "pegylation" is based on grafting a dense layer of polymer on the surface of nanoparticles, thereby reducing particle aggregation[90] and inhibiting serum protein interactions, which otherwise leads to opsonization and clearance by the reticuloendothelial system.[91] The main factors which influence this are the surface density of the compound and the PEG chain length. Pegylation induces a concentration dependent decrease in serum protein binding and increase in particle circulation time. This is due to the conformation of PEG into a mushroom (0–4 mol%) or a mushroom/brush state (5–9 mol%) on the particle surface. The higher density of PEG pushes more polymer chains into a brush conformation, thereby providing more steric hindrance for serum protein interaction.[92] However, incorporating ≥10 mol% PEG will not further decrease protein adsorption, will not fully incorporate into liposomes and can reduce drug encapsulation.[56,93,94] Moreover, pegylation beyond 5 mol% on TSL has been shown to induce membrane integrity collapse and premature drug leakage.[95] As a result, pegylation of TSL formulations usually does not reach much beyond 5 mol%. Increasing the PEG chain length (molecular weight 350–45 000 gram mole$^{-1}$) has a positive effect on reducing serum protein adsorption and improving circulation time.[93,94] However, increasing PEG chain length reduces membrane permeability and thereby hampers drug loading and release from pegylated particles.[96] Due to these reasons, frequently a PEG chain with a molecular weight of 2000 ($PEG_{2000}$) is selected as a consensus between membrane permeability and particle circulation time. Ganglioside $GM_1$ has also been investigated as a surface modification for TSL by Maruyama and colleagues.[30] Introducing this surface modification resulted in a reduction of blood clearance and uptake by the reticuloendothelial system (RES), which increased drug accumulation in heated tumors by two-fold. Nevertheless, the results for $GM_1$ as a liposome surface modification were inconsistent between animal species[97] and comparative studies between $GM_1$ and PEG showed that the latter performed better in all the previously mentioned aspects.[98,99] When follow up research also showed that PEG assists in improving drug release at mild HT (Section 6.5) whereas $GM_1$ did not, there was no need to further investigate $GM_1$ as a surface modification for TSL.[100] Stearylpolyoxyethylene (POE) as a surface modification for TSL showed similar results as $GM_1$ and therefore also no consecutive research was conducted using POE.[31]

One interesting candidate in this field is 1,2-dipalmitoyl-*sn*-glycero-3-phosphodiglycerol ($DPPG_2$), an anionic, synthetic phospholipid which could be incorporated in TSL up to 70 mol%. In contrast to DSPE-$PEG_{2000}$,[101] $DPPG_2$ lipids do not form micelles when used above 10 mol% and because of these reasons, high quantities of this lipid can be incorporated in the TSL membrane. The incorporation of 30 mol% $DPPG_2$ in an DPPC : DSPC TSL resulted in a high circulation half-life of encapsulated carboxyfluorescein (CF) of 9.6 ± 1.3 h in hamsters and 5.0 ± 0.9 h in rats, while maintaining a rapid drug release at 42 °C[42,56] (Figure 6.4). This combination of high circulation half-life and rapid drug release makes the $DPPG_2$-based TSL formulations exceptionally promising candidates for reliable clinical treatments. How this lipid exactly functions in the TSL membrane mechanistically is currently investigated.

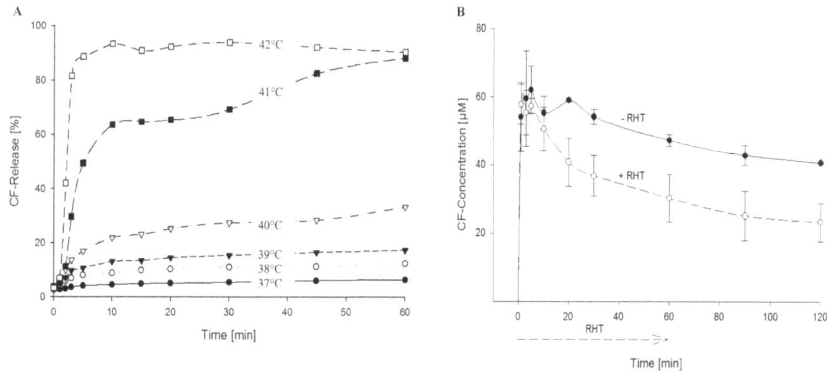


**Figure 6.4** Carboxyfluorescein (CF) release (A) and pharmacokinetic profiles (B) of DPPC:DSPC:DPPG$_2$ based thermosensitive liposomes (TSLs) with molar ratio 50:20:30. The formulation showed negligible CF leakage at 37 °C and fast, triggered release at 42 °C in bovine serum (A). CF had a long circulation time in hamsters with a CF $t_{1/2}$ of 9.6 ± 1.3 h and regional hyperthermia (RHT) of a tumor in a window chamber caused the $t_{1/2}$ to significantly drop as CF is highly released at the tumor site. Figures reused with permission from L. H. Lindner, M. E. Eichhorn, H. Eibl, N. Teichert, M. Schmitt-Sody, R. D. Issels and M. Dellian, *Clin. Cancer Res.*, 2004, **10**, 2168–2178.[42]

As stated before, several surface modifications have been introduced into the TSL field, which are intended to improve release. Han and coworkers investigated the poly(*N*-isopropylacrylamide-*co*-acrylamide) (PNIPAM-AAM) surface modification and found that next to the release improving capacities of the polymer, protein adsorption onto the particles was reduced.[43,102] However, additional pegylation caused a significant reduction in protein binding and induced further drug release improvement. Similar results were reported for poly[2-(2-ethoxy)ethoxyethyl vinyl ether (EOEOVE)] modified TSLs with the additional findings that indicated that pegylation was an absolute necessity for improved circulation time and reduced particle clearance *in vivo.*[50] Others that made use of poly(*N*-(2-hydroxypropyl)methacrylamide mono/dilactate) (pHPMA mono/dilactate) or elastin like polypeptide (ELP) have also emphasized the use PEG next to the release improving surface modification for adequate TSL circulation time.[59]

Taken together, there have been many surface modifications described for TSL formulations that have an influence on circulation time and drug release. What is important to consider is the fine balance between drug release and circulation time. Not all surface modifications have been comparatively tested *in vivo* for analysis of the fate of TSLs once they enter the blood stream and how this might affect drug delivery. These studies are required in order to make better judgment on optimal forms of surface modification for TSLs. PEG is still most commonly used as a surface modification to improve circulation time. However, there is room for improvement as pegylated particles are more effectively cleared after repeated injections[103,104] and anaphylactic responses have been reported after i.v. administration.[105]

## 6.5 Release Improvement

In the previous sections, we introduced an increasing interest in release improvement of TSL in the recent decades. The pegylated, cholesterol containing, traditional TSL (TTSL) has a $T_m$ around 42–45 °C,[9] which is rather high when considering clinical HT.[106] Temperatures in this range can induce significant damage to healthy tissue surrounding the tumor and causes vascular shutdown, which could hamper effective drug delivery.[107] Pegylation of liposomes can assist in improving release in high cholesterol containing TSL formulations,[9] but in cholesterol free TSL, there is a specific quantitative range of %mol PEG to use for optimal TSL release and stability. Hossann and co-workers observed a concentration dependent effect of DSPE-PEG$_{2000}$ on CF release from DPPC/DSPC liposomes.[56] Because incorporation of 10 mol% of DSPE-PEG$_{2000}$ resulted in reduced CF loading efficacy and 5 mol% of DSPE-PEG$_{2000}$ is sufficient to ensure long circulation of TSL, only 1 mol% and 5 mol% of DSPE-PEG$_{2000}$ were intensively investigated in this study. In comparison to pure DPPC/DSPC liposomes, 1 mol% of DSPE-PEG$_{2000}$ slightly reduced CF release, whereas TSL with 5 mol% of DSPE-PEG$_{2000}$ demonstrated a 1.3 fold increase of CF release. More recently, Li *et al.* confirmed these results for the same base lipid composition using a 1–10 mol% DSPE-PEG$_{2000}$ concentration range. Although the release profile for the pure DPPC/DSPC formulation was not shown, higher pegylation resulted in higher CF release. However, ≥5 mol% of DSPE-PEG$_{2000}$ induced considerable drug leakage at body temperature.[95] They concluded that 5 mol% of DSPE-PEG$_{2000}$ was optimal for ensuring fast release from DPPC/DSPC liposomes and maintaining stability at body temperature. Others emphasized that a pegylated TSL solely made from DPPC as a base lipid needs an additional compound for enhancing drug release.[108] Dewhirst, Needham and co-workers pioneered this field with the introduction of lysolipids in pegylated TSL (LTSL).[39] They hypothesized that the TSL formulation had to be able to ensure full release of the drug payload during the transition time of the particle travelling through the heated tumor, and therefore a TSL with full drug release within seconds had to be designed.[5] Adding 10 mol% of monopalmitoyl-phosphatidylcholine (lyso-PC16) to a DPPC liposome decreased the $T_m$ of the bilayer from 41.9 °C to 40.0 °C, resulting in an increase of CF release from 15–20% to 80% at 41 °C in saline.[109] Adding PEG to the formulation and using DOX as the drug showed similar results. The LTSL released 45% DOX in 20 sec in fetal bovine serum (FBS) at 42 °C, whereas the non-lysolipid containing TSL and TTSL released only 20% and 40% after 30 min, respectively.[39] More recently, the lyso-PC16 was exchanged for monostearoylphosphatidylcholine (lyso-PC18) to slightly enhance liposomal stability at 37 °C.[110] Therefore, the LTSL formulation Thermodox®, which is currently in clinical trials, is composed of DPPC : lyso-PC18 : DSPE-PEG$_{2000}$ at a molar ratio of 86.5 : 9.7 : 3.8.[14–16,108] There was also considerable effort spent on analysis of the release mechanism from LTSL. Lysolipids tend to accumulate at the grain boundaries during melting of the base lipids, establishing a pore in the membrane due to the micelle forming nature of the lysolipid.[110,111] The additional PEG in the LTSL further sterically stabilizes the pore, thereby assisting effective rapid drug release.[108,112] Despite these promising results, there are some reports that give some doubt on using lysolipids in TSL formulations. Firstly, the lysolipid can dissociate from the membrane when incubated with acceptor vesicles *in vitro*[111] or in circulation.[113] Secondly, the circulation time of these lysolipid containing vesicles are remarkably lower than for TSLs that lack lysolipid, possibly reducing DOX delivery by LTSL.[113] Tagami *et al.* developed an alternative LTSL formulation composed of DPPC : Brij78 (96 : 4), named HaT (hyperthermia-activated cytotoxic).[53,114] Brij78, a surfactant in the lipid bilayer of HaT, is responsible for the ultrafast DOX release from HaT based on its chemical structure, a PEGylated ($PEG_{880}$) single acyl chain that combines the chemical structures of lyso-PC and a PEG-lipid in a single molecule. The HaT liposomes released DOX significantly faster than LTSL, an improvement which resulted in a factor 1.4 increase of DOX delivery to murine EMT6 mammary carcinoma while maintaining a similar pharmacokinetic profile.[53] By optimizing the DOX loading methodology and post-inserting Brij78 solely in the outer leaflet of TSL, the DOX release rate, circulation half-life, intratumoral DOX delivery and therapeutic efficacy were further enhanced.[115] However, 1H-NMR analysis on acceptor vesicles incubated with HaT liposomes showed extraction of Brij78 similarly as described above for lysolipids.[53] This extraction could cause a loss of stability of HaT in circulation and enhanced clearance of the particles *in vivo*, thereby reducing delivered drug to the heated tumor. Lindner *et al.* synthesized a novel $DPPG_2$ lipid which not only prolongs the circulation time of TSL (see previous section), but also provides an ultra-fast temperature-triggered drug release.[42] The $DPPG_2$-based formulation composed of DPPC : DSPC : $DPPG_2$ (5 : 2 : 3 mol%) showed 70% DOX release within the first 20 sec of heating at 42 °C in FBS and required 2 min for total release.[56] Furthermore, there was also less than 5% DOX leakage observed within one hour of incubation at 37 °C, giving the formulation promising properties for further *in vivo* investigation. Follow up work was conducted by addition of miltefosine to the formulation.[46] Miltefosine, a cytotoxic drug from the class of alkylphosphocholines, is closely related to lysolipids, but lacks the glycerol backbone and therefore the hydrolysable ester bond present in lysolipids. The incorporation of miltefosine in the TSL membrane did not only yield TSL with high heat-triggered and cytotoxic potency, but also demonstrated an increased release of CF by a mechanism which is currently under investigation.

There are also other release improvement strategies that go beyond the use of specific lipids or surfactants. One example is the use of thermosensitive polymers which above the lower critical solution temperature (LCST) switch from a hydrophilic to hydrophobic, aggregated state. The precipitation of the hydrophobic polymer on the liposome surface increases the interaction with the lipid bilayer and leads to the temperature-induced liposome destabilization, thereby triggering drug release.[116] Kono and co-workers have pioneered this field with poly-*N*-isopropylacrylamide (poly-NIPAM), a thermosensitive polymer with a LCST of 32 °C.[117] NIPAM was copolymerized with octadecyl acrylate (ODA) which served as an anchor to fix the polymer in the lipid bilayer. Incorporation of copoly-(NIPAM-ODA) into Egg-PC, DPPC or DOPE membranes resulted in improved drug release at 41 °C of CF or calcein.[117,118] However, at temperatures around the LCST of the copolymer (27 °C) there was considerable drug release observed, thereby confirming the membrane destabilizing effect of the polymer. This resulted in follow up research where NIPAM was covalently linked to acryloylpyrrolidine (APr), a polymer with a LCST of 52 °C to establish a net LCST of 40 °C.[119] Not only could the stability of the TSL be improved below 30 °C, it also could optimize the release by a terminal-anchor design of the polymer. Pegylation of these polymer enriched TSLs had a similar effect on drug release as for LTSL, stabilizing the formulation below LCST and enhancing drug release rates above.[43,120] An alternative polymer-modified TSL formulation established by Kono and colleagues included the use of a block copolymer of the thermosensitive (2-ethoxy)ethoxyethyl vinyl ether (EOEOVE) and the anchor octadecyl vinyl ether (ODVE).[50,121] The resulting copolymer EOEOVE-*block*-ODVE that was incorporated into a DOX loaded TSL showed a stronger temperature sensitizing effect than the traditional NIPAM based TSL and significantly supressed tumor growth in a subcutaneous C-26 colon cancer model. Alternatively, poly(*N*-(2-hydroxypropyl)methacrylamide mono/dilactate) (pHPMA mono/dilactate) thermosensitive polymers have also been investigated for TSL based drug delivery.[45,64]

More recently, significant effort has been put into TSL formulations that make use of biodegradable peptides for release improvement of drugs. Park and co-workers succeeded in improving release by using an elastin-like polypeptide (ELP) of a Valine-Proline-Glycine-Valine-Glycine peptide repeat.[26,59] The principle is similar to the previously mentioned mechanism of action of surface linked polymers. Each amino acid in the peptide has an inverse transition temperature ($T_t$), which makes the peptide water soluble below the net $T_t$ and undergoes a conformational change to a gel-like coascervate phase above this temperature, thereby establishing membrane instability and faster drug release.[122,123] A detailed comparison of ELP-TSL *versus* LTSL showed that the prior has higher stability at body temperature, giving the particle a higher circulation time (41% *vs.* 14% injected DOX dose 5 h after i.v. injection in mice, respectively) which improved therapeutic efficacy in a murine squamous cell carcinoma model.[59] In two studies by Al-Ahmady *et al.*, the introduction of a leucine zipper peptide into the TSL membrane was described for enhancement of drug release.[70,124] Upon HT exposure, the established α-helix peptide bundle would undergo a phase change to a disordered conformation of two monomers forming a pore which would ensure effective drug release. The leucine zipper TSL showed higher stability than LTSL *in vitro* and *in vivo*, but the introduction of the peptides into the membrane did not result in release enhancement at 42 °C. Although leucine zipper TSL treated tumors accumulated more DOX than LTSL treated tumors, the therapeutic efficacy was similar when an intravascular drug release protocol was applied.

In order to obtain a TSL with a rapid drug release profile, some release improvement to base lipid composition is required. However, this is usually connected to a reduction in TSL stability at physiological temperature and enhanced blood clearance of the particles. Due to these reasons it is important to fine tune formulations with optimal stability in the blood stream and fast release at the heated tumor. There is currently a lack of comparative studies that describe TSL formulations with the focus on fastest drug release *versus* TSLs that focus on increased circulation time. These studies would give more insight into how an optimal TSL should be designed for the most effective drug delivery to solid tumors.

## 6.6 Encapsulated Compounds

As shown in Table 6.1, various types of drugs have been encapsulated into TSL formulations. Hydrophilic drugs are encapsulated in the aqueous core of the liposome, possibly also interacting with the membrane, while hydrophobic drugs are embedded in the lipid bilayer. Many factors like liposome composition, surface charge, method of preparation, drug loading protocol and drug characteristics influence the drug loading efficacy and liposome stability, therefore making an optimal liposome/drug loading preparation process a challenging task.[125] Frequently, hydrophobic compounds tend to leak more easily from liposomes than hydrophilic compounds entrapped in the aqueous core,[126] a process which is highly enhanced in the presence of serum proteins or cell membranes,[127] resulting in low improvement of *in vivo* circulation time *versus* free drug.[128]

In several studies, the principle of TSL based drug delivery is first shown with a fluorescent and highly hydrophilic drug like CF or calcein.[35,42,45,95] These model drugs, which are practical to use, non-toxic and ideal for formulation comparisons are encapsulated into liposomes by passive encapsulation, the hydration of a lipid film with a water solution containing these compounds, resulting in an encapsulation efficiency of 1–10%. The low encapsulation and the hydrophilic nature of these compounds causes a relatively slow diffusion across the bilayer, thereby giving a low premature drug leakage at 37 °C and release at 42 °C. When these model drugs are replaced with the most commonly used anthracycline DOX, which is actively loaded up to 100% and less hydrophilic, the drug leakage and release is more profound[43,56,64,129] (Figure 6.5). By which method the active loading is performed also plays a role in drug release and stability.[25,130]

Other actively loaded anthracyclines showed increasing membrane association with increasing hydrophobicity Idarubicin > Daunorubicin > Doxorubicin > Epirubicin.[131] Idarubicin is an example of a more hydrophobic anthracycline that has been investigated in TSL based drug delivery.[77] The various TSLs described in this study all showed a rapid premature burst release of 10–40% within the first minutes of incubation in 37 °C FBS, which is likely caused by a membrane-bound fraction of the drug that dissociates from the liposome upon serum protein interaction (Section 6.8). These findings were

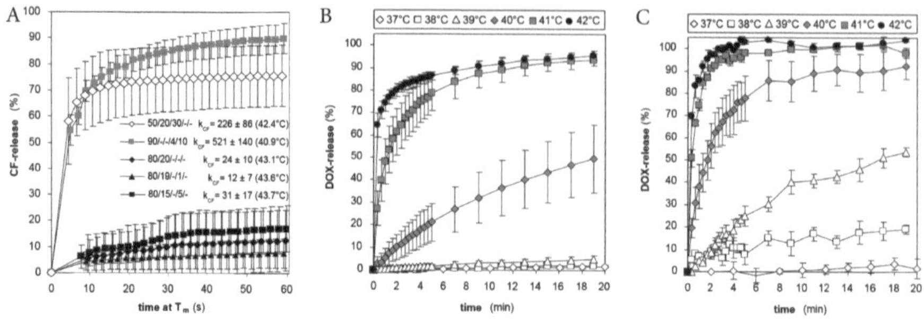


**Figure 6.5** Carboxyfluorescein (CF) release kinetics from thermosensitive liposomes (TSLs) at their specific phase transition temperature ($T_m$) with different DPPC:DSPC:$DPPG_2$:DSPE-$PEG_{2000}$:Lyso-PC lipid molar ratios in bovine serum (A) and Doxorubicin (DOX) release profiles from DPPC:DSPC:$DPPG_2$ based (TSLs) with molar ratio 50:20:30 in HEPES-buffered saline (B) or bovine serum (C). The lysolipid based TSL showed the fastest temperature-induced CF release (A). However, the $DPPG_2$-based TSL showed fast and complete release of the chemotherapeutic DOX (C). Release of DOX in serum deprived media reduced effective drug release from these TSLs (B). Reprinted from *Biochimica et Biophysica Acta (BBA) – Biomembranes*, **1768**, M. Hossann, M. Wiggenhorn, A. Schwerdt, K. Wachholz, N. Teichert, H. Eibl, R. D. Issels and L. H. Lindner, *In vitro* stability and content release properties of phosphatidylglyceroglycerol containing thermosensitive liposomes, 2491–2499, Copyright 2007, with permission from Elsevier.

in contrast with less hydrophobic Doxorubicin or Epirubicin loaded TSLs.[25,52] However, there are more extreme hydrophobic drugs like plumbagin and paclitaxel which are tightly embedded in the TSL bilayer and dissociate from the membrane rather inefficiently *in vitro*.[41,57] Lindner and colleagues also tested the highly lipophilic miltefosine engrafted TSL on cells, of which is still unknown whether the compound fully dissociates from the membrane upon HT exposure or if the cellular uptake of miltefosine was induced by HT-enhanced binding of the particles to the cells.[46] Another example of how various TSL incorporated compounds show differences in release and stability is described by May *et al.*[60] In this study, the two water soluble, with low membrane permeability drugs, Oxaliplatin and Gemcitabine were compared in a Brij-containing TSL. Despite the chemical similarities of these drugs and being incorporated in a Brij-TSL with the same lipid composition, the drug leakage at 37 °C, stability in circulation and intratumoral drug levels were significantly lower for Oxaliplatin TSL. In other cases, the effective release of high molecular weight proteins from TSL has been reported, but still more consecutive research needs to be performed to describe the mechanism of action and *in vivo* effectivity.[38,55]

Some studies have also described the encapsulation of MRI-contrast agents in TSL to represent the release of encapsulated drugs in an image-guided way. This so called "dose painting" by encapsulation of *e.g.* $MnSO_4$ or Gadolinium-based contrast agents into TSL next to DOX, showed the heterogeneous drug delivery of the contrast agent throughout a tumor *in vivo* by $T_1$-weighed MRI.[132,133] The principle relies on the contrast agent being introduced in a large aqueous volume after release, thereby inducing a shift in $T_1/T_2$ relaxivity. Although the changes in relaxivity and DOX release do not follow the exact same kinetics *in vitro*,[134] the change in intratumoral $T_1$ relaxivity imaged by MRI *in vivo* correlated adequately with the DOX quantities obtained by HPLC *ex vivo*, a result which was found for LTSL[135] and $DPPG_2$-based TSL.[136]

This short compilation of TSL research where encapsulated compounds with different chemical properties are compared indicates that every TSL-drug combination is unique and cannot always be represented by *e.g.* model drugs in terms of release and stability kinetics. The exact interaction and dynamics of drugs with liposomal bilayers is complex and should be investigated very thoroughly for each new TSL formulation. Furthermore, the complexity of these membrane-drug dynamics will rise when interactions with serum components are taken into account (Section 6.8). Although image-guided drug delivery for TSL by MRI contrast agents is promising, it is important to keep in mind that the translation to cellular uptake in the tumor is an area which remains to be investigated.

## 6.7 Targeting

In order to establish higher drug delivery values, some have spent significant effort on molecular targeting of TSL to the tumor cell or cells in the tumor microenvironment. For TSL, two targeting strategies have been investigated, targeting the tumor cell or targeting the neovasculature. Active targeting can be achieved by coupling ligands such as monoclonal antibodies, antibody fragments (Fab), proteins, peptides, carbohydrates or glycoproteins onto the liposome surface.[137,138] Another method is by using lipids with a cationic head group.

For targeting of the tumor cell, the TSL needs to extravasate from the blood stream into the interstitial space *via* the earlier mentioned EPR effect. The targeting agent can then bind overexpressed proteins on the tumor cell surface and drug release can take place prior or after endocytosis of the particle.[139] The first targeted TSL formulation was described by Sullivan and colleagues in 1986.[27] They successfully encapsulated [$^3$H]uridine into TSL with coupled anti-H2K$^k$ antibodies on the liposome surface, which target the major histocompatibility complex (MHC) class I expressed on murine T lymphoma (RDM4) cells. The targeted TSL gave three times higher uridine uptake by RDM4 cells compared to non-targeted TSL when the cells were incubated with the liposomes for 1 h, followed by a 41 °C incubation for drug release.[27] Folate-targeting of TSL could even improve drug uptake in multi drug resistant cancer cell lines and significantly reduce drug $IC_{50}$ values.[40] Similar *in vitro* findings were described for HER2 and pamidronate targeted TSL.[54,73] However, *in vivo* results on TSL targeting to tumor cells are lacking. The only published work that has shown potential for targeting TSL to tumor cells *in vivo* was obtained by Al-Ahmady *et al.*[61] and Yang *et al.*[65] The prior study focused on targeting a TTSL[9] to MDA-MB-435 cells by a linked anti-MUC-1 antibody. The MUC-1-targeted TTSL showed higher intratumoral DOX levels than the non-targeted TTSL, depending on the HT protocol used in a s.c. murine xenograft model. However, no change in therapeutic potential could be observed. The latter study investigated simultaneous targeting of overexpressed CD13 on tumor cells and vascular endothelium by an Asparagine-Glycine-Arginine (NGR) peptide. The therapeutic efficacy in a fibrosarcoma xenograft model was improved by the targeting relatively to the non-targeted counterpart. Nevertheless, the inconsistency in administered dose between experimental groups makes statistically relevant comparison impossible.

More research has been performed on targeting of TSL to the tumor vasculature. One site for targeting is the overexpression of negatively charged proteins on the surface of endothelial cells of the neovasculature.[140] Although a cationic TSL (CTSL) was already described in 1999,[36] using it for targeting tumor vasculature was only recently described.[141] CF and DOX loaded CTSL could enhance drug delivery compared to non-targeted TSL to endothelial cells in a similar fashion as described above, but also *in vivo* the enhanced delivery to the endothelial cell was imaged by dorsal skinfold window chamber models.[141,142] In follow up experiments, it became clear that although CTSL was slightly faster cleared from circulation than non-targeted TSL (47% *versus* 52% injected DOX dose 1 h after i.v. injection in mice, respectively), the intratumoral DOX levels for the CTSL were 1.7 fold higher than for non-targeted TSL when tumors were pre-heated prior to administration of the particles. Nevertheless, there was no therapeutic difference reported between CTSL and non-targeted TSL in a murine B16BL6 melanoma model.[76] Another described option for vascular targeting of TSL is overexpressed $\alpha_v\beta_3$ integrins on the endothelial cell by an Arginine-Glycine-Aspartic acid (RGD) peptide.[62,143] *In vitro*, the targeting ensured higher particle endocytosis, drug delivery to endothelial cells and *in vivo* imaging modalities also suggested higher particle accumulation into tumors. However, the *in vivo* therapeutic effect of these formulations still has to be investigated. Finally, Wang *et al.* described targeting of excessively clotted plasma proteins in solid tumors by a Cysteine-Arginine-Glutamine-Lysine-Alanine (CREKA) peptide.[72] Currently, this work is the only described case where the targeting of TSL resulted in a significant improvement of *in vivo* solid tumor treatment over non-targeted TSL.

In conclusion, we can state that the principle of targeting of TSL to the tumor cell or the tumor neovasculature has not been very convincing so far. The explanation for this can be found in the broader understanding of how particles function in circulation and the tumor microenvironment. For targeting of the tumor cell, one has to consider that the EPR effect is very variable between tumor types[144] and even within a tumor.[145] The EPR effect is often overestimated and is a questionable process considering clinical translation.[146,147] Liposomes with an average hydrodynamic diameter of 100 nm are also relatively large nanoparticles in the field of nanomedicine, which reduces the odds for effective extravasation from the blood stream, deep into the tumor interstitium.[148] A study by Lokerse and colleagues has also shown that DPPC : DSPC : DSPE-$PEG_{2000}$ TSLs are too unstable within the time required for accumulation and extravasate in too small amounts into heated tumors, thereby failing to elicit a therapeutic effect.[149] Another important factor which might hamper targeting by the surface conjugated ligand is proteins and cells in circulation that interact with the liposomes. It has been shown that a so called "protein corona" that forms onto a nanoparticle surface can greatly reduce the targeting capacities,[150,151] thereby providing an explanation for the lack of convincing data on targeted TSL studies *in vivo*.

## 6.8 Testing Release

As has already been indicated to some extent in the previous chapters, the environment of the TSL greatly determines the stability and drug retention. This means that to test drug release and stability of TSL formulations *in vitro*, physiologically representative conditions have to be chosen. Most *in vitro* release and stability studies make use of the fluorescence quenching and dequenching of a (model) drug during loading and release, respectively. The most important factor when performing these assays is in which solution the release is determined. Firstly, the solution needs to have an osmotic pressure similar to where the liposomes will be introduced *in vivo*, which is usually the bloodstream. Varying osmolality will result in different ion interactions with the liposome surface[152] and can induce liposomes to swell or shrink, inducing possible drug leakage.[153] Therefore, a physiological saline or HEPES buffered saline with a fixed pH would be a suitable medium for testing stability and release from TSL. However, the *in vivo* blood stream is much more complex and there are many other factors that need to be considered. Serum proteins interact with liposomes and form coronas on the liposome surface.[154] Depending on the lipids used, different types and different quantities of proteins can bind the liposome surface,[151] which results in drug leakage and shorter circulation half-lives.[155,156] Because of these reasons, we recommend that *in vitro* release and stability tests of TSLs are performed in serum.[157] We have seen on previous occasions that the release and leakage of drugs from TSL formulations is higher and faster in FBS *versus* saline[9,56] (Figure 6.5). For TSL containing lysolipid, the release at HT is quite similar between saline and FBS, because this TSL type forms pores in the membrane for drug release.[24] TSL formulations that do not make use of release improving agents are much more dependent on the protein interactions for effective drug release at HT. Hossann and colleagues investigated the role of serum components and acceptor vesicles on release from several TSL formulations extensively.[157] They showed that drug release from $DPPG_2$-containing TSL is remarkably different in serum of different animal species and followed up with investigating the role of human serum albumin (HSA), human immunoglobulin G (IgG) and cholesterol containing DPPC vesicles (DPPC/Chol-LLV) on drug release. HSA interaction with $DPPG_2$-, PEG- or PEG/Lysolipid-based TSL induced a reduction in drug leakage at 37 °C but had a concentration dependent increase on drug release at 42 °C. IgG had a high destabilizing effect on LTSL, while for $DPPG_2$-based TSL it was a major factor for drug release mostly in the mild HT range (40–43 °C). Incubations with DPPC/Chol-LLV were performed to illustrate if natural lipid exchange that occurs in circulation would influence TSL stability. LTSL and PEG-TSL showed significant premature drug release at physiological temperature, whereas $DPPG_2$-based TSL was much less affected. Another study by Shibata *et al.* showed similar findings for *e.g.* Haptoglobin, Clusterin, Fibronectin, Apolipoprotein and complement protein interactions affecting drug release from NTSL.[158] These findings illustrate how relevant it is to test TSL release and stability in the right media which represents the *in vivo* blood stream as best as possible. Nevertheless, *in vitro* release and stability studies for TSL formulations have to be critically interpreted as the blood circulation is more complex and difficult to test in a model setting.[25]

## 6.9 Conclusions

Using TSLs for optimal local solid tumor chemotherapy is a field which is currently expanding into a broad variety of TSL formulations described to fulfil this goal. By fine tuning the factors discussed above, one can design a TSL which fits the desired features. However, it is important to consider clinical relevance of each described formulation. Targeting of TSL for example, makes a formulation novel, but is likely not effective *in vivo*. Other examples which focus on release improvement by novel moieties have yet to be proven more effective than the relatively simple TSL formulations that solely make use of a mixture of phospholipids with a surface modification like PEG, especially in extensive *in vivo* experiments. What is now required are more relevant *in vivo* experiments that further explain the mode of action of different TSL formulations in circulation and whether it is relevant to continue to optimize formulations or whether the difference between formulations is negligible. Once this information is available, the TSL based drug delivery methodology will be far better established and will make the introduction into clinical trials more efficient.

## References

1. R. Soloman and A. A. Gabizon, *Clin. Lymphoma Myeloma*, 2008, **8**, 21–32.
2. H. Hashizume, P. Baluk, S. Morikawa, J. W. McLean, G. Thurston, S. Roberge, R. K. Jain and D. M. McDonald, *Am. J. Pathol.*, 2000, **156**, 1363–1380.
3. H. Maeda, J. Wu, T. Sawa, Y. Matsumura and K. Hori, *J. Controlled Release*, 2000, **65**, 271–284.
4. K. M. Laginha, S. Verwoert, G. J. Charrois and T. M. Allen, *Clin. Cancer Res.*, 2005, **11**, 6944–6949.

5. G. Kong, G. Anyarambhatla, W. P. Petros, R. D. Braun, O. M. Colvin, D. Needham and M. W. Dewhirst, *Cancer Res.*, 2000, **60**, 6950–6957.
6. J. N. Weinstein, R. L. Magin, M. B. Yatvin and D. S. Zaharko, *Science*, 1979, **204**, 188–191.
7. M. B. Yatvin, J. N. Weinstein, W. H. Dennis and R. Blumenthal, *Science*, 1978, **202**, 1290–1293.
8. S. Unezaki, K. Maruyama, N. Takahashi, M. Koyama, T. Yuda, A. Suginaka and M. Iwatsuru, *Pharm. Res.*, 1994, **11**, 1180–1185.
9. M. H. Gaber, K. Hong, S. K. Huang and D. Papahadjopoulos, *Pharm. Res.*, 1995, **12**, 1407–1416.
10. A. A. Manzoor, L. H. Lindner, C. D. Landon, J. Y. Park, A. J. Simnick, M. R. Dreher, S. Das, G. Hanna, W. Park, A. Chilkoti, G. A. Koning, T. L. ten Hagen, D. Needham and M. W. Dewhirst, *Cancer Res.*, 2012, **72**, 5566–5575.
11. P. S. Yarmolenko, Y. Zhao, C. Landon, I. Spasojevic, F. Yuan, D. Needham, B. L. Viglianti and M. W. Dewhirst, *Int. J. Hyperthermia*, 2010, **26**, 485–498.
12. R. M. Staruch, M. Ganguly, I. F. Tannock, K. Hynynen and R. Chopra, *Int. J. Hyperthermia*, 2012, **28**, 776–787.
13. M. L. Hauck, S. M. LaRue, W. P. Petros, J. M. Poulson, D. Yu, I. Spasojevic, A. F. Pruitt, A. Klein, B. Case, D. E. Thrall, D. Needham and M. W. Dewhirst, *Clin. Cancer Res.*, 2006, **12**, 4004–4010.
14. Celsion, *Phase 1/2 Study of ThermoDox with Approved Hyperthermia in Treatment of Breast Cancer Recurrence at the Chest Wall (DIGNITY)*, 2009.
15. Celsion, *Study of ThermoDox with Standardized Radiofrequency Ablation (RFA) for Treatment of Hepatocellular Carcinoma (HCC) (OPTIMA)*, 2014.
16. Celsion, *Targeted Chemotherapy Using Focused Ultrasound for Liver Tumours (TARDOX)*, 2014.
17. S. Mura, J. Nicolas and P. Couvreur, *Nat. Mater.*, 2013, **12**, 991–1003.
18. D. Papahadjopoulos, K. Jacobson, S. Nir and T. Isac, *Biochim. Biophys. Acta*, 1973, **311**, 330–348.
19. D. Marsh, A. Watts and P. F. Knowles, *Biochemistry*, 1976, **15**, 3570–3578.
20. D. H. Kim, M. J. Costello, P. B. Duncan and D. Needham, *Langmuir*, 2003, **19**, 8455–8466.
21. L. Cruzeiro-Hansson and O. G. Mouritsen, *Biochim. Biophys. Acta*, 1988, **944**, 63–72.
22. W. Curatolo, B. Sears and L. J. Neuringer, *Biochim. Biophys. Acta*, 1985, **817**, 261–270.
23. T. Kaasgaard, C. Leidy, J. H. Crowe, O. G. Mouritsen and K. Jorgensen, *Biophys. J.*, 2003, **85**, 350–360.
24. C. D. Landon, J. Y. Park, D. Needham and M. W. Dewhirst, *Open Nanomed. J.*, 2011, **3**, 38–64.
25. W. J. Lokerse, E. C. Kneepkens, T. L. ten Hagen, A. M. Eggermont, H. Grull and G. A. Koning, *Biomaterials*, 2016, **82**, 138–150.
26. S. M. Park, J. M. Cha, J. Nam, M. S. Kim, S. J. Park, E. S. Park, H. Lee and H. R. Kim, *PLoS One*, 2014, **9**, e103116.

27. S. M. Sullivan and L. Huang, *Proc. Natl. Acad. Sci. U. S. A.*, 1986, **83**, 6117–6121.
28. K. Iga, N. Hamaguchi, Y. Igari, Y. Ogawa, H. Toguchi and T. Shimamoto, *J. Pharm. Sci.*, 1991, **80**, 522–525.
29. J. L. Merlin, *Eur. J. Cancer*, 1991, **27**, 1026–1030.
30. K. Maruyama, S. Unezaki, N. Takahashi and M. Iwatsuru, *Biochim. Biophys. Acta*, 1993, **1149**, 209–216.
31. K. Iga, K. Ohkouchi, Y. Ogawa and H. Toguchi, *J. Drug Targeting*, 1994, **2**, 259–267.
32. T. P. Chelvi, S. K. Jain and R. Ralhan, *Oncol. Res.*, 1995, **7**, 393–398.
33. J. L. Merlin, S. Marchal, C. Ramacci, D. Notter and C. Vigneron, *Int. J. Hyperthermia*, 1995, **11**, 855–865.
34. D. Sharma, T. P. Chelvi, J. Kaur and R. Ralhan, *Melanoma Res.*, 1998, **8**, 240–244.
35. H. Hayashi, K. Kono and T. Takagishi, *Bioconjugate Chem.*, 1999, **10**, 412–418.
36. K. Kono, A. Henmi and T. Takagishi, *Biochim. Biophys. Acta*, 1999, **1421**, 183–197.
37. O. Ishida, K. Maruyama, H. Yanagie, M. Eriguchi and M. Iwatsuru, *Jpn. J. Cancer Res.*, 2000, **91**, 118–126.
38. Y. Yuyama, M. Tsujimoto, Y. Fujimoto and N. Oku, *Cancer Lett.*, 2000, **155**, 71–77.
39. D. Needham, G. Anyarambhatla, G. Kong and M. W. Dewhirst, *Cancer Res.*, 2000, **60**, 1197–1201.
40. M. H. Gaber, *J. Biochem., Mol. Biol. Biophys.*, 2002, **6**, 309–314.
41. S. B. Tiwari, R. M. Pai and N. Udupa, *J. Drug Targeting*, 2002, **10**, 585–591.
42. L. H. Lindner, M. E. Eichhorn, H. Eibl, N. Teichert, M. Schmitt-Sody, R. D. Issels and M. Dellian, *Clin. Cancer Res.*, 2004, **10**, 2168–2178.
43. H. D. Han, B. C. Shin and H. S. Choi, *Eur. J. Pharm. Biopharm.*, 2006, **62**, 110–116.
44. J. Woo, G. N. Chiu, G. Karlsson, E. Wasan, L. Ickenstein, K. Edwards and M. B. Bally, *Int. J. Pharm.*, 2008, **349**, 38–46.
45. L. Paasonen, B. Romberg, G. Storm, M. Yliperttula, A. Urtti and W. E. Hennink, *Bioconjugate Chem.*, 2007, **18**, 2131–2136.
46. L. H. Lindner, M. Hossann, M. Vogeser, N. Teichert, K. Wachholz, H. Eibl, W. Hiddemann and R. D. Issels, *J. Controlled Release*, 2008, **125**, 112–120.
47. A. Puri, G. Kramer-Marek, R. Campbell-Massa, A. Yavlovich, S. C. Tele, S. B. Lee, J. D. Clogston, A. K. Patri, R. Blumenthal and J. Capala, *J. Liposome Res.*, 2008, **18**, 293–307.
48. H. Zhang, Z. Y. Wang, W. Gong, Z. P. Li, X. G. Mei and W. L. Lv, *Int. J. Pharm.*, 2011, **414**, 56–62.
49. N. D. Winter, R. K. Murphy, T. V. O'Halloran and G. C. Schatz, *J. Liposome Res.*, 2011, **21**, 106–115.
50. K. Kono, T. Ozawa, T. Yoshida, F. Ozaki, Y. Ishizaka, K. Maruyama, C. Kojima, A. Harada and S. Aoshima, *Biomaterials*, 2010, **31**, 7096–7105.

51. T. Ta, A. J. Convertine, C. R. Reyes, P. S. Stayton and T. M. Porter, *Biomacromolecules*, 2010, **11**, 1915–1920.
52. Y. Wu, Y. Yang, F. C. Zhang, C. Wu, W. L. Lu and X. G. Mei, *J. Liposome Res.*, 2011, **21**, 221–228.
53. T. Tagami, M. J. Ernsting and S. D. Li, *J. Controlled Release*, 2011, **152**, 303–309.
54. B. Smith, I. Lyakhov, K. Loomis, D. Needle, U. Baxa, A. Yavlovich, J. Capala, R. Blumenthal and A. Puri, *J. Controlled Release*, 2011, **153**, 187–194.
55. X. Zhang, P. F. Luckham, A. D. Hughes, S. Thom and X. Y. Xu, *Int. J. Pharm.*, 2011, **421**, 291–292.
56. M. Hossann, M. Wiggenhorn, A. Schwerdt, K. Wachholz, N. Teichert, H. Eibl, R. D. Issels and L. H. Lindner, *Biochim. Biophys. Acta, Biomembr.*, 2007, **1768**, 2491–2499.
57. P. Kulshrestha, M. Gogoi, D. Bahadur and R. Banerjee, *Colloids Surf., B*, 2012, **96**, 1–7.
58. A. Yudina, M. Lepetit-Coiffe, M. De Smet, S. Langereis, H. Grull and C. Moonen, *J. Controlled Release*, 2012, **161**, 90–97.
59. S. M. Park, M. S. Kim, S. J. Park, E. S. Park, K. S. Choi, Y. S. Kim and H. R. Kim, *J. Controlled Release*, 2013, **170**, 373–379.
60. J. P. May, M. J. Ernsting, E. Undzys and S. D. Li, *Mol. Pharmaceutics*, 2013, **10**, 4499–4508.
61. Z. S. Al-Ahmady, O. Chaloin and K. Kostarelos, *J. Controlled Release*, 2014, **196**, 332–343.
62. M. S. Kim, D. W. Lee, K. Park, S. J. Park, E. J. Choi, E. S. Park and H. R. Kim, *Colloids Surf., B*, 2014, **116**, 17–25.
63. Y. N. Dou, J. Zheng, W. D. Foltz, R. Weersink, N. Chaudary, D. A. Jaffray and C. Allen, *J. Controlled Release*, 2014, **178**, 69–78.
64. M. van Elk, R. Deckers, C. Oerlemans, Y. Shi, G. Storm, T. Vermonden and W. E. Hennink, *Biomacromolecules*, 2014, **15**, 1002–1009.
65. Y. Yang, Y. Yang, X. Xie, X. Cai, H. Zhang, W. Gong, Z. Wang and X. Mei, *Biomaterials*, 2014, **35**, 4368–4381.
66. S. Limmer, J. Hahn, R. Schmidt, K. Wachholz, A. Zengerle, K. Lechner, H. Eibl, R. D. Issels, M. Hossann and L. H. Lindner, *Pharm. Res.*, 2014, **31**, 2276–2286.
67. C. S. Shemesh, C. W. Hardy, D. S. Yu, B. Fernandez and H. Zhang, *Photodiagn. Photodyn. Ther.*, 2014, **11**, 193–203.
68. C. Al Sabbagh, N. Tsapis, A. Novell, P. Calleja-Gonzalez, J. M. Escoffre, A. Bouakaz, H. Chacun, S. Denis, J. Vergnaud, C. Gueutin and E. Fattal, *Pharm. Res.*, 2015, **32**, 1585–1603.
69. X. Liang, J. Gao, L. Jiang, J. Luo, L. Jing, X. Li, Y. Jin and Z. Dai, *ACS Nano*, 2015, **9**, 1280–1293.
70. Z. S. Al-Ahmady, C. L. Scudamore and K. Kostarelos, *Int. J. Cancer*, 2015, **137**, 731–743.
71. E. V. Rosca, M. Wright, R. Gonitel, W. Gedroyc, A. D. Miller and M. Thanou, *Mol. Pharmaceutics*, 2015, **12**, 1335–1346.

72. C. Wang, X. Wang, T. Zhong, Y. Zhao, W. Q. Zhang, W. Ren, D. Huang, S. Zhang, Y. Guo, X. Yao, Y. Q. Tang, X. Zhang and Q. Zhang, *Int. J. Nanomed.*, 2015, **10**, 2229–2248.
73. H. Song, J. Zhang, X. Liu, T. Deng, P. Yao, S. Zhou and W. Yan, *Pharm. Dev. Technol.*, 2016, **21**, 680–687.
74. T. Tagami, M. Kubota and T. Ozeki, *J. Pharm. Sci.*, 2015, **104**, 3824–3832.
75. Z. Peng, E. Fang, C. Wang, X. Lu, G. Wang and Q. Tong, *J. Nanosci. Nanotechnol.*, 2015, **15**, 3823–3833.
76. B. M. Dicheva, A. L. Seynhaeve, T. Soulie, A. M. Eggermont, T. L. Ten Hagen and G. A. Koning, *Pharm. Res.*, 2016, **33**, 627–638.
77. T. Lu, W. J. Lokerse, A. L. Seynhaeve, G. A. Koning and T. L. ten Hagen, *J. Controlled Release*, 2015, **220**, 425–437.
78. T. P. W. McMullen, R. N. A. H. Lewis and R. N. McElhaney, *Curr. Opin. Colloid Interface Sci.*, 2004, **8**, 459–468.
79. V. Ben-Yashar and Y. Barenholz, *Biochim. Biophys. Acta*, 1989, **985**, 271–278.
80. J. de Gier, J. G. Mandersloot and L. L. van Deenen, *Biochim. Biophys. Acta*, 1969, **173**, 143–145.
81. P. A. Hyslop, B. Morel and R. D. Sauerheber, *Biochemistry*, 1990, **29**, 1025–1038.
82. E. Oldfield and D. Chapman, *FEBS Lett.*, 1972, **23**, 285–297.
83. L. Redondo-Morata, M. I. Giannotti and F. Sanz, *Langmuir*, 2012, **28**, 12851–12860.
84. L. S. Guo, R. L. Hamilton, J. Goerke, J. N. Weinstein and R. J. Havel, *J. Lipid Res.*, 1980, **21**, 993–1003.
85. K. Maruyama, T. Yuda, A. Okamoto, S. Kojima, A. Suginaka and M. Iwatsuru, *Biochim. Biophys. Acta*, 1992, **1128**, 44–49.
86. E. J. Luna and H. M. McConnell, *Biochim. Biophys. Acta*, 1977, **470**, 303–316.
87. D. Papahadjopoulos, S. Nir and S. Oki, *Biochim. Biophys. Acta*, 1972, **266**, 561–583.
88. T. M. Allen, C. Hansen, F. Martin, C. Redemann and A. Yau-Young, *Biochim. Biophys. Acta*, 1991, **1066**, 29–36.
89. A. Gabizon, R. Catane, B. Uziely, B. Kaufman, T. Safra, R. Cohen, F. Martin, A. Huang and Y. Barenholz, *Cancer Res.*, 1994, **54**, 987–992.
90. D. Needham, T. J. McIntosh and D. D. Lasic, *Biochim. Biophys. Acta*, 1992, **1108**, 40–48.
91. G. Blume, G. Cevc, M. D. Crommelin, I. A. Bakker-Woudenberg, C. Kluft and G. Storm, *Biochim. Biophys. Acta*, 1993, **1149**, 180–184.
92. S. M. Moghimi and J. Szebeni, *Prog. Lipid Res.*, 2003, **42**, 463–478.
93. R. Gref, M. Luck, P. Quellec, M. Marchand, E. Dellacherie, S. Harnisch, T. Blunk and R. H. Muller, *Colloids Surf., B*, 2000, **18**, 301–313.
94. N. Dos Santos, C. Allen, A. M. Doppen, M. Anantha, K. A. Cox, R. C. Gallagher, G. Karlsson, K. Edwards, G. Kenner, L. Samuels, M. S. Webb and M. B. Bally, *Biochim. Biophys. Acta*, 2007, **1768**, 1367–1377.

95. L. Li, T. L. ten Hagen, D. Schipper, T. M. Wijnberg, G. C. van Rhoon, A. M. Eggermont, L. H. Lindner and G. A. Koning, *J. Controlled Release*, 2010, **143**, 274–279.
96. K. Hashizaki, H. Taguchi, C. Itoh, H. Sakai, M. Abe, Y. Saito and N. Ogawa, *Chem. Pharm. Bull.*, 2003, **51**, 815–820.
97. D. Liu, F. Liu and Y. K. Song, *Pharm. Res.*, 1995, **12**, 508–512.
98. A. Mori, A. L. Klibanov, V. P. Torchilin and L. Huang, *FEBS Lett.*, 1991, **284**, 263–266.
99. A. L. Klibanov, K. Maruyama, V. P. Torchilin and L. Huang, *FEBS Lett.*, 1990, **268**, 235–237.
100. K. Maruyama, S. Unezaki, T. Yuda, O. Ishida, N. Takahashi, A. Suginaka, L. Huang and M. Iwatsuru, *J. Liposome Res.*, 1994, **4**, 143–165.
101. B. Ashok, L. Arleth, R. P. Hjelm, I. Rubinstein and H. Onyuksel, *J. Pharm. Sci.*, 2004, **93**, 2476–2487.
102. H. D. Han, M. S. Choi, T. Hwang, C. K. Song, H. Seong, T. W. Kim, H. S. Choi and B. C. Shin, *J. Pharm. Sci.*, 2006, **95**, 1909–1917.
103. E. T. Dams, P. Laverman, W. J. Oyen, G. Storm, G. L. Scherphof, J. W. van Der Meer, F. H. Corstens and O. C. Boerman, *J. Pharmacol. Exp. Ther.*, 2000, **292**, 1071–1079.
104. T. Ishida, R. Maeda, M. Ichihara, K. Irimura and H. Kiwada, *J. Controlled Release*, 2003, **88**, 35–42.
105. J. Szebeni, *Mol. Immunol.*, 2014, **61**, 163–173.
106. F. K. Storm, W. H. Harrison, R. S. Elliott and D. L. Morton, *Cancer Res.*, 1979, **39**, 2245–2251.
107. J. C. Lin and C. W. Song, *Cancer Res.*, 1993, **53**, 2076–2080.
108. D. Needham, J. Y. Park, A. M. Wright and J. Tong, *Faraday Discuss.*, 2013, **161**, 515–534, discussion 563–589.
109. G. R. Anyarambhatla and D. Needham, *J. Liposome Res.*, 1999, **9**, 491–506.
110. J. K. Mills and D. Needham, *Biochim. Biophys. Acta*, 2005, **1716**, 77–96.
111. M. C. Sandstrom, L. M. Ickenstein, L. D. Mayer and K. Edwards, *J. Controlled Release*, 2005, **107**, 131–142.
112. L. M. Ickenstein, M. C. Arfvidsson, D. Needham, L. D. Mayer and K. Edwards, *Biochim. Biophys. Acta*, 2003, **1614**, 135–138.
113. B. Banno, L. M. Ickenstein, G. N. Chiu, M. B. Bally, J. Thewalt, E. Brief and E. K. Wasan, *J. Pharm. Sci.*, 2010, **99**, 2295–2308.
114. T. Tagami, M. J. Ernsting and S. D. Li, *J. Controlled Release*, 2011, **154**, 290–297.
115. T. Tagami, J. P. May, M. J. Ernsting and S.-D. Li, *J. Controlled Release*, 2012, **161**, 142–149.
116. K. Kono, *Adv. Drug Delivery Rev.*, 2001, **53**, 307–319.
117. K. Kono, H. Hayashi and T. Takagishi, *J. Controlled Release*, 1994, **30**, 69–75.
118. H. Hayashi, K. Kono and T. Takagishi, *Biochim. Biophys. Acta, Biomembr.*, 1996, **1280**, 127–134.
119. K. Kono, R. Nakai, K. Morimoto and T. Takagishi, *Biochim. Biophys. Acta, Biomembr.*, 1999, **1416**, 239–250.

120. K. Kono, K. Yoshino and T. Takagishi, *J. Controlled Release*, 2002, **80**, 321–332.
121. K. Kono, T. Murakami, T. Yoshida, Y. Haba, S. Kanaoka, T. Takagishi and S. Aoshima, *Bioconjugate Chem.*, 2005, **16**, 1367–1374.
122. D. W. Urry, *J. Phys. Chem. B*, 1997, **101**, 11007–11028.
123. D. E. Meyer and A. Chilkoti, *Biomacromolecules*, 2004, **5**, 846–851.
124. Z. S. Al-Ahmady, W. T. Al-Jamal, J. V. Bossche, T. T. Bui, A. F. Drake, A. J. Mason and K. Kostarelos, *ACS Nano*, 2012, **6**, 9335–9346.
125. S. B. Kulkarni, G. V. Betageri and M. Singh, *J. Microencapsulation*, 1995, **12**, 229–246.
126. D. R. Khan, E. M. Rezler, J. Lauer-Fields and G. B. Fields, *Chem. Biol. Drug Des.*, 2008, **71**, 3–7.
127. R. K. Chowdhary, I. Shariff and D. Dolphin, *J. Pharm. Pharm. Sci.*, 2003, **6**, 13–19.
128. A. Cabanes, K. E. Briggs, P. C. Gokhale, J. A. Treat and A. Rahman, *Int. J. Oncol.*, 1998, **12**, 1035–1040.
129. L. Li, T. L. ten Hagen, M. Hossann, R. Suss, G. C. van Rhoon, A. M. Eggermont, D. Haemmerich and G. A. Koning, *J. Controlled Release*, 2013, **168**, 142–150.
130. A. Fritze, F. Hens, A. Kimpfler, R. Schubert and R. Peschka-Suss, *Biochim. Biophys. Acta*, 2006, **1758**, 1633–1640.
131. N. Dos Santos, K. A. Cox, C. A. McKenzie, F. van Baarda, R. C. Gallagher, G. Karlsson, K. Edwards, L. D. Mayer, C. Allen and M. B. Bally, *Biochim. Biophys. Acta, Biomembr.*, 2004, **1661**, 47–60.
132. B. L. Viglianti, S. A. Abraham, C. R. Michelich, P. S. Yarmolenko, J. R. MacFall, M. B. Bally and M. W. Dewhirst, *Magn. Reson. Med.*, 2004, **51**, 1153–1162.
133. A. M. Ponce, B. L. Viglianti, D. Yu, P. S. Yarmolenko, C. R. Michelich, J. Woo, M. B. Bally and M. W. Dewhirst, *J. Natl. Cancer Inst.*, 2007, **99**, 53–63.
134. M. de Smet, S. Langereis, S. van den Bosch and H. Grull, *J. Controlled Release*, 2010, **143**, 120–127.
135. B. L. Viglianti, A. M. Ponce, C. R. Michelich, D. Yu, S. A. Abraham, L. Sanders, P. S. Yarmolenko, T. Schroeder, J. R. MacFall, D. P. Barboriak, O. M. Colvin, M. B. Bally and M. W. Dewhirst, *Magn. Reson. Med.*, 2006, **56**, 1011–1018.
136. M. Peller, L. Willerding, S. Limmer, M. Hossann, O. Dietrich, M. Ingrisch, R. Sroka and L. H. Lindner, *J. Controlled Release*, 2016, **237**, 138–146.
137. P. Sapra and T. M. Allen, *Prog. Lipid Res.*, 2003, **42**, 439–462.
138. E. Ruoslahti, *Adv. Mater.*, 2012, **24**, 3747–3756.
139. R. A. Petros and J. M. DeSimone, *Nat. Rev. Drug Discovery*, 2010, **9**, 615–627.
140. R. B. Campbell, D. Fukumura, E. B. Brown, L. M. Mazzola, Y. Izumi, R. K. Jain, V. P. Torchilin and L. L. Munn, *Cancer Res.*, 2002, **62**, 6831–6836.
141. B. M. Dicheva, T. L. ten Hagen, L. Li, D. Schipper, A. L. Seynhaeve, G. C. van Rhoon, A. M. Eggermont, L. H. Lindner and G. A. Koning, *Nano Lett.*, 2013, **13**, 2324–2331.

142. B. M. Dicheva, T. L. ten Hagen, D. Schipper, A. L. Seynhaeve, G. C. van Rhoon, A. M. Eggermont and G. A. Koning, *J. Controlled Release*, 2014, **195**, 37–48.
143. B. M. Dicheva, T. L. ten Hagen, A. L. Seynhaeve, M. Amin, A. M. Eggermont and G. A. Koning, *Pharm. Res.*, 2015, **32**, 3862–3876.
144. M. Bolkestein, E. de Blois, S. J. Koelewijn, A. M. Eggermont, F. Grosveld, M. de Jong and G. A. Koning, *J. Nucl. Med.*, 2016, **57**, 601–607.
145. S. N. Ekdawi, J. M. Stewart, M. Dunne, S. Stapleton, N. Mitsakakis, Y. N. Dou, D. A. Jaffray and C. Allen, *J. Controlled Release*, 2015, **207**, 101–111.
146. T. Lammers, F. Kiessling, W. E. Hennink and G. Storm, *J. Controlled Release*, 2012, **161**, 175–187.
147. J. W. Nichols and Y. H. Bae, *J. Controlled Release*, 2014, **190**, 451–464.
148. Z. Popovic, W. Liu, V. P. Chauhan, J. Lee, C. Wong, A. B. Greytak, N. Insin, D. G. Nocera, D. Fukumura, R. K. Jain and M. G. Bawendi, *Angew. Chem., Int. Ed. Engl.*, 2010, **49**, 8649–8652.
149. W. J. Lokerse, M. Bolkestein, T. L. Ten Hagen, M. de Jong, A. M. Eggermont, H. Grull and G. A. Koning, *Theranostics*, 2016, **6**, 1717–1731.
150. A. Salvati, A. S. Pitek, M. P. Monopoli, K. Prapainop, F. B. Bombelli, D. R. Hristov, P. M. Kelly, C. Aberg, E. Mahon and K. A. Dawson, *Nat. Nano*, 2013, **8**, 137–143.
151. M. Hadjidemetriou, Z. Al-Ahmady, M. Mazza, R. F. Collins, K. Dawson and K. Kostarelos, *ACS Nano*, 2015, **9**, 8142–8156.
152. J. Sabin, G. Prieto, J. M. Ruso, R. Hidalgo-Alvarez and F. Sarmiento, *Eur. Phys. J. E: Soft Matter*, 2006, **20**, 401–408.
153. E. Nagamachi, Y. Hirai, K. Tomochika and Y. Kanemasa, *Microbiol. Immunol.*, 1992, **36**, 231–234.
154. G. Caracciolo, *Nanomedicine*, 2015, **11**, 543–557.
155. T. M. Allen and L. G. Cleland, *Biochim. Biophys. Acta*, 1980, **597**, 418–426.
156. A. Chonn, S. C. Semple and P. R. Cullis, *J. Biol. Chem.*, 1992, **267**, 18759–18765.
157. M. Hossann, Z. Syunyaeva, R. Schmidt, A. Zengerle, H. Eibl, R. D. Issels and L. H. Lindner, *J. Controlled Release*, 2012, **162**, 400–406.
158. H. Shibata, H. Yoshida, K. Izutsu, Y. Haishima, T. Kawanishi, H. Okuda and Y. Goda, *Int. J. Pharm.*, 2015, **495**, 827–839.

CHAPTER 7

# Targeted Delivery with Ultrasound Activated Nano-encapsulated Drugs

D. GOUREVICH*[a] AND S. COCHRAN[b]

[a]Pharmaseed Ltd, Nes-Zyona, Israel; [b]Glasgow University, UK
*E-mail: Dana.gourevich@gmail.com

## 7.1 Introduction

Targeted drug delivery (TDD) has been a subject of investigation for many research groups worldwide[1] with imaging technologies such as magnetic resonance imaging (MRI), ultrasound (US) and computed tomography (CT) introducing the possibility for real-time guidance. The potential of targeting a drug at a specific location in the body while maintaining a toxin-free environment in the vital organs is a possibility that is now becoming more likely[2,3] but many steps remain in taking what is a promising idea through a suitable model to a fully functioning and reliable remedy.

The various methods of TDD shown in Figure 7.1, both physical, *e.g.* focused ultrasound (FUS), laser *etc.*, and chemical, *e.g.* various drug carriers, have their own benefits and drawbacks. Hence, this chapter describes an example of a combination of physical and chemical methodologies and shows how this may exploit the best features of each approach to create a better drug delivery tool.[4]

Drug Discovery Series No. 63
Theranostics and Image Guided Drug Delivery
Edited by Maya Thanou

Published by the Royal Society of Chemistry, www.rsc.org

**Figure 7.1** A schematic outline of possible drug delivery systems. To provide a localized, more efficient site-specific therapy and to decrease severe side-effects, research groups worldwide are investigating mechanisms for targeted drug delivery. As presented in this figure, there are various mechanisms to target drugs to specific locations. These include chemically modified drug carriers such as polymers and the widely known liposomes. Additional methods include biological navigation systems such as enzymes and viruses for site-specific delivery and others employ physical phenomena such as radio-frequency electric fields, laser radiation, microwave radiation and ultrasound. The mechanism described in this chapter combines chemically modified cyclodextrin-based molecular carriers for safe drug distribution with heat and pressure by focused ultrasound for site-specific drug delivery.

In clinical practice, US is conventionally divided into two major streams: diagnostic and therapeutic. The diagnostic stream mainly refers to imaging, while the therapeutic stream covers a wider spectrum including lithotripsy, phacoemulsification, FUS and assorted TDD. Ultrasound-mediated therapeutic drug delivery (USmTDD) can be achieved *via* two different mechanisms, which can be used separately or together, the first being locally increased temperature[5] and the second the application of mechanical force, *via e.g.* cavitation or micro-streaming. Both of these have the potential to increase cell membrane permeability by different routes.

Heat treatment is known to increase *in situ* drug uptake in cancer cells[6] and can be generated with FUS relatively straightforwardly through the absorption by tissue and conversion of kinetic energy into heat. Cavitation, on the other hand, is a more problematic tool in USmTDD. Acoustic cavitation can be defined as acoustically-driven bubble oscillation and collapse.[7] In general, natural occurring cavitation bubbles and their oscillations and collapse form a spontaneous and rather unpredictable phenomenon, which is presently difficult to control. Nonetheless, it is possible to generate stable, controllable cavitation using engineered cavitation nuclei, *e.g.* microbubbles (MBs). MBs are widely utilised in investigations of USmTDD under specific acoustic conditions[8] to generate cavitation to increase cell membrane permeability by sonoporation[9] and cellular massage.[10] One of the accelerating research areas in cavitation is the utilization of FUS to facilitate *in situ* drug release from various carriers.[11]

In the work on which this chapter is based, the source of FUS was a clinical MRI-guided focused ultrasound surgery (MRgFUS) system, CE-marked and FDA-approved for treatment of uterine fibroids. The use of such a system is beneficial not only in providing a research tool with integrated monitoring but also in providing an easier transition to pre-clinical *in vivo* studies[12] and potentially on to clinical trials.

From the point of view of chemistry, novel nanocarriers can be synthesized from a cyclodextrin (CD) unit, for complexation with the drug, covalently bonded to an amino-acid-based functional group. CDs are natural, water soluble, sugar based, cone structured molecules with lipophilic inner cavities.[13] Their ability to form complexes with a wide variety of molecules has been studied for over a century[14] and they have been shown to enhance solubility of non-soluble drugs, contribute to the stability of guest molecules, reduce volatility, and even mask odours. Another important aspect of CDs, especially for TDD, is that the equilibrium of the drug–CD complex is dynamic[15] and it must thus be properly targeted to reach the intended treatment location. This is usually achieved by creating a chemical derivative, with a functional group as the targeting tool.[16] Here, a newer approach is described, in which FUS is introduced to a cell monolayer in order to enhance membrane permeability, through both mechanical and thermal mechanisms, in conjunction with a chemically modified CD.

The drug considered in this chapter is doxorubicin (Dox), an antibiotic substance discovered over fifty years ago, which is currently classified as the most potent treatment for many malignancies and especially for solid tumours.[17] Its main drawback as a chemotherapeutic agent is the high cardiotoxicity associated with the cleavage of the Dox molecule and creation of free radicals.[18] Although many efforts have been made to decrease its unwanted side effects, *e.g.* by encapsulating it in liposomes,[19] the challenge of efficient Dox delivery without harming the surrounding tissues, still exists. It was recently shown that CDs can potentially aid in the prevention of the Dox cleavage and hence decrease their cytotoxicity.[20,21]

The CD-based carrier considered here was chemically modified in a way that resulted in a successful Dox delivery vehicle for testing *in vitro*, with FUS used as the mechanism for drug release. From the data obtained so far, it is proposed that the combination of FUS, applied with a clinical system, with a nanocarrier based on CD technology will not only allow research into a wide range of encapsulated drugs in an *in vitro* research environment but also a faster track to use of site-specific drug delivery with non-invasive application in patients.

## 7.2 Development and Characterization of a Cyclodextrin-based Drug Carrier

### 7.2.1 Chemical Modification and Cyclodextrin Derivatives

The motivation for chemical modification of CDs varies and may include improvement of their solubility, increase in their stereo-selectivity for specific guest molecules in host-guest complexes, and the possibility to explore enzyme-catalyzed reaction mechanisms.[22] When carrying out any modification, two important factors must be considered. First, one must consider the hydroxyl groups' nucleophilicity, as the initial reaction of modification will be an electrophilic attack on these groups, and second, it is important to recognise the CD's ability to form complexes with the reagents in use. If a strong complex is formed, the product will be defined by the reagent, whereas, for a weak complex, the product is guided by the hydroxyl groups' nucleophilicity. Additional important considerations are the cavity size of the CD and the solvent used for the reaction. For example, with dimethylformamide, a sulphonation reaction occurs non-selectively on all of the hydroxyl groups, whereas the same reaction in pyridine will be directed to the hydroxyl groups on the 6th position.[23,24]

As presented in Figure 7.2A, the CD's glucopyranose ring has primary hydroxyl groups on the sixth position and secondary hydroxyl groups on the second and third positions, with the hydroxyl groups on the second position being most acidic and the ones on the third position the most difficult to access. The hydroxyl groups on the sixth position are the most nucleophilic and will react selectively with less reactive reagents. The intramolecular hydrogen bonds created between the hydroxyl groups on the second and third positions restrict their rotation while the hydroxyl groups on the sixth position rotation are free, which reduces the accessibility to the cavity from the primary face.[23]

Sole substitution of the CD's hydroxyl groups on either the second, third or sixth position, also known as monosubstitution, usually increases the CD's aqueous solubility compared to the natural one and hence is used, along with other materials, to enhance drug delivery to specific sites.[25] However, it is a challenge to synthesize such molecules because of difficulties in selectively modifying the three different hydroxyl groups on the outer surface of the CD.[26] Indeed, in general, the higher the number and variety of groups to be substituted on the CD, the harder will be the synthesis.[27] A way proposed to avoid a mixture of unwanted products is by protection of certain hydroxyl groups and thereby directing of the reaction to the desired yield.[28] An alternative, unconventional method for selective substitution on CDs is to utilize power US with microwaves, suggested not only to yield higher amounts of a specific product but also to dramatically shorten the reaction time.[29] In the study reported in this chapter, a γCD was covalently bonded to an amino acid derivative with a naphthalene moiety to increase its potential for encapsulation of the target molecule (Figure 7.2B).

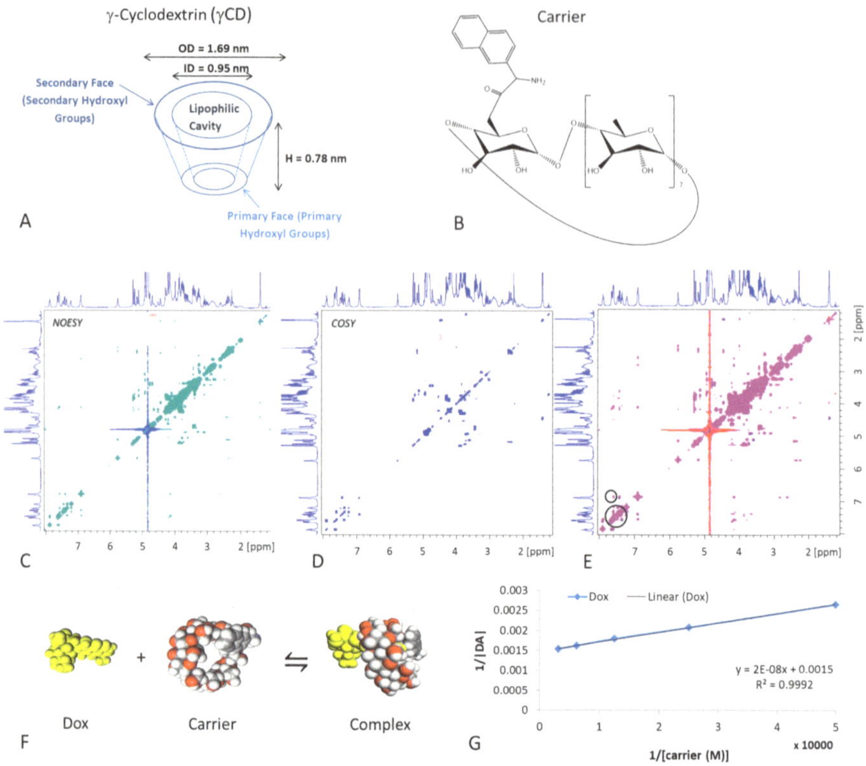


**Figure 7.2** (A) γ-CD parameters and 3D structure and (B) a modified nanocarrier used to encapsulate Doxorubicin. 1H-NMR (C) NOESY and (D) COSY spectra of 1 : 1 Dox : carrier complex in D2O. (E) Subtraction of the COSY and NOESY spectra. Evidence for the 3D interaction between the carrier and the Dox is shown by the peaks marked in black circles. (F) Carrier and Dox in a suggested encapsulated form in a space-filling style side view, (G) Benesi–Hildebrand plot of Dox fluorescence, where the ratio of the intercept and the slope (equation presented bottom right) gives $K_{binding}$ for the complex.

### 7.2.2 Doxorubicin as a Guest Molecule in a Cyclodextrin-based Complex

Doxorubicin belongs to the anthracycline family, a well-known antibiotic drug that is widely used in anti-cancer treatments, with anthracycline-type drugs ranked among the most effective oncological treatments discovered so far. The cytotoxicity mechanism of Dox is *via* interaction and modification of the DNA structure by disturbance of various DNA-related processes such as replication, transcription, and repair mechanisms.[17] However, anthracycline-type drugs are also known to create reactive oxygen species (ROS)[18] which are highly cytotoxic and can result in adverse side effects including major cardiac damage.[30] A possible way to prevent the creation of ROS is by protection of the quinone group by complexation with a CD.[21]

Dox has also been shown to interact with other widely consumed aromatic molecules, such as caffeine, which lower its pharmacological activities.[31] Consequently, encapsulation of Dox inside a CD can significantly increase its medico-biological activity within the body by prevention of unwanted encapsulation by various molecules, yet can also maintain its cytotoxicity while simultaneously lowering the potential creation of free radicals by maintaining the minimum dose and the protection of the quinine group from hydrolysis.[20,21] Complexation of Dox, as a lipophilic guest molecule, within a CD in aqueous solution is usually a spontaneous interaction in which the Dox escapes the surrounding media into the cavity of the CD.

### 7.2.3 Characterization of Cyclodextrins and Their Complexes

The high density of electrons inside the CD's cavity can change the local environmental polarity, affecting the electronic transitions and hence the optical behaviour of both the guest and the CD. Accordingly, inclusion complexes are widely studied by various optical methods such as UV-visible spectrophotometry, fluorescence, and infrared imaging, along with various nuclear magnetic resonance (NMR) techniques,[13] with the latter considered to be the most conclusive in identifying CD-based inclusion complexes.[32] Milestones were established in the early 1970s for the use of NMR as a method to monitor complexation. Throughout this period, it was observed by $^1$H-NMR that, due to shielding of the inner cavity protons (3H, 5H) by the guest molecule, a significant upfield shift occurs. On the other hand, the hydrogen atoms located on the exterior of the cavity (2H, 4H, 6H) are generally not affected.[13] Although it was initially discovered using $^1$H-NMR, nowadays $^{13}$C-NMR, $^{15}$N-NMR, $^{31}$P-NMR and $^{19}$F-NMR are also in use for studies of inclusion complexes, even with solid samples.[33]

In the work described here, information on the 3D structure of the complex was achieved by NOESY and COSY measurements. Subtraction of the COSY (Figure 7.2D) and NOESY (Figure 7.2C) full spectra provided information on the non-covalently bonded protons present in proximity to each other in the 3D structure of the molecule. The subtraction image, *i.e.* Figure 7.2E, shows that a number of peaks appearing in the NOESY spectrum are absent in the COSY spectra, marked by black circles. The presence of these peaks does not necessarily indicate the complexation reaction, since they could represent the 3D structure of the Dox or carrier on its own, but their amplitude and nature strengthen the assumption that some interactions do occur. By magnifying the subtraction of the COSY spectrum from the NOESY one at 6–8 ppm, a clear spatial correlation can be seen between the peaks at 6.9 ppm and 7.6 ppm (marked with black circles). As the peak at 6.9 ppm most likely belongs to Dox 3H, which is located on the first phenyl ring next to the methyl group, and the peak at 7.6 ppm can be attributed to the carrier's naphthalene protons, this is a strong indication of $\pi$–$\pi^*$ interaction between the Dox aromatic rings and the naphthalene conjugate on the carrier. Taking into account the overall $^1$H-NMR data, it can be suggested that the Dox encapsulation occurs as suggested in Figure 7.2F, with the Dox entering the γCD from the secondary face with the aromatic rings forward and forming π–π* interactions with the naphthalene group.

It has been reported that addition of natural γCD to Dox at a ratio of up to 5000 : 1 can significantly stabilize the Dox in aqueous solution *via* complex formation.[34] In the work described in this chapter, fluorescence measurements were conducted at 592 nm to investigate the strength of the binding of Dox as a guest molecule in the host–guest complex with the carrier. The signal decay with the addition of the carrier suggested that good coupling between the Dox and the γCD-naphthalene derivative is formed due to encapsulation, allowing energy transfer between them. Therefore, the Dox excited state decays non-radiatively leading to lower fluorescence signals. According to the fluorescence studies presented in Figure 7.2G, the main Dox fluorescence peak at 592 nm is decreased by a factor of eight as a result of the addition of the carrier at the maximum concentration of 320 μM, with significant encapsulation observed at carrier : Dox ratios as low as 1 : 1 and satisfactory levels at 20 : 1; the latter was therefore the ratio applied *in vitro*.

The observed decrease in the intensity of Dox fluorescence peaks is in good correspondence with literature which reports that encapsulation within a CD can significantly lower the fluorescence and absorbance signals of the guest molecule.[21,35] In addition, it has been shown that, for these changes, the Benesi–Hildebrand method[36] of calculation of the binding constant, $K_{\text{binding}}$, can be applied.[21,35,37] According to these calculations, $K_{\text{binding}} = 61944 \pm 5592.2\ \text{M}^{-1}$. This value is a factor more than 300 time higher than that established by Anand *et al.* (2009)[20] for the parental γCD with Dox in 1 : 1 complex formation. The higher value of $K_{\text{binding}}$ can be attributed to the modification of the γCD with the naphthalene group, potentially exhibiting π–π* interactions with the Dox's aromatic rings.

To conclude this section, the overall data suggest that interaction can be achieved between a γCD host and Dox as a guest molecule, providing an indication of complex formation which can potentially aid prevention of the Dox cleavage and generation of dangerous ROS, and thus reduce harmful side-effects.

## 7.3 Adaptation of Clinical MRgFUS System for *in vitro* Application of FUS

Ultrasound is a well-known diagnostic and therapeutic modality which has also been a topic of intensive research for TDD. Energy from FUS can propagate easily through soft tissue, allowing mechanical and thermal effects to be achieved on cells in a focal zone remote from the FUS source. Hence, USmTDD has potential as a tool for both drug release *in situ* and membrane permeability enhancement.[38] Although it has been widely investigated, there is still a major shortfall in proof of the repeatability and reproducibility of the results generated by USmTDD in different research groups.

An additional difficulty is the need to differentiate cavitation effects from those deriving from increased temperature and verification that any heating produced unavoidably does not contribute to observed biological effects. Hence, in this chapter, an experimental system is described, based on a clinical MRgFUS system, which allows differentiation between these mechanisms correlated to FUS application.[39,40] This system presents the possibility for researchers with access to MRgFUS systems to exploit the same arrangements for USmTDD-related studies at different sites, thus representing one possible route to the standardization needed ultimately for confidence in clinical translation.

Figure 7.3A depicts sonication using a clinical MRgFUS system. It was tested using a thermal camera and wired thermocouple measurements as well as MRI water proton resonance frequency (PRF) phase mapping thermometry

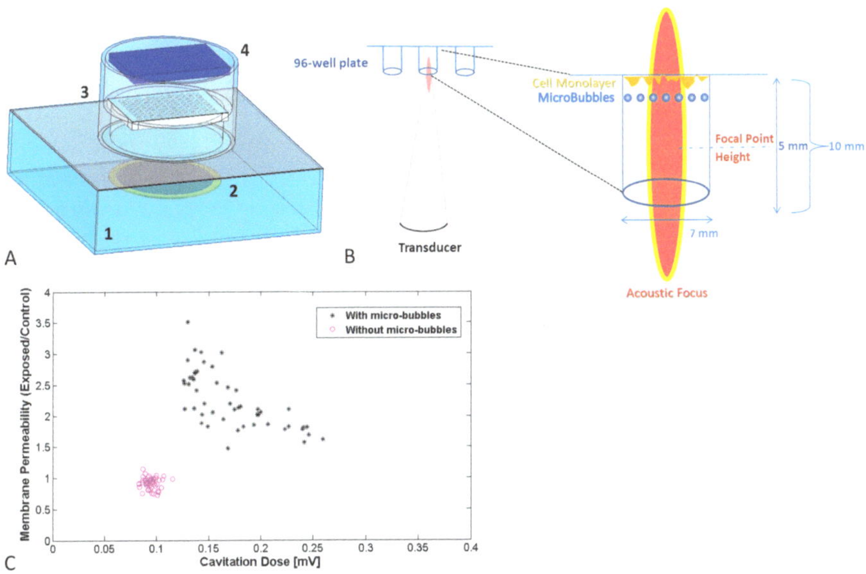


**Figure 7.3** (A) MRgFUS based sonication arrangement with: 1. Water tank filled with water maintained at 30 °C; 2. Ultrasonic transducer with half the elements active, providing a total aperture of 60 mm; 3. Inverted 96-well plate placed 92 mm above the transducer; 4. Absorbing material placed 50 mm above the 96-well plate. (B) Schematic representation of ultrasonic wave propagation through a single well within the 96-well tissue culture plate. (C) Induced membrane permeability following continuous sonication for 10 s with the following parameters: acoustic power level $P = 1W$, focused at the centre of each well in all axes; peak negative pressure, PNP = 0.53 MPa; intensity, $I = 18.77\ W\ cm^{-2}$; and frequency, $f = 0.95$ MHz, applied to cultures of MCF-7 human breast cancer cells.

to ensure that no significant heating occurred during sonication at the power and energy levels corresponding to those used *in vitro*. The importance of this is that, during FUS application for sonoporation, any effects observed *in vitro* correlate solely with the physical effects of FUS on cells. Additionally, acoustic measurements were made with a needle hydrophone to evaluate the FUS beam pattern within a single well in the 96-well plate, with a schematic representation shown in Figure 7.3B, and acoustic measurements confirmed that the sealing film and base material of the 96-well culture plate are suitable for *in vitro* USmTDD research. Furthermore, following acoustic evaluation, a specific beam pattern and focal point position were chosen for the *in vitro* studies. Figure 7.3B is a schematic representation of FUS propagation within a single well in a 96-well plate. The width and length of the focal zone of the FUS beam were selected to ensure satisfactory coverage of the interior of the well whilst avoiding unwanted heating in the plastic perimeter.

To gain knowledge of the mechanism of uptake enhancement by MBs with FUS, membrane integrity tests were performed to quantify the level of lactate dehydrogenase (LDH) present in the medium due to cell membrane leakage. As can be seen in Figure 7.3C, there is an increase in the membrane permeability by a factor of up to 3.5 due to FUS in the presence of MBs and no significant change without MBs. From these results it can be concluded that the mechanism of enhanced cellular uptake *via* sonoporation is achieved specifically by the presence of MBs, with their role promoting continuous stable cavitation. A suggested pathway by which the MBs may increase membrane permeability is called cellular massage, in which oscillations of the MBs in proximity to cells influences membrane permeability.

A possible threshold is reported to be between stable and transient cavitation.[41,42] As explained previously, in stable cavitation, MBs undergo steady state oscillation during the application of ultrasound, expanding during negative pressure peaks and shrinking during positive pressure peaks. The oscillations can be either linear, *i.e.* the bubble size changes equally in every direction, or non-linear, *i.e.* the bubble may take different shapes and is no longer spherical during oscillation. The presence of stable cavitation in proximity to cells also causes micro-streaming, producing large shear forces on the cells and contributing to increased membrane permeability. Stable cavitation generates an acoustic signature that depends on the type of oscillation and which contains harmonic and sub-harmonic peaks of the applied FUS frequency. In contrast, transient cavitation is a process in which, during the MB expansion stage under negative pressure, the MB can no longer sustain itself and violently implodes. During such an event, if there is a nearby surface such as a cell membrane then a jet is formed towards it from the fluid surrounding the MB. This jet can be considered analogous to a micro-needle and may rupture the cell membrane. Unlike stable cavitation, transient cavitation has a broadband acoustical signature in the frequency domain.

In Figure 7.3C, an increasing trend of membrane permeability with decreasing cavitation dose (CaD) can be seen in the presence of MBs. This correlation between CaD and the membrane integrity indicates that greater membrane permeability is achieved for smaller values of CaD (0.12 mV < CaD < 0.19 mV) suggesting that stable, nonlinear cavitation with a peak at one-half the FUS frequency is preferable for increased drug uptake.

When applying FUS to cells, especially into a cell monolayer, the design of the sonication chamber is very important. It must be transparent to US propagation, without inherent attenuation, with direct coupling to the source of the FUS, preferably through degassed water. Moreover, the excess ultrasonic energy propagating outside the chamber is also important since it may be reflected back into the chamber unless deliberate measures are taken to prevent this. Any reflection from the chamber walls or other structures should be considered in the construction and evaluation of the system. From a biological point of view, the chamber should be sterile to maintain healthy cells. It should be sealed for the time of FUS exposure, yet accessible for cell handling and, most important, should allow control groups isolated from acoustic or other stimuli. Hence most studies in USmTDD have been carried out in cell suspensions[43] or in an Opti-cell environment.[44,45]

Although the Opti-cell single-chamber environment is designed for work with cells and is US transparent, it is particularly difficult to achieve isolated control, positioning this approach purely as an optical observation tool, with the outcomes presented below. Figure 7.4A depicts the effects of sonication

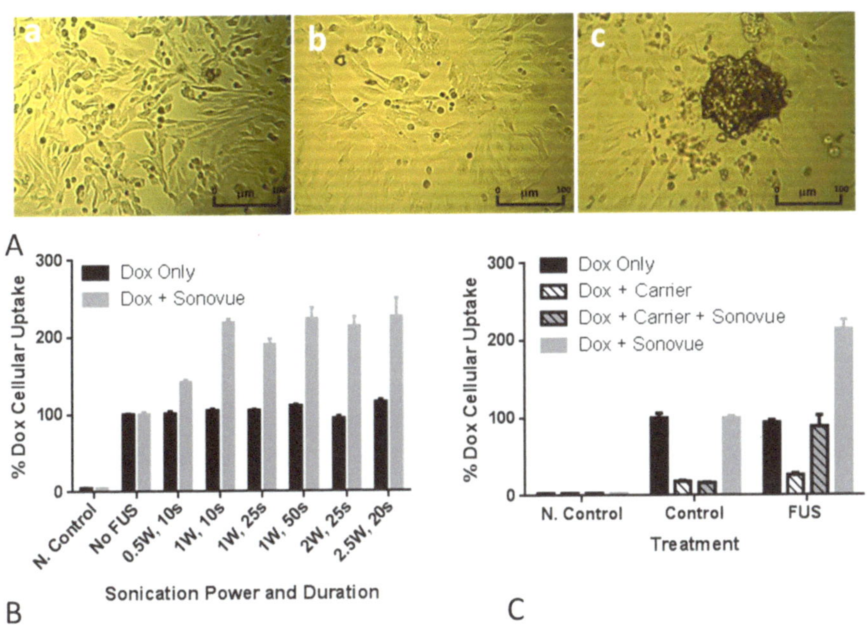


**Figure 7.4** (A) 5 Wa continuous sonication for 20 s (a) prior to sonication, (b) after sonication, and (c) after 20 Wa continuous sonication. (B) Cellular uptake of Dox into MCF-7 cells following various FUS applications. (C) Cellular uptake of Dox into MCF-7 cells with and without carrier following sonication at 2.5 Wa for 20 s.

by the MRgFUS system on A375m human breast cancer cells observed in an Opti-cell chamber at different FUS power levels. As can be seen, at 5 Wa (Watts acoustic power) no visible change can be seen because of sonication whereas, at 20 Wa, there is a detached cell agglomerate. Although it is clear from Figure 7.4A that the cell monolayer has been damaged, there is no possibility to assess the scale of damage since the Opti-cell is a unitary chamber. In the work described here, this led to the development of a different arrangement to accommodate both visual and quantitative evaluation of the effects of FUS, presented in Figure 7.3A and B.

A major part of the establishment of an *in vitro* sonoporation arrangement is to carry out a baseline study, including tuning of the parameters and definition of the optimal sonication patterns. Initial measurements are necessary to identify the optimal parameters by scanning possible sonication powers, times and energy levels. Since the half-life of the MBs during sonication at 1 MHz has been shown to be as low as 41 s,[46] sonication times longer than 50 s were considered irrelevant for the work described here. As can be seen in Figure 7.4B, all power levels except the lowest approximately doubled Dox uptake. The lack of major differences between the higher power levels appears to confirm that MB activity is a threshold-dependent phenomenon.

To evaluate the efficiency of FUS as a stimulus for drug release from the nanocarrier, both sonoporation and heating experiments were conducted. During sonoporation, the cells were exposed to FUS in the presence of encapsulated drug with and without MBs. According to the uptake results presented in Figure 7.4C, the presence of the nanocarrier in the Dox solution reduces the uptake of the drug by about 80%. This figure does not change when FUS is applied to the cells with the carrier plus drug combination but without MBs as a cavitation nuclei. However, in the presence of MBs, the uptake of Dox into the cells is increased by a factor of more than five, resulting in almost as large quantities of Dox penetrating the cells as in the control experiment with free Dox. This presents the possibility of localized release under specified *in situ* FUS conditions.

To explore the safety of the application of FUS to the cell monolayers, it is also necessary to verify that FUS in the presence of MBs does not cause irreversible degradation of the viability of the sonicated cells. In the work described here, MTT-based viability tests were performed which showed that, although there was cell loss of 27% ± 2% immediately after application of FUS, after a recovery period of 24 h, cell loss had halved to 13% ± 1%, indicating that the cells remaining after sonication are behaving approximately normally.

An additional effect associated with the application of FUS is temporary temperature increase *in situ*, with FUS heating in tissues the best understood mechanism of FUS behaviour *in vivo*.[47] The thermal effects of FUS are presently used mainly in tissue ablation in MRgFUS[48] with controlled heating achieved *via* the application of a high intensity focused ultrasound (HIFU) beam.[49] Utilization of the same technique but with a lower temperature rise allows moderate hyperthermia which is known for its ability to increase cellular uptake of anticancer drugs.[12] Hyperthermia in conjunction with MBs has also been considered recently as a means to enhance cellular uptake of anticancer agents *in vivo*.[6]

To study thermal effects on the carrier-drug combination and their interaction with the cell layer described in this chapter, an additional MRgFUS arrangement was developed, closely based on the arrangement in Figure 7.3A but with the crucial difference being placement of tissue-mimicking phantom material above and below the 96-well plate deliberately to induce controlled heating in proximity to the cells. Heating for 2 min at exactly 43 °C, or up to 4 min for a temperature in the range 42.5–43 °C has been described as 'the optimal thermal dose' for cancer therapy.[50] Hence, in the arrangement discussed here, the FUS-induced heating exposure time was set to 3 min, with a heating pattern comprising temperature rise to approximately 43 °C in around 1 min followed by sustained moderate heat treatment for 3 min continuously as shown in Figure 7.5A. This profile was monitored with MRI thermometry to ensure it was uniform for every sonication performed.

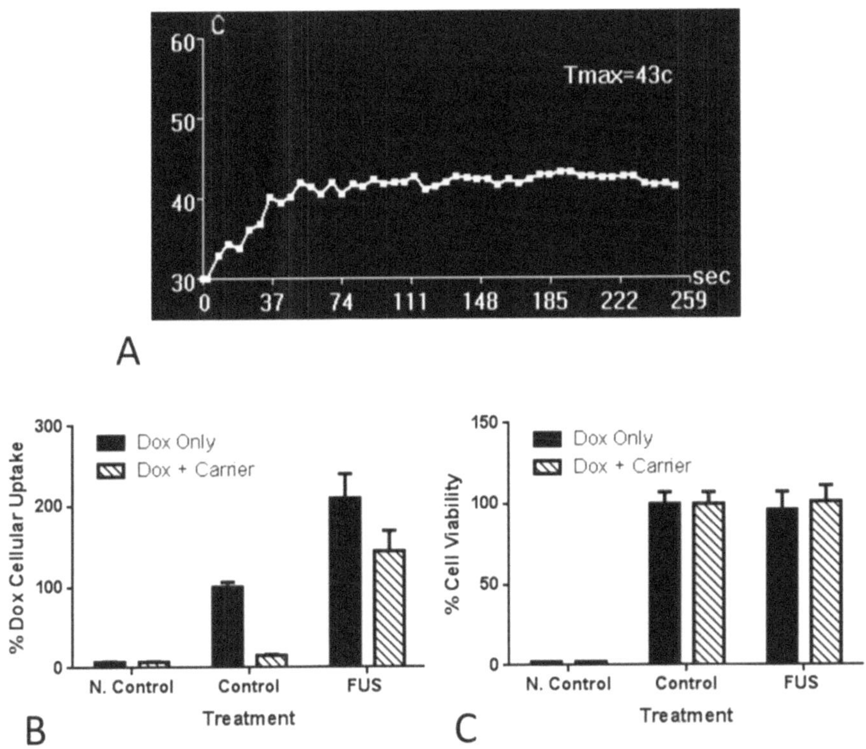


**Figure 7.5** (A) PRF temperature measurement with MRI during continuous sonication of 30 W for 30 s followed by pulsed sonication of 30 W for 180 s, 0.5 s on and 3 s off. (B) Dox uptake into MCF7 cells and (C) viability of MCF7 cells following this temperature protocol.

Simultaneously with this arrangement, cavitation monitoring was also possible using the MRgFUS system to ensure that the drug uptake effects shown in Figure 7.5B were associated solely with heating. As can be seen, establishment of an elevated temperature of 43 °C for 3 min can double the uptake of Dox into the cells. Furthermore, release and uptake of the encapsulated drug also results, to a level similar to that of the free non-encapsulated drug. As shown in Figure 7.5C, no impact was visible on the viability of the cells by FUS induced heating with the protocol applied here. This data thus indicates the possibility to introduce systemically a safer version of a drug and then by moderate local heating *in situ* to target it at specific tissue using remote FUS application.

The observed increase in the amount of Dox present within the cells following heating of the carrier plus drug complex may result from two possible mechanisms, the first being Dox release from encapsulation because of the thermal sensitivity of the carriers and the second that the encapsulated drug complex penetrates the cells because of heat-enhanced membrane permeability. The latter would also explain the increase in the uptake of free Dox. To test these pathways, heating without FUS was performed in a separate experiment using contact heating, with the temperature increase within the wells monitored in real time *via* thermocouples. With this arrangement, it was found that the increase in uptake of free Dox was a factor 2.5× higher than with FUS-induced heating. However, the increase in uptake of the encapsulated drug was similar or slightly higher with FUS-induced heating rather than with contact heating. Although it is most likely that both mechanisms apply, the results suggest that the possible mechanisms of uptake increase with contact heating are heat-induced membrane permeabilisation, as indicated by the high free Dox uptake values, with potential drug release, whereas with FUS induced heating, the first mechanism has more effect, with the drug released from the CD because of the temperature increase, indicated by the higher encapsulated Dox uptake results.

Throughout these studies, the presence of nanocarriers did not negatively influence the viability of cells. Indeed, a slight increase in viability was observed; although not statistically significant, this may be caused by stimulation of the dehydrogenase cell activity by the CDs.[51] Most importantly, though, the results indicate that the CD nanocarrier has no cytotoxicity at concentrations up to 500 μM.

## 7.4 Discussion and Conclusions

This chapter has illustrated the concept of USmTDD, combining the synthesis and evaluation of a novel molecular carrier and its implementation *in vitro* with arrangements for specific investigation of thermal and mechanical effects of FUS.

The drug carrier used as an illustration is based on CD technology. It consists of a truncated-cone-shaped sugar molecule, γCD, covalently bonded to an amino acid derivative with a naphthalene moiety. Spectroscopic evaluation by $^{1}H$- and $^{13}C$-NMR provided evidence supporting the realisation of the intended structure, which aimed to be a vehicle for delivery of Dox *via* a complexation reaction. The examination of Dox in this reaction with the carrier was investigated with fluorescence imaging and NMR. According to the fluorescence studies, a significant decrease in the Dox signal was recorded, with direct dependence on the carrier concentration, $K_{binding}$ of approximately 62 000 $M^{-1}$, a value significantly higher than $K_{binding}$ of Dox to the unmodified γCD[20]. This could be attributed to the formation of $\pi$–$\pi$* interactions between the Dox and the naphthalene moiety of the carrier. The same effect was also evident in 3D NMR studies, where correlation was suggested between the protons of the aromatic rings of the Dox and those of the naphthalene. This data, together with other 3D and 2D NMR studies, suggest the encapsulation conformation presented in Figure 7.2. The overall chemical analysis of the carrier and Dox, on their own as well as in a complex, aid understanding of their behaviour in various media, as well as validating their chemical structures and their roles in complex formation. The carrier itself was verified to be non-toxic at tested concentrations of 500 μM, as confirmed by viability studies.

To examine the potential of the CD-based drug carrier to aid delivery of Dox into targeted cancer cells *via* USmTDD, several *in vitro* arrangements were explored in the work. As the main goal of the *in vitro* studies was to evaluate individually the mechanical and thermal effects of FUS during USmTDD, two different types of FUS application arrangements were designed. The first, dealing with mechanical effects on cells, utilised low power FUS in the presence of MBs to generate controlled stable cavitation and the second, utilising a temperature rise *in situ*, produced moderate hyperthermia in proximity to the cells. To validate the impact of FUS-induced heating on the cell monolayer; an additional contact-heating arrangement was investigated.

The common basis of all the USmTDD arrangements was a clinical MRg-FUS system. To validate the accuracy and proper functioning of these arrangements, thorough thermal and acoustical evaluation was conducted. Thermal evaluation of the sonoporation experiments using both a thermal camera and thermocouples confirmed that there is no significant temperature rise as a result of sonication at the energy levels applied *in vitro*, in turn confirming that the effects of FUS observed during sonication to achieve sonoporation were mechanical rather than thermal. According to these studies, in the presence of MBs, the uptake of Dox into human cancer cells more than doubled, while the viability of the cells was above 85% after 24 h recovery. Membrane integrity test results were in line with the uptake values, suggesting that the increased Dox uptake is due to increased cell membrane permeability caused by the MBs.

Alongside the sonoporation arrangements, FUS-induced heating was also considered, together with a control experiment without FUS. Both were evaluated for application of temperature-induced TDD. MRI thermometry as well as thermocouples were used to define the optimal sonication protocol for FUS-induced heating *in vitro*, comprising continuous sonication to produce a rapid temperature rise, followed by 3 min stable heating in a pulsed mode of 0.5 s on, 3 s off. Heating at 'the optimal thermal dose' for cancer therapy,[50] *i.e.* 2–4 min at 43 °C, resulted in a doubling in the cellular uptake of Dox, with cell viability following heating using FUS being above 96%.

In addition to the *in vitro* investigation of increased Dox uptake because of mechanical and thermal effects of FUS, the ability of the carrier to encapsulate and release the Dox, *in vitro*, *via* the same mechanisms, was established. According to the *in vitro* evaluation of the complex, the carrier successfully encapsulated the Dox, resulting in 80% reduction in Dox penetration within the cancer cells. The release mechanisms using mechanical and thermal effects were found to be successful in achieving close to 100% release of Dox as compared to the control. *Via* the sonoporation mechanism, the uptake of encapsulated Dox in the presence of MBs was increased by a factor higher than five, resulting in Dox uptake into the cells almost matching the free Dox control. A suggested mechanism for the increase in cell uptake is *via* penetration of the complete CD plus Dox complex into the cells due to increased cell membrane permeability achieved by the MBs, followed by dissociation within the cell itself.

Thermal effects of FUS resulted in an increase by up to a factor of 8.5 in the uptake of the encapsulated Dox into the cells. This was validated using contact heating, where an increase by a factor of 5.7 was recorded. These increases suggest two possible mechanisms. The first is the release from the encapsulation due to the heat sensitive feature of the carriers and the second is that the encapsulated drug translocates across the cell membrane due to heat-enhanced membrane permeability. Although it is likely that both mechanisms act simultaneously, the results suggest that the mechanism of increased uptake in heating without FUS is heat induced membrane permeability, as indicated by the free Dox uptake values, with potential drug release, whereas in FUS-induced heating it is likely to relate to the first mechanism, with the drug released from encapsulation, as indicated by the higher encapsulated Dox uptake results.

In conclusion, the arrangements for investigation of USmTDD *in vitro* proved to be successful in achieving the desired goals of a well monitored, reliable research environment with high repeatability, able to differentiate between the thermal and mechanical effects of FUS. In addition, the results confirmed the CD nanocarrier as a promising drug delivery vehicle for USmTDD for controlled release utilising both thermal and cavitation effects. The potential of MBs as a tool to enhance membrane permeability by sonoporation has been confirmed by increased drug uptake with and without the CD nanocarrier. The significant increase in uptake, together with relatively high cell viability, has confirmed that membrane permeabilisation is a reversible and non-permanent process[52] which does not lead directly to cell necrosis. In addition, moderate heating was shown to be highly beneficial in drug release and cellular uptake increase of free and encapsulated drug.

While the fundamental objective of the work described here was the establishment of a sonication protocol and related equipment for *in vitro* studies, development toward the *in vivo* stage and clinical trials was kept in mind.

As previously explained, all the arrangements involving FUS were based on a commercially available, clinically approved system for MRgFUS. Although the translation from work *in vitro* to *in vivo* research is not straightforward, a strong background *in vitro* can contribute greatly to the success of outcomes *in vivo*.

Prior to the work described in this chapter, no properly controllable research environment existed for *in vitro* studies. In principle, the adaptation of a clinical MRgFUS system for *in vitro* research that is described allows the same system to be used for all stages of USmTDD development, from *in vitro* studies, through the pre-clinical stage to clinical trials. This allows physicians to be a part of a wider USmTDD research group, from the beginning of the product definition, bringing real meaning to the term "from bench to bedside".

## Acknowledgements

The financial support from the European Community's Seventh Framework Programme (FP7/2007–2013) under grant agreement 230674 (Nanoporation project) is acknowledged, as is the work of the authors' co-authors on papers emerging from this grant.

## References

1. L. Brannon-Peppas and J. O. Blanchette, *Adv. Drug Delivery Rev.*, 2004, **56**, 1649–1659.
2. A. Corti, F. Pastorino, F. Curnis, W. Arap, M. Ponzoni and R. Pasqualini, *Med. Res. Rev.*, 2012, **32**, 1078–1091.
3. A. Rösler, G. W. M. Vandermeulen and H.-A. Klok, *Adv. Drug Delivery Rev.*, 2001, **53**, 95–108.
4. Z. Gao, H. D. Fain and N. Rapoport, *Mol. Pharmaceutics*, 2004, **1**, 317–330.
5. H. Grull and S. Langereis, *J. Controlled Release*, 2012, **161**, 317–327.
6. A. Yudina, M. Lepetit-Coiffe, M. De Smet, S. Langereis, H. Grull and C. Moonen, *J. Controlled Release*, 2012, **161**, 90–97.
7. C. L. Morfey, *Dictionary of Acoustics*, Academic Press, 2001.
8. A. Yudina, M. Lepetit-Coiffe and C. T. Moonen, *Mol. Imaging Biol.*, 2011, **13**, 239–249.
9. W. G. Pitt, G. A. Husseini and B. J. Staples, *Expert Opin. Drug Delivery*, 2004, **1**, 37–56.
10. J. Wu and W. L. Nyborg, *Adv. Drug Delivery Rev.*, 2008, **60**, 1103–1116.
11. A. L. Klibanov, T. I. Shevchenko, B. I. Raju, R. Seip and C. T. Chin, *J. Controlled Release*, 2010, **148**, 13–17.
12. L. Chen, Z. Mu, P. Hachem, C. Ma, A. Wallentine and A. Pollack, *Phys. Med. Biol.*, 2010, **55**, 7399–7410.
13. R. Singh, N. Bharti, J. Madan and S. Hiremath, *J. Pharm. Sci. Technol*, 2010, **2**, 171–183.
14. G. Schmid, *Trends Biotechnol.*, 1989, **7**, 244–248.

15. M. E. Brewster and T. Loftsson, *Adv. Drug Delivery Rev.*, 2007, **59**, 645–666.
16. F. van de Manakker, T. Vermonden, C. F. van Nostrum and W. E. Hennink, *Biomacromolecules*, 2009, **10**, 3157–3175.
17. C. A. Frederick, L. D. Williams, G. Ughetto, G. A. Van der Marel, J. H. Van Boom, A. Rich and A. H. Wang, *Biochemistry*, 1990, **29**, 2538–2549.
18. T. Šimůnek, M. Štěrba, O. Popelová, M. Adamcová, R. Hrdina and V. Geršl, *Pharmacol. Rep.*, 2009, **61**, 154–171.
19. M. O'brien, N. Wigler, M. Inbar, R. Rosso, E. Grischke, A. Santoro, R. Catane, D. Kieback, P. Tomczak and S. Ackland, *Ann. Oncol.*, 2004, **15**, 440–449.
20. R. Anand, S. Ottani, F. Manoli, I. Manet and S. Monti, *RSC Adv.*, 2012, **2**, 2346–2357.
21. O. Swiech, A. Mieczkowska, K. Chmurski and R. Bilewicz, *J. Phys. Chem. B*, 2012, **116**, 1765–1771.
22. K. Uekama, *Chem. Pharm. Bull.*, 2004, **52**, 900–915.
23. A. R. Khan, P. Forgo, K. J. Stine and V. T. D'Souza, *Chem. Rev.*, 1998, **98**, 1977–1996.
24. J. Szejtli, *Pure Appl. Chem.*, 2004, **76**, 1825–1845.
25. W. Tang, I. W. Muderawan, T.-T. Ong and S.-C. Ng, *Tetrahedron: Asymmetry*, 2007, **18**, 1548–1553.
26. I. W. Muderawan, T. T. Ong, T. C. Lee, D. J. Young, C. B. Ching and S. C. Ng, *Tetrahedron Lett.*, 2005, **46**, 7905–7907.
27. Y. Himeno, A. Miyagawa, M. Kawai and H. Yamamura, *Tetrahedron*, 2009, **65**, 9474–9480.
28. N. Badi, N. Jarroux and P. Guégan, *Tetrahedron Lett.*, 2006, **47**, 8925–8927.
29. F. Trotta, K. Martina, B. Robaldo, A. Barge and G. Cravotto, *J. Inclusion Phenom. Macrocyclic Chem.*, 2007, **57**, 3–7.
30. P. S. Green and C. Leeuwenburgh, *Biochim. Biophys. Acta, Mol. Basis Dis.*, 2002, **1588**, 94–101.
31. J. Piosik, M. Zdunek and J. Kapuscinski, *Biochem. Pharmacol.*, 2002, **63**, 635–646.
32. C. M. Fernandes, R. A. Carvalho, S. P. da Costa and F. J. Veiga, *Eur. J. Pharm. Sci.*, 2003, **18**, 285–296.
33. H. M. C. Marques, *Flavour Fragrance J.*, 2010, **25**, 313–326.
34. S. Emara, I. Morita, K. Tamura, S. Razee, T. Masujima, H. A. Mohamed, S. M. El-Gizawy and N. A. El-Rabbat, *Talanta*, 2000, **51**, 359–364.
35. M. Mukhopadhyay, D. Banerjee, A. Koll, A. Mandal, A. Filarowski, D. Fitzmaurice, R. Das and S. Mukherjee, *J. Photochem. Photobiol., A*, 2005, **175**, 94–99.
36. H. A. Benesi and J. Hildebrand, *J. Am. Chem. Soc.*, 1949, **71**, 2703–2707.
37. X.-J. Dang, M.-Y. Nie, J. Tong and H.-L. Li, *J. Electroanal. Chem.*, 1997, **437**, 53–59.
38. D. Gourevich, O. Dogadkin, A. Volovick, L. Wang, J. Gnaim, S. Cochran and A. Melzer, *J. Controlled Release*, 2013, **170**, 316–324.
39. D. Gourevich, Y. Hertzberg, A. Volovick, Y. Shafran, G. Navon, S. Cochran and A. Melzer, *Ultrasound Med. Biol.*, 2013, **39**, 507–514.

40. D. Gourevich, A. Volovick, O. Dogadkin, L. Wang, H. Mulvana, Y. Medan, A. Melzer and S. Cochran, *Ultrasound Med. Biol.*, 2015, **41**, 1853–1864.
41. C. E. Brennen, *Cavitation and Bubble Dynamics*, Cambridge University Press, 2013.
42. T. Leighton, *The Acoustic Bubble*, Academic press, 2012.
43. L. B. Feril, T. Kondo, Q.-L. Zhao and R. Ogawa, *Cancer Lett.*, 2002, **178**, 63–70.
44. J. Eisenbrey, P. Huang, J. Hsu and M. Wheatley, *Ultrasonics*, 2009, **49**, 628–633.
45. L. B. Feril Jr, K. Tachibana, T. Kondo, R. Ogawa, Q.-L. Zhao, K. Yamaguchi, K. Ogawa, H. Endo, Y. Irie and Y. Harada, *J. Med. Ultrason.*, 2010, **37**, 3–8.
46. S. Tinkov, R. Bekeredjian, G. Winter and C. Coester, *Appl. Acoust.*, 2009, **70**, 1323–1329.
47. V. Frenkel, *Adv. Drug Delivery Rev.*, 2008, **60**, 1193–1208.
48. T. Peters and K. Cleary, *Image-guided Interventions: Technology and Applications*, Springer Science & Business Media, 2008.
49. G. Ter Haar, *Ultrasound Med. Biol.*, 1995, **21**, 1089–1100.
50. S. A. Sapareto and W. C. Dewey, *Int. J. Radiat. Oncol., Biol., Phys.*, 1984, **10**, 787–800.
51. A. Al-Omar, S. Abdou, L. De Robertis, A. Marsura and C. Finance, *Bioorg. Med. Chem. Lett.*, 1999, **9**, 1115–1120.
52. M. R. Böhmer, A. L. Klibanov, K. Tiemann, C. S. Hall, H. Gruell and O. C. Steinbach, *Eur. J. Radiol.*, 2009, **70**, 242–253.

CHAPTER 8

# *Theranostics in the Gut*

FRASER STEWART[a], HELEN MULVANA[b], INKE NÄTHKE[a] AND SANDY COCHRAN*[b]

[a]School of Life Sciences, University of Dundee, Dundee, DD1 4HN, UK;
[b]School of Engineering, University of Glasgow, Glasgow, G12 8QQ, UK
*E-mail: sandy.cochran@glasgow.ac.uk

## 8.1 Introduction

This chapter outlines a strategy to apply theranostics and image-guided drug delivery to the gastrointestinal (GI) tract. Oral and, less commonly, rectal delivery of drugs are very well-established. However, as with intravenous delivery they share the limitation to systemic application. Awareness of video capsule endoscopy (VCE) and related technologies suggests the potential for targeting for two reasons. First, the enormous global investment in microscale technology, including electronics, communications, and microfluidics, and the related business models for applications of capsule endoscopy (CE) are opening up many new possibilities. Second, the widespread acceptance of VCE as a diagnostic modality is encouraging research into other applications in the GI tract, including diagnosis with wireless communication and therapy.

VCE is based on optical imaging of the surface of the gut and, just as ultrasound (US) was a major step forward in conventional endoscopy following its establishment with optics, so a similar step may be expected in CE, with work already under way in that regard. Additionally, though, US carries the potential to combine diagnosis and therapy, particularly in the form of

Drug Discovery Series No. 63
Theranostics and Image Guided Drug Delivery
Edited by Maya Thanou

Published by the Royal Society of Chemistry, www.rsc.org

US-mediated targeted drug delivery (UmTDD), a subject of active contemporary research interest. Such a combination has the potential to improve early detection and treatment of a range of GI pathologies, notably including inflammatory bowel disease (IBD) in the small intestine.

The detailed content of this chapter begins in Section 8.2 with a review of the GI tract, its organisation and structure and diseases of common susceptibility. Basic concepts in CE are introduced in Section 8.3, including diagnostic modalities and therapeutic potential. A detailed description of UmTDD is given in Section 8.4, including the important phenomenon of cavitation and the related topic of microbubbles (MBs). Section 8.5 links UmTDD and MB technology with CE specifically then Section 8.6 returns to the topic of GI diseases, with a particular focus on current treatments and potential theranostics related to UmTDD. The chapter covers a wide range of contemporary questions, many of which remain unanswered; a small selection of the most important are considered in Section 8.7.

## 8.2 The Gastrointestinal Tract

Inflammatory diseases and cancer of the GI tract are a major health problem, with 2.2 million IBD patients in Europe[1] and 447,000 new colon cancer cases in 2012.[2] Tools to detect these diseases early, deliver treatment locally, and measure response to therapy could vastly improve clinical outcomes. Whilst endoscopes can access the upper and lower sections of the GI tract for visual examination and sampling of tissue, they are uncomfortable and cannot easily access the middle section of the GI tract where IBD is prevalent. Thus, the ability to produce endoscopic capsules constitutes an important advance. Adding functions such as diagnostic US imaging and US-mediated therapy to the existing visual diagnosis possible with endoscopic capsules may allow measurement of tissue structure and related image-guided theranostic treatment.[3,4]

### 8.2.1 Organisation and Structure of the Gastrointestinal Tract

The GI tract connects the oral cavity with the anus and comprises the oesophagus, stomach, small intestine, and large intestine (colon), as shown in Figure 8.1(a). Its main functions are the breakdown of food, absorption of nutrients, and packaging of waste product for elimination. This involves physical breakdown facilitated by chewing; chemical breakdown and processing by acids, enzymes and bile products in the stomach and upper small intestine; and further breakdown into molecular components by the colonic microbiome. Once food is digested into compounds that can be transported into the body, they are absorbed across the epithelial layer that lines the small and large intestine. In the latter, water is also absorbed and undigested waste products, together with cells shed from the small intestine, are packaged for excretion.[5]

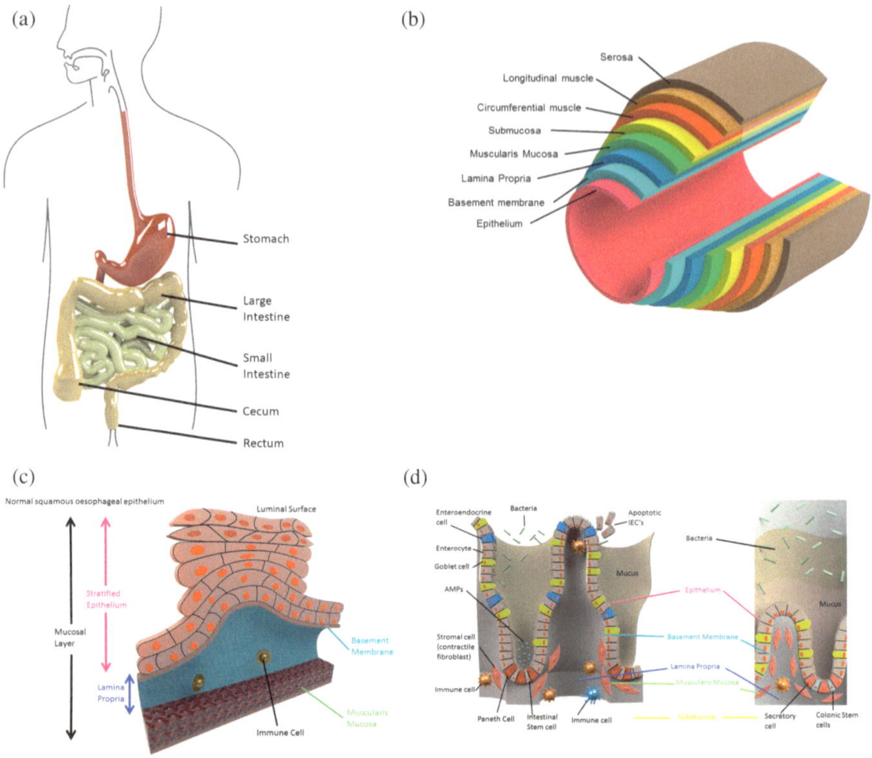


**Figure 8.1** (a) Schematic of the overall GI tract in the human body including the oral cavity, oesophagus, stomach, small and large intestine (colon). (b) Schematic of the relative order of the layers surrounding the lumen of the GI tract. (c) Details of the oesophageal and (d) small intestinal wall showing these different layers in more detail. (c) The oesophagus is lined by a stratified epithelium. In this case, epithelial cells are produced by the cell layer just above the basement membrane. Their proliferation pushes differentiating cells towards the lumen. During this process they become increasingly flattened and their nuclei deteriorate. Immune cells are normally present in the Lamina Propria. (d) In the small and large intestine, the epithelial layer is formed by a columnar epithelium that contains a number of different cell types. Mucus is secreted by Goblet cells to form a protective layer that also houses many different types of bacteria. In the crypt base, Paneth cells in the small intestine and similar, related cells in the colon, reside in the bottom of crypts. They secrete antimicrobial peptides (AMPs) that provide a protective environment for neighbouring stem cells (intestinal stem cells, ISC, and colonic stem cells, CSC). Underneath the basement membrane, stromal fibroblasts form a contractile structure. Different types of immune cells are present in close association with the epithelial layer and throughout the Lamina Propria. Proliferation in the crypt, coupled with unknown mechanisms, facilitate movement of cells upwards towards the villus (the region beyond the crypt) in the intestine or the top of crypts in the colon, where they are extruded into the lumen and undergo apoptosis.

The lumen of the GI tract, which is external to the body, contains not only bacteria but also many other materials that would be detrimental to the body if allowed to access it. This makes the barrier function of the lining crucial for normal health. It also means that defence systems exist to protect against inappropriate invasion of digestive tract contents. In addition to the specialized tissue layers that constitute the wall of the GI tract, many different types of immune cells provide surveillance mechanisms for protection.

The tissue forming the GI tract changes from oesophagus to colon to reflect the function of each region. Common to all segments are four general layers that form the tube, as shown in Figure 8.1(b). The serosa is the outermost layer and is a thin protective sheath. Next is the muscularis, which comprises two orthogonally aligned layers of muscle and provides contractility. It is connected to the submucosa, a layer of stromal cells and connective tissue. The mucosa is innermost, forming the lining of the lumen and supported by a basement membrane connecting it to the submucosa. This innermost layer is an epithelium. Immune cells are found mostly in the submucosa and specialised immune cells can also be found directly associated with the epithelium.

In the mouth and oesophagus, the epithelium is stratified similarly to the skin. In stratified epithelia, multiple layers of cells are continually generated from stem cells that reside near the basement membrane and cells in the uppermost layer are continually shed into the lumen. At the gastro-oesophageal junction, the layered arrangement changes to a simple, cuboidal epithelium comprising a single layer of epithelial cells that continues through the rest of the digestive tract. In the stomach, this layer is highly invaginated into deep 'pits'. These U-shaped invaginations house specialized secretory epithelial cells that produce acid and other digestive reagents. Figure 8.1(c) shows the stratified epithelium of the oesophagus.

The duodenum is the beginning of the small intestine, connected to the stomach. Here the shape of the surface of the lining changes and invaginated folds, called crypts, are arranged around finger-like protrusions called villi, Figure 8.1(d). The duodenum also houses glands that secrete alkaline mucus to neutralize the highly acidic stomach contents. This mucus protects the epithelium and ensures that digestive enzymes in the small intestine are active and can chemically digest food further so that it can be absorbed. The next section is the jejunum. Its main function is nutrient absorption. The structure of the epithelium in the jejunum is similar to that in the duodenum, with prominent villi and higher order tissue folds that together maximise the surface area for efficient nutrient absorption. The final section of the small intestine is the ileum. Its function is to absorb vitamins, bile salts and any products not absorbed in the jejunum.

The ileum is connected to the colon at the cecum. The epithelium changes again in the cecum and villi are no longer present from this region onwards. Instead, crypts become deeper to create a smoother surface to aid the packaging of indigestible waste products, absorption of water and preparation of the waste for excretion. These functions are assisted by abundant gut flora comprising mostly bacteria that ferment residual waste products further and protect against overgrowth of harmful bacteria.

### 8.2.2 Diseases of the Gastrointestinal Tract

The most common diseases of the GI tract are cancer and IBD. These emerge from defects in the biological processes that normally protect the GI tract. The lining of the GI tract is constantly renewed, with stem cells residing in the bases of crypts driving the rapid cell turnover that underpins this process. In humans, 20–50 million cells are shed every minute from the small intestine,[6] ensuring that cells do not persist once damaged by toxic agents such as by-products of digestion and bacterial fermentation. The renewal process also allows the epithelial lining to react to and remodel itself if instructed by signals it receives, facilitating rapid repair in response to physical, chemical, or biological insults.

The stem cell-driven, rapid and continuous turnover of the epithelium plays a role in cancer and may be responsible for its relatively high incidence in the GI tract, particularly the colon. Even the normally extremely low probability of genetic mutations becomes significant when considering that, over the average human lifetime, approximately $2.9 \times 10^{11}$ and $2 \times 10^{12}$ stem cell divisions occur in the linings of the small intestine and colon respectively.[7] This means the likelihood of a transforming mutation occurring that could initiate a tumour is high in this tissue. Indeed, cancer of the bowel is the second most common cause of cancer deaths in the developed world.[8]

IBD is the other common disorder of the intestinal tract. Normally, the active immune surveillance in the GI tract ensures rapid, efficient detection of foreign, invading pathogens, prompting the activation of immune responses to eliminate them. However, hyper-stimulation or inappropriate activation of the immune response can cause tissue damage. Minor defects in the barrier function of the epithelium or chronic exposure to damaging reagents can activate the immune system, causing recruitment of inflammatory cytokines and other factors that are detrimental when present in excess and/or for prolonged times.

Specific examples of common IBDs that affect the GI tract are Barrett's oesophagus, Crohn's disease (CD) and ulcerative colitis. In Barrett's oesophagus, reflux of acidic stomach contents into the lower end of the oesophagus causes chronic damage and inflammation of the epithelium. One consequence is a change from the normally stratified structure of the epithelium to a simple, columnar epithelium mimicking that of the intestines. CD is manifest as tissue inflammation and can occur in any region of the GI tract, whereas ulcerative colitis is restricted to the colon. In both cases, persistent inflammation leads to damage of several tissue layers. In CD, this extends beyond the submucosa to the serosa whereas, in ulcerative colitis, it is restricted to the mucosa and submucosa. Both are chronic, responding only intermittently to therapy and requiring careful, usually life-long clinical management. Symptoms include abdominal pain, diarrhoea, and blood in the stool. Importantly, the chronic tissue inflammation in both of these diseases predisposes to cancer. Current treatments include anti-inflammatory agents and antibiotics to reduce bacteria and immune-suppressants for more severe and progressed cases. Surgery is also used to remove affected regions.

## 8.3 Basic Concepts of Capsule Endoscopy

Conventional endoscopy is an important clinical tool for diagnosis of pathology in the GI tract. A minimally invasive approach with a relatively long technological history, endoscopy is controlled entirely from outside the body by a highly skilled clinician and usually requires at least mild patient sedation. Whilst endoscopy in the form of double-balloon enteroscopy (DBE) can be used to penetrate the entire GI tract, this is not an easy procedure. In contrast, CE relies on devices small enough for oral ingestion that then make their own journey through the full length of the GI tract. Device dimensions are generally reported in terms of diameter x length, *e.g.* 11 mm × 31 mm being the largest clinically available, ingestible device, and this is how size is defined here. CE does not require sedation, is easily administered and leaves the patient free to move around. Its potentially routine nature and the possibility for frequent use offer a different paradigm for diagnosis and therapy in the GI tract, with the potential for an integrated, theranostic approach.

### 8.3.1 Capsule Endoscopy for Diagnosis

The first CE device was developed in early 2000,[9] with optical video imaging as the diagnostic modality. Since then, VCE has provided a minimally-invasive technique to image the entire GI tract, eliminating the gap between (upper) gastroduodenoscopy and (lower) colonoscopy[10] and enabling routine visualisation of the entire small intestine. Current VCE devices include Endo-Capsule (Olympus America Inc., PA, USA), MicroCam (IntroMedic Co Ltd., Seoul, Korea) and OMOM Capsule (Chongqing Jinshan Science & Technology (Group) Co Ltd., Chongqing, China). Since its original development, VCE has been used in more than 1.5 million patients.[11] Three different types are now available: (1) small bowel (SB), the most common, allowing imaging of the entire small intestine; (2) colon, with two cameras for improved visualisation and longer battery life to accommodate the longer journey to the diagnostic site; and (3) oesophageal, with a high frame rate to capture video during the short passage of the oesophagus.

#### *8.3.1.1 Video Capsule Endoscopy*

Following wide-ranging clinical exploration, the main use of VCE has emerged in detection of obscure GI bleeding (OGIB), with 60–70% of all VCE diagnoses relating to this disorder.[12] This is based on the superiority of VCE over other relevant diagnostic modalities including computed tomography (CT), magnetic resonance imaging (MRI), and push enteroscopy.[13] A second important use for VCE is in the diagnosis and evaluation of IBD, specifically CD, which is detected more reliably with VCE than other diagnostic modalities.[14] However, strictures caused by CD are a contraindication since they risk capsule retention. This can be taken into account by the patient taking a soluble capsule to assess patency of the GI tract before CE.

Several other uses of VCE are noteworthy. VCE has been proposed as a means for diagnosing oesophageal diseases such as Barrett's oesophagus and oesophagitis. However, it was found to be inferior to endoscopy[15] and this is likely due to the fast transit in the oesophagus and reduced insufflation.[16] Another use is in diagnosis of small bowel tumours.[16] However, a study comparing VCE to DBE showed that VCE detected only 5 of 15 tumours identified by DBE.[17] Hence, though VCE has diagnostic potential, a negative result should still be followed by DBE if tumours are highly suspected. It has also been reported that VCE could be useful for diagnosis of Coeliac disease. In a study with 166 patients positively diagnosed by biopsy,[18] VCE identified 89% with a specificity of 95%. In another study, two-thirds of patients showed mucosal changes with VCE beyond the proximal small bowel, the usual site for biopsy.[19]

#### 8.3.1.2 Fluorescence Imaging

In addition to the common use of video imaging in CE, fluorescence imaging capsule endoscopy (FICE) could aid in early detection of cancer.[20] The basis for this approach is that fluorophores present in healthy tissue such as flavin adenine dinucleotide (FAD) are often significantly reduced in cancerous tissue.[21,22] These fluorophores emit light with wavelengths in the range 490–590 nm when excited using light with wavelengths 380–500 nm. FICE can thus be used to detect cancerous regions by measuring the presence of endogenous fluorophores along the GI tract.

Currently, the availability of FICE devices is limited. One example is a wireless FI-based capsule[23] designed as a proof of concept. However, its size is 20 mm × 100 mm, preventing ingestion. Another, smaller example, with lower power consumption, is the wireless FI capsule reported by Al-Rawhani *et al.*[24] This device improved over FI endoscopes and previous FICE devices, but with a size of 16 mm × 48 mm, it is still not ingestible. Work is under way to develop ingestible FICE devices by reducing the size of the optical components which presently limit the overall size.

#### 8.3.1.3 pH Measurement

CE can also be used to monitor other physiological parameters that can report disease states. For instance, pH monitoring has been successful with CE for improving the diagnosis of oesophageal reflux disease (ORD).[25] In this case, a wireless pH capsule can be fixed in the oesophagus to transmit data wirelessly to a receiver for up to 96 hours. The performance of the pH capsule is at least equal to other means of measuring oesophageal pH. Other CEs containing pH sensors also exist, such as the microelectronic pill by Johannessen *et al.* In this case, numerous sensors including pH, aim to detect pathological conditions such as IBD and pancreatic disease.[26] Additionally, pH monitoring can be used to aid device localization, based on different pH values in different regions of the GI tract.[27] Adding pH sensing to CE may thus be useful and complement other modalities.

#### *8.3.1.4 Ultrasound Imaging*

Unlike the diagnostic imaging modalities discussed so far, ultrasound (US) can image beneath the surface mucosa, similarly to endoscopic ultrasound (EUS).[28] US imaging is safe, inexpensive and has the potential to be miniaturised for deployment in small devices. At present, ultrasound CE (USCE) is not available clinically; however, work is under way to develop ingestible USCE devices. Memon *et al.*[29] are designing CE with a cylindrical US array for 360° imaging in the GI tract, and Lee *et al.*[30] and Liu *et al.*[31] have published early work on USCE with mechanically-scanned US devices, including pre-clinical demonstration. Another example is the so-called Sonopill,[32] a device aiming to combine USCE with other diagnostic and therapeutic modalities; it is discussed in detail in Section 8.3.3.

#### *8.3.1.5 Other Capsule Endoscopy Devices*

Other CE devices relating to theranostics must also be mentioned. The Crosby capsule is a small device attached to a flexible tube to take biopsies of the mucosa of the small intestine.[33] It has now been largely superseded by conventional endoscopy for taking biopsies in the upper GI tract and its main use now is in paediatric diagnosis.[34] Additionally, a prototype wireless system has been proposed for a capsule to take biopsies in the GI tract.[35] However, only *ex-vivo* testing has been performed with this capsule so far. Finally, work is under way to develop an extremely low-cost chemomechanical capsule specialised to sample the microbiome of the small intestine.[36]

### 8.3.2 Capsule Endoscopy for Therapeutic Use

Capsule endoscopy, realised as VCE, has become a well-established diagnostic modality for the GI tract. CE also has potential as a vehicle for targeted delivery of therapeutics, *i.e.* in the right place at the right time, and therapeutic CE (TCE) is an important future option.

#### *8.3.2.1 Targeted Drug Delivery*

The advantage of integrating targeted drug delivery (TDD) with CE in a theranostic configuration is that a drug can be delivered to a specific location, permitting a higher dose there whilst sparing healthy tissue. Examples include the Enterion capsule[37] and the Intelisite capsule.[38] Both contain reservoirs of around 1 ml of drug that can be released by a remote trigger. However, a number of challenges remain to be met before such capsules can be fully functional: accurate localisation of the capsule is required to ensure the correct delivery site; a means of preferential delivery into or through the mucosa will be beneficial to maximize efficacy; and a means to maintain position in the GI tract may be required, in opposition to physiological motion, for the TDD to be more than momentary.

A TCE device has been reported to overcome some of these challenges,[39] including a mechanism to hold the capsule in place and a microneedle for injecting drugs into the mucosa. However, testing has not yet been reported so its effectiveness is unclear. Additionally, there is a risk that the microneedle could perforate the mucosa. A minimally-invasive means to permeabilise the mucosa may be preferable and focused US is a possible mechanism. A proof-of-concept UmTDD capsule has been created, called SonoCAIT[4] (Sonication capsule for autonomous imaging and therapy). The SonoCAIT concept is to use focused US both to permeabilise and physically push therapeutics into the mucosa. Its principles are based on the extensive work in other areas of UmTDD described in Section 8.4. Early results using cells *in vitro* and animal tissue *ex vivo* are promising and it is discussed in greater detail in Section 8.5.

#### *8.3.2.2 Other*

Just as with diagnosis, the potential benefits of TCE are leading to other exploration. For example, a prototype TCE device has been developed to deliver a surgical clip to aid natural orifice transluminal endoscopic surgery (NOTES) procedures.[40] *In-vivo* experiments using this TCE were reported to be successful.

### 8.3.3 Sonopill

The Sonopill programme (reference UK EPSRC EP/K034537) integrates research in theranostic activity in CE by combining the expertise of engineers, biologists and clinicians to develop and test a fully functional device. The specific target is to combine microultrasound (μUS) with other modalities such as VCE, FICE, and pH measurement in an ingestible capsule for detection and/or treatment of GI conditions such as Barrett's oesophagus, Coeliac disease, Crohn's disease and colon cancer as outlined in Section 8.2. Figure 8.2 shows the range of CE capabilities under consideration. The use of μUS as the main diagnostic modality will provide imaging beneath the mucosa to measure tissue changes[3] and FICE is also being considered; the potential for these to be combined in a single device with VCE is illustrated in Figure 8.3.

The ability of CE to deliver therapeutics is also being explored within Sonopill. For instance, the UmTDD capsule, SonoCAIT uses focused US to promote drug uptake intracellularly by sonoporation and/or intercellularly by opening tight junctions to permeabilize the mucosal barrier. Facilitating

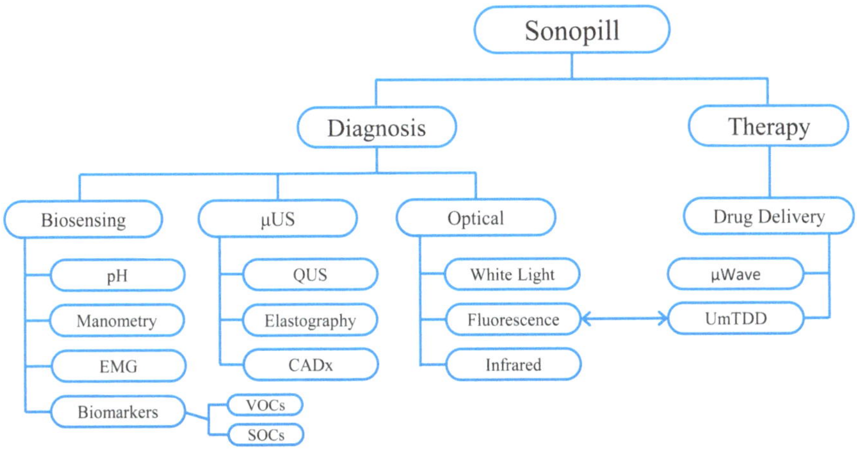


**Figure 8.2** Sonopill taxonomy outlining a range of possible modalities. CADx: computer-assisted diagnosis, EMG: electromyography, VOCs: volatile organic compounds, SOCs: soluble organic compounds, µWave: microwave, and FUS: focused-US surgery. Fluorescence imaging and UmTDD connected to highlight their importance for theranostic delivery.

increased drug uptake would provide a further advantage over oral and systemic intravenous administration. Further details are discussed in Sections 8.4, focusing on UmTDD, and 8.5, on TCE technology.

## 8.4 Ultrasound-mediated Targeted Drug Delivery

Ultrasound-mediated Targeted Drug Delivery shares many of the attractions of imaging with US as a well-established, low-cost, portable technology not reliant on ionising radiation. US can provide a safe, non-invasive means to deliver drugs or genes to cells *in vivo* while providing concurrent imaging in a theranostic configuration to identify and target specific organs or tissues accurately.

First demonstrated in the 1980s[41] as a way to target and enhance drug delivery, particularly in the presence of ultrasound contrast agent (USCA) MBs, UmTDD research has been progressing for over two decades, generating a wealth of literature on applications to increase uptake of a wide range of drugs and genes.[42,43] Relative to other available alternatives, UmTDD is the only viable, non-invasive way to repeat treatment to increase therapeutic index for disease management without the limitations of viral vectors, direct injection or electroporation and interesting preclinical results have been achieved *in vivo*. This is particularly important when considering its combination with CE. However, delivery rates have remained low and unpredictable, limiting progress towards clinical application. In the absence of an in-depth understanding of the mechanisms that contribute to cellular uptake during UmTDD, a means to improve local delivery is highly desirable.

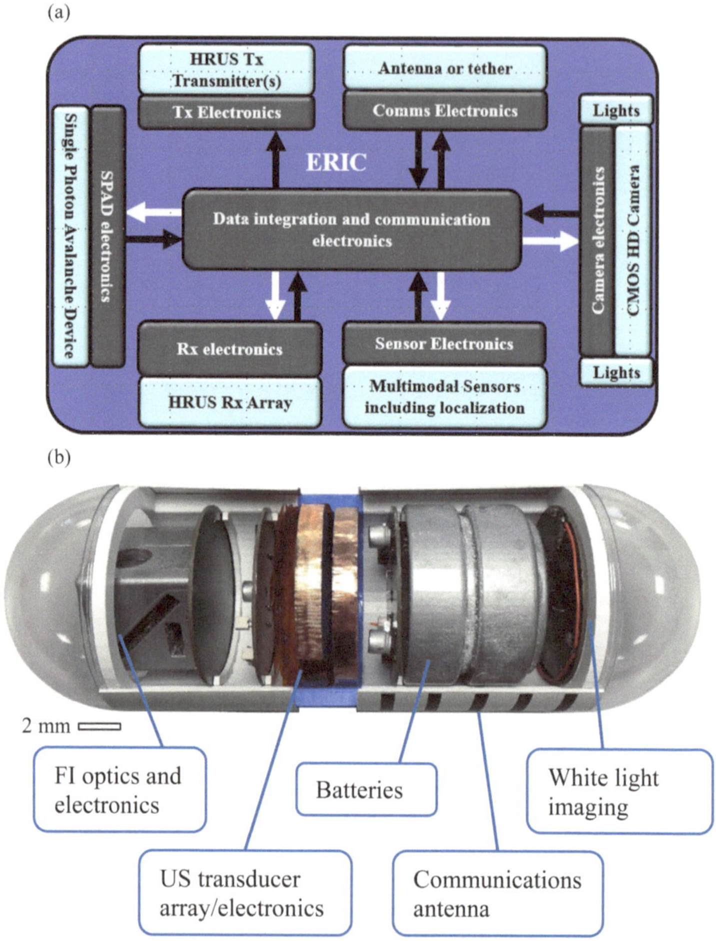


**Figure 8.3** (a) Schematic showing the components in a multimodality capsule including μUS and (b) photograph of a 10x scale model showing the layout of the components within the shell.

In this section, the main principles of UmTDD are discussed, with particular emphasis on the relevance of UmTDD realised with CE.

### 8.4.1 Ultrasound Delivery

Sonoporation is defined as the temporary permeabilization of a cell membrane and related intracellular uptake of molecules with the assistance of US, often in the presence of MBs. It can be applied to target and enhance uptake of drugs and genes into specific cells and tissues with drug or DNA associated with the MBs. The MBs may act as carrier, or the drug may be incorporated within the MB.[44] MBs can also enhance drug uptake when administered alongside the drug in suspension; while it is not necessary to load the drug into the MBs, it is necessary that the drug is in the immediate vicinity of the insonated MBs.[45] In general, the plasma membrane pore formed as a result of the insonated MB cell interaction is non-lethal, with pore diameters ranging from several tens to a few hundreds of nm, or even up to 2–3 μm,[46] allowing the cell to reseal and recover.[47] Most studies affirm that cell membrane resealing occurs within milliseconds to seconds[47] but can take as long as 24 hours.[46]

Sonoporation is generally considered to occur as a result of acoustic mechanisms, but thermal and chemical effects of US can also increase cell membrane permeability and molecular uptake.[47] Rupture of the cell membrane allows molecules, including drugs and ions, to enter by passive diffusion. In particular, if the concentration of $Ca^{2+}$ ions is higher outside the cell, the massive influx which can result can increase endocytosis and subsequent molecular uptake.[48,49] Several studies have shown endocytosis induced by US with MBs, with a smaller number demonstrating similar effects using US alone. One example is the caveolae-mediated endocytosis reported by Lioneiti *et al.*[50] Other studies have shown that mechanical forces induced by US can stimulate endocytosis; where cell plasma membrane deformation occurs, it may be sensed by mechanosensitive structures in the cell leading to both endocytosis and exocytosis.[47]

#### 8.4.1.1 *Hyperthermia*

US at high frequency and intensity can be used to heat tissues. If the US beam is focused into a small volume, sufficiently high temperature gradients can be achieved to induce useful thermal effects for therapy. These include tissue ablation, activation of liposomes containing drugs and heating of specifically targeted tissue to affect cell membrane integrity. Indeed, US can lead to sufficient local temperature increases to change the physicochemical state of the phospholipid bilayer and alter the permeability of cell membrane.[42]

#### 8.4.1.2 *Cavitation*

Cavitation refers to the oscillation of MBs of endogenous gas under activation by an US pressure field. In medical US, gas present in tissues and blood is forced to aggregate during the low pressure rarefaction of the oscillating pressure wave and subsequently undergoes radial oscillation. 'Stable cavitation' indicates that the gas bubbles remain nominally intact and are not ruptured during oscillation and 'inertial caviation' refers to the forced implosion of bubbles resulting in a massive local deposition of energy. Both forms can generate local effects in cells and tissues, including micro-streaming, local shear forces and membrane rupture. In the absence of exogenous agents to seed the MBs, high US intensities are needed to generate cavitation in tissues and this can cause significant, and more importantly, not highly controlled tissue damage.

Microstreaming induced by stable cavitation sets up liquid flow and if this occurs in proximity to a cell, shear stresses can be generated in the range 100–1000 Pa, depending on the US parameters.[47] For context, the shear stresses associated with blood flow are estimated at 0.1–4 Pa. The relatively high shear stress generated by cavitation may then lead to membrane rupture and can also result in bulk streaming, where fluid motion occurs in the direction of US propagation. In this case, MBs may be pushed in the direction of the US, a potential drug targeting mechanism which can also lead to membrane disruption.[51] Finally, MB oscillation close to cells can cause pushing and pulling forces during the expansion and compression phases respectively. Hence processes associated with stable cavitation do not need MB disruption to provoke membrane rupture.[52]

Inertial cavitation occurs at higher insonation pressures and provokes violent MB collapse capable of generating shock waves or liquid microjets, the former by violent symmetrical MB contraction and the latter as a result of aspherical collapse due to proximity to a boundary such as a cell membrane.[45] Both generate highly concentrated forces capable of perforating cell membranes. Inertially-cavitating MBs do not need to be in direct contact with the cell membrane to produce pores though several studies confirm that the maximum distance between cell and MBs should not exceed the microbubble diameter.[47]

Several membrane repair processes following MB pore formation have been identified. In general, small pores are self-sealed, while the sealing of larger pores involves exocytosis.[47] This may be employed by the cell either to decrease the membrane tension to facilitate spontaneous resealing, or to form patches that fuse with the plasma membrane and reseal the disruption.

### 8.4.2 Ultrasound-driven Microbubble Delivery

USCA MBs were first produced commercially to enhance diagnostic imaging in the late 1990s. More recently, their ability to modulate and enhance US fields has led to their investigation for US therapy. MBs allow the bioeffects of ultrasound to be generated in tissue at reduced intensities, within the range of standard clinical imaging systems. This has important implications for TCE with US, where the maximum intensity will be low because of the limited energy and volume for the US device.

#### *8.4.2.1 The Physics of Microbubbles*

Use of US and MBs for UmTDD is an active area of research and significant progress has been made in understanding the physics of MB oscillations and their interaction with tissues and cells. USCAs range in size from 1–10 μm in diameter and consist of a core of biologically inert gas, commonly a fluorocarbon, stabilised by a lipid, protein or polymer shell. At approximately the same as a typical red blood cell (<8 μm), microbubbles remain within the vasculature, though, in some cases, able to traverse capillaries, and do not carry the risk of embolism presented by adding gas bubbles to the blood stream. As a result of their use to enhance contrast in diagnostic US imaging, several commercial agents exist[51,53] and many research groups have developed their own formulations tailored to specific applications.

When driven to oscillate at relatively low US intensities, MBs can enhance the generation of local bioeffects by inducing poration and microstreaming[42,54] causing porosity in the microvasculature[55,56] and cell membranes.[57] The effects generated do not cause long-term damage according to present knowledge and, in each case, the vessel or cell recovers as noted previously, trapping material which may have entered passively or been swept into the cell.[57,58]

Significant advances have been made to improve understanding of the physics of MB oscillation. For example, models to describe and predict MB oscillation and shell behaviour have been developed, in many cases based on and validated through high speed optical observation.[59–61] A useful review of theoretical shell models can be found in Doinikov *et al.*[62] In turn, this has advanced understanding of how such oscillations, in close proximity to cells, may generate biological effects. In addition to inertial cavitation,[63] development of shear stresses and microstreaming[64,65] are now known to influence cell membrane behaviour and uptake mechanisms.

## 8.4.3 Targeting

Ultrasound is itself a form of targeting since only tissues exposed to the US field will exhibit enhanced uptake. This is particularly relevant for TCE where tissues can, in theory, be targeted precisely because of their proximity to the device.

Acoustic radiation force refers to the non-oscillatory, time-averaged force exerted by US.[66] It can be used to manipulate small particles, pushing them in the direction of propagation or translating them against the direction of propagation, as has been demonstrated recently.[67] The ability to manipulate particles to position them in proximity to tissue or a cell surface has obvious advantages in UmTDD since particles to be delivered must be in close proximity to the cell membrane. Additionally, the significant acoustic impedance contrast between their gas cores and the surrounding blood pool makes MBs particularly susceptible to acoustic radiation forces, though the magnitude of the effect is highly dependent on the MB diameter.[68]

The addition of molecular probes such as antibodies to specific proteins offers further potential for UmTDD to be applied to specific tissues or regions of disease. There are several ways to add such molecules to MB shells. Whilst this is possible with protein and polymer shells, work to date has been dominated by methods suited to lipid-shelled MBs. Mulvana *et al.* have produced a review of the adaptation of MBs to accept such payloads.[69]

### 8.4.4 Bioeffects and Delivery Mechanisms Using Ultrasound

As a result of early work demonstrating that MBs driven to implosion could generate pores in cell membranes,[70] most studies which followed focused on generating inertial cavitation for drug delivery. Improved delivery rates were achieved through parametric investigation of the US and MB parameters, with lower US frequencies and higher intensities shown to provide small improvements.[71] However, the therapeutic potential of increasing US intensity on MB dose is limited; higher intensities may cause undesirable tissue damage and reduce the persistence in the circulation of MBs capable of generating effects on cells.[44] Neither can the MB dose be raised indefinitely due to the increasing risk of embolism.

Recently, and much less obviously, it has been demonstrated that the mode of therapeutic agent delivery is influenced by the US intensity. Cock *et al.* have shown that uptake is mostly characterised by a diffuse presence of the agent within the cytoplasm at low intensities, while delivery is seen to be punctuate at higher intensities.[72] It is surmised that low intensity leads to uptake dominated by endocytosis and delivery at higher intensities is through inertial cavitation. Several other groups have provided further detail, indicating that clathrin-dependent endocytosis is particularly vulnerable to UmTDD.[47,73]

One innovative approach to improving UmTDD efficiency has been to develop means of delivering short-lived MBs directly to the treatment site using intravascular devices,[74] establishing a precedent for localised delivery of MBs, US and therapeutics to realize UmTDD.

Utilisation of endocytosis as a potential delivery route represents an opportunity not addressed so far in UmTDD. However, it is clear that the intended delivery route has significance for the design of the USCE device, not least in terms of the intensity and duration of the US exposure, thus the persistence of the device, but also in terms of the possible payload and therefore the restrictions on drug formulation.

### 8.4.5 Carriers, Agents and Their Uses

In addition to the practice of delivering a drug alongside MBs in a combined suspension, a large body of work has demonstrated incorporation of the drug directly into the MBs or packaged such that it can be attached directly. Attachment can improve co-localisation of the drug to the MB and packaging and, with a liposome[75] or polymer shell,[76] can reduce off target delivery and systemic toxicity. There are several ways to add drugs to MBs, including direct incorporation, electrostatic incorporation, and chemical conjugation of the drug or packaged drug, Figure 8.4. Direct incorporation of the drug into the shell is simplest[77] and albumin-shelled MBs are particularly well suited to this.[78] By selecting appropriately charged shell materials, hydrophilic drugs can be incorporated directly within the shell of the MB *via* electrostatic force.[79,80] Finally, drugs packaged within liposomes and as nanoparticles can be conjugated directly to the shell either electrostatically or chemically,

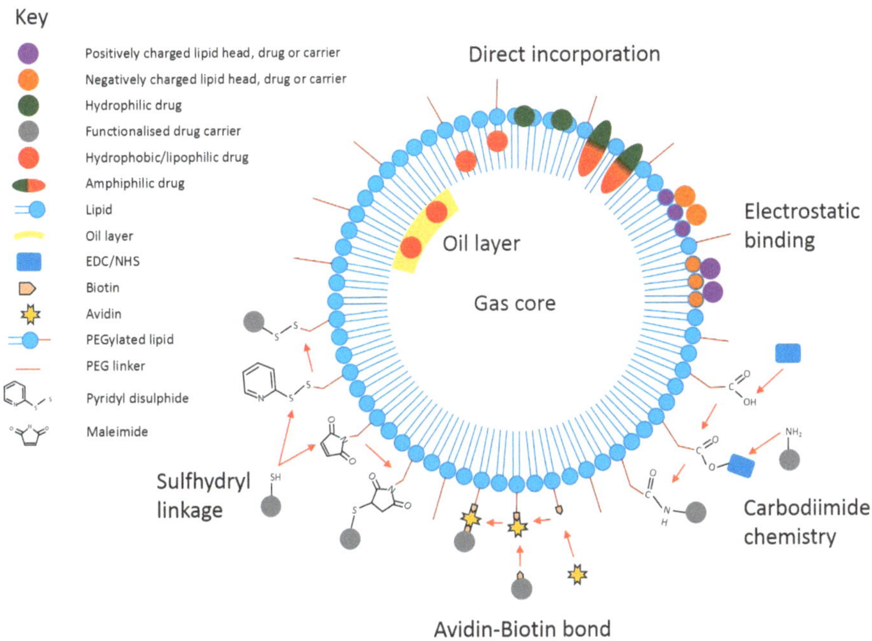


**Figure 8.4** Potential loading techniques for lipid microbubbles shown schematically, applicable to both protein and polymer microbubbles. Hydrophobic/lipophilic, hydrophilic or amphiphilic interactions between the drug and shell allow direct incorporation. Charge differences between the shell molecules and the drug or drug carrier allow electrostatic binding. Finally, various chemical binding processes are possible with functionalisation of the drug carrier and/or the shell. Abbreviations: EDC: 1-Ethyl-3-(3-dimethylaminopropyl) carbodiimide, NHS: *N*-hydroxysuccinimide, PEG: polyethylene glycol. © 2016 IEEE. Reprinted, with permission from H. Mulvana, R. J. Browning, Y. Luan, N. de Jong, M. X. Tang, R. J. Eckersley and E. Stride, Characterization of Contrast Agent Microbubbles for Ultrasound Imaging and Therapy Research, *IEEE Trans. Ultrason. Ferroelectr. Freq. Control.*, 2017, **64**(1), 232–251, DOI: 10.1109/TUFFC.2016.2613991.

where the nature of the chemical linkage depends on the chemistry of the shell and the drug or drug packaging. For example, a commonly used chemical link is avidin-biotin[81] and as such the attachment chemistry for use with both lipid and protein shelled MBs is well described.[82]

Selection of the appropriate carrier for a drug will depend on its chemistry, size and requirements for eventual bioavailability upon delivery. For example, using chemical attachment *via* avidin-biotin to attach liposomes to MBs is considered to improve the loading capacity of microbubbles through a 35-fold increase in loading surface area.[83] It is also important to consider how the addition of materials to the MB shell will influence its acoustic characteristics and stability within and outside the body, and how any attachment chemistry will influence drug delivery or biological safety.[69]

## 8.5 Theranostic Ultrasound Capsule Endoscopy

Capsule endoscopy has emerged in the past 15 years as a diagnostic modality for the GI tract. However, despite the many options available for TDD, including the UmTDD described in Section 8.4, its potential for therapy has been limited to the proof-of-concept studies outlined in Section 8.3.2. The present section considers how UmTDD can be applied to theranostic CE and its potential to overcoming outstanding challenges. Work carried out in the field is reviewed and SonoCAIT is used as an example to illustrate potential future capabilities for theranostic USCE.

### 8.5.1 Ultrasound Approaches for UmTDD in Capsule Endoscopy

Ultrasound is pivotal in therapeutic treatments such as thermal ablation and UmTDD, as described in Section 8.4. There are two reasons why thermal ablation is not presently viewed as relevant in the GI tract. First, the dimensions of the ellipsoidal lesions generated in hyperthermic focused ultrasound surgery (FUS) carry a major risk of perforating the wall of the GI tract. Additionally, the necessary US source is likely to be both too large and to need too much power to be viable. Thus, only UmTDD is considered here.

As noted in Section 8.4, US methods were first proposed for TDD some decades ago, with the phenomena of cavitation associated with a range of non-thermal biological effects, such as sonoporation and changes in membrane permeability.[84] An advantage of UmTDD over conventional means of drug delivery is its ability to target delivery of therapeutics locally, allowing a higher dose to be administered, hence increasing treatment efficacy while reducing toxicity to healthy tissue.[85] Another advantage is that focused US can enhance therapeutic effects, as originally discussed by Miller *et al.*, allowing therapeutics to be delivered deep into tissue.[84]

#### *8.5.1.1 Extracorporeal and Intracorporeal Ultrasound*

The implementation of US in UmTDD can be achieved extra- or intracorporeally, with the former presently most widely used. US can produce mechanical and thermal effects deep within tissues non-invasively and can be focused to a region approximately 2 mm in diameter.[86] This has been applied in FUS to ablate tumours in different organs hyperthermally.[87] Applying US similarly in TDD could also improve treatment, as suggested by the finding that mild hyperthermia can release therapeutic agents from thermo-sensitive carriers.[88] Using US in this manner would involve encapsulating highly potent drugs in a thermosensitive carrier for local release at a target site by applying focused US. Whilst thermal ablation requires high intensity focused US (HIFU) to produce sufficiently high temperatures, mild hyperthermia for TDD requires much lower intensities, making it viable to include an appropriate US source in an ingestible capsule.

Currently, UmTDD systems typically comprise an US transducer placed extracorporeally, coupled to the skin with US gel or water and directed at the target.[89] The main advantages of extracorporeal transducers are that they are minimally invasive and do not have size constraints or require power budgeting. However, this approach also has several drawbacks. The US beam can be adversely affected by tissue interfaces including bone, which may interfere with the beam and produce hotspots.[90] Patient movement must also be considered, since this could cause focussing of the transducer into an unwanted area. These problems can be overcome by complex arrangements that permit continually adjustment of the physical position of the US transducer or a multi-element array with transmission from each element defined to ensure constant focus on the target.

Some of the difficulties of extracorporeal US are overcome by intracorporeal US, such as an US source within a CE device to provide access to the entire length of the GI tract. The need to miniaturise the US source to fit into an ingestible capsule is made possible by the low power requirements of UmTDD. The capsule configuration also eliminates issues relating to patient movement and potential obstruction by bone. Remaining challenges include the fact that the digestive process generates gas, making satisfactory US coupling potentially difficult, the possibility that robotics may be required to position and localize the capsule, and the need for expensive miniaturised bespoke electronics.

An added advantage of US-based TCE is that an ingestible focused-US capsule could also incorporate imaging modalities, *e.g.* high-definition video or US imaging, to localize and/or monitor drug delivery in a truly theranostic configuration. As already demonstrated, capsules can be configured to include a reservoir of drugs for release at a target location. Furthermore, the US source could enhance therapeutic effects by using cavitation and acoustic radiation forces to push therapeutics into the bowel wall, as outlined in Section 8.4.

This approach to theranostic CE is exemplified by the SonoCAIT capsule[4] which contains a miniature US transducer, video camera and therapeutics channel, Figure 8.5, combined in a capsule with ingestible dimensions, 10 mm × 30 mm. Its purpose is to prove the concept of an ingestible, theranostic capsule with potential clinical uses. The concave, spherically-focused US source has an outer diameter of 6 mm and length of 4 mm, allowing it to fit within a capsule. Testing has shown that it can produce US pressures suitable for reversible sonoporation and acoustic radiation forces sufficient to direct MBs physically.

## 8.5.2 Imaging for Ultrasound-mediated Targeted Drug Delivery

As mentioned in the previous section, features such as gas-filled pockets in the GI tract are highly echogenic, significantly affecting the US beam. This leads to a requirement to image the target regions and to monitor UmTDD to

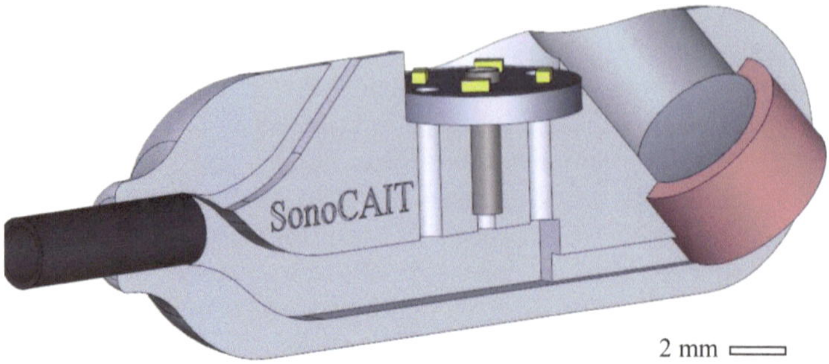


**Figure 8.5** Computer-aided rendering of SonoCAIT (cabling omitted for clarity). Components from right to left: Focused-US transducer, illumination source with four light-emitting diodes, CMOS camera, therapeutics channel and tether. The capsule measures 30 mm in length and 10 mm in diameter, similar in size to commercial capsule endoscopes.

minimise potential adverse effects. Imaging also provides an opportunity to monitor immediate local effects such as cavitation and can be used to identify the position of the capsule and the US focus, as outlined below.

#### *8.5.2.1 Localisation*

As discussed in Section 8.3, VCE is a promising diagnostic tool for GI conditions. Combining it with an US source for UmTDD would allow identification of a target site and monitoring of therapeutic delivery using a combination of forward and side facing cameras. Both configurations are already available in VCE and SonoCAIT is designed to demonstrate the usefulness of a confocal side-facing camera to identify the correct mucosal region for treatment, Figure 8.5.

Another method to monitor the position of a CE device is by measuring pH as it moves through the GI tract, with pH changing from 1.0–2.5 in the stomach to 6.6 in the small intestine and 7.0 in the colon.[27] Although, pH alone will not suffice to report capsule position with the accuracy needed for UmTDD, it may provide sufficient detail when combined with an imaging modality such as video or μUS.

μUS refers to US imaging using frequencies greater than 20 MHz, the common upper limit for conventional diagnostic ultrasound. Such frequencies provide higher spatial resolution but reduced penetration depth. Nonetheless, compared with video imaging, μUS can provide images of the interior of tissues and the relatively small thickness of GI tract wall noted in Section 1.2 means it is fully accessible for imaging by μUS. Furthermore, μUS has been shown to identify small changes in tissue structure in murine models of colon cancer.[3] Thus, combining μUS with UmTDD is promising for theranostics in the GI tract.

Another option for co-localizing the capsule and pathology in the GI tract is provided by MRI or CT. For example, a theranostic pathway could be identified in which the patient ingests a capsule, pH monitoring is used to determine when it leaves the stomach, the patient is then monitored with MRI or CT, and treatment is initiated when the capsule reaches the relevant region.

#### *8.5.2.2 Treatment Monitoring*

Magnetic resonance imaging-guided FUS (MRgFUS) refers to FUS taking place in an MRI scanner. The use of MRI in therapeutic US applications allows better control and monitoring of the pharmacokinetics and pharmacodynamics of a drug and of its biodistribution.[91] Specifically, MRI can report therapeutic success accurately.[92] MRI can also measure temperature variation directly, which can report on the release of therapeutic agents from thermosensitive carriers. This is promising for theranostic USCE. However, a technical issue is the need for the USCE to be MRI-compatible. Additionally, the use of an MRI scanner, with associated skilled staff, is a major expense.

In contrast to MRI, US imaging systems are highly portable and much less expensive. During treatment, it is possible to use US imaging to detect cavitation and the location where it occurs.[93] On the other hand, direct temperature measurement is not yet possible with US but reports of US thermometry based on changes in attenuation or speed of sound have provided early results.[94]

μUS imaging has been introduced as a means to diagnose and localise areas of disease using CE. It also presents a treatment monitoring option using the capsule itself, rather than an extracorporeal imaging system. Furthermore, it could be used to monitor therapeutic delivery to a target location. Quantitative US (QUS) may also be used to monitor treatment effects. Quantifying the properties of tissue before and after treatment, QUS could report changes in tissue structure indicating if treatment has been effective. An advantage of μUS and QUS is that imaging is delivered by the capsule itself, thus freeing patients to continue their daily lives in the same manner as existing VCE.

Infrared (IR) imaging is a further option.[95] It can be used to detect temperature changes in the GI tract, such as in areas of the bowel affected by IBD. It thus holds promise to aid the identification of sites requiring treatment and for post-treatment monitoring of alterations in inflammation in response to therapy.[96] IR cameras can be miniaturised for CE deployment, making it possible to achieve such imaging in the GI tract itself.

## 8.6 Realising UmTDD in the Gastrointestinal Tract

Having discussed mechanisms and tools, this section now considers current detection and treatment methods for the major diseases of the GI tract to illustrate the potential impact of UmTDD with theranostic CE.

IBD can lead to cancer, and itself carries significant morbidity and some mortality, so both are important targets for new clinical interventions. Patient outcome in inflammatory and neoplastic diseases of the GI tract greatly benefits from early detection and localized treatment, which is the target of theranostic USCE with UmTDD. Early detection allows treatment before significant tissue damage has occurred, making success much more likely and local delivery of drugs reduces side effects associated with the currently common systemic administration. It also allows much larger and thus potentially more effective doses to be delivered directly to affected areas.

### 8.6.1 Barrett's Oesophagus

Barrett's oesophagus is usually confirmed visually using upper GI endoscopy together with detailed histological examination of endoscopically-obtained biopsies from areas representing the entire circumference of the oesophagus.[97] However, early stage inflammation is difficult to detect visually and small lesions are easily missed in the circumferential sampling approach, illustrating the need for more reliable detection methods that can measure tissue changes more comprehensively. The conversion of the acid- and inflammation-induced tissue changes associated with Barrett's oesophagus to cancer in 10–15% of cases[98] means there is a clear need not only to detect the disease but also for reliable identification of patients with the potential to develop cancer.

The current approach to treat Barrett's oesophagus is to reduce acid reflux using proton pump inhibitors and dietary control and/or to remove affected areas using radio ablation or surgery. Recognized risk factors for Barrett's oesophagus include gender, age, body weight and distribution, and family history.[98] The same factors, together with the size and grade of dysplasia are also associated with the risk of developing cancer and are used to decide intervals for regular surveillance. Nonetheless, many patients still develop oesophageal cancer, a disease associated with a five-year survival of no more than 15%.[99]

The use of CE in Barrett's oesophagus appears most promising for diagnosis, particularly with μUS, but may also be indicated for localized drug delivery, with both potentially amenable to the use of tethered capsules.[100]

### 8.6.2 IBD: Crohn's Disease and Ulcerative Colitis

CD disease is an idiopathic inflammation that can affect any region of the GI tract from the mouth to the anus. Patients most commonly experience abdominal pain and diarrhoea, but can have other symptoms including rectal bleeding, weight loss, nausea, vomiting and vitamin deficiencies. It is thought to be the result of an imbalance between pro- and anti-inflammatory mediators but the underlying cause is unknown, making it difficult to diagnose. The abundance of inflammatory markers in the blood and a variety of imaging techniques such as CT, MRI, barium contrast studies, and abdominal radiography can help detection. Endoscopy is another feasible approach, hindered by its inability to access the small intestine easily. Importantly, only overt and progressed disease can be detected by these means. Therapeutic intervention aims to induce sustained remission and covers a wide range including corticosteroids, anti-inflammatories, anti-tumour necrosis factor (TNF) agents, diet, antibiotics, immune suppressants, probiotics, parasites that can divert the immune system, and surgery. However, these treatments rarely achieve permanent cure and the inflammation can recur anywhere.

It is highly likely that successful treatment would be easier if inflammatory changes could be detected during disease development. Thus there is great need to detect local inflammation early before extensive tissue damage has occurred. In addition, treating inflammation locally would allow delivery of higher doses of therapeutics while greatly reducing systemic side effects. Both these issues well match the potential of USCE and UmTDD outlined here.

Ulcerative colitis is characterized by inflammatory changes in the colon associated with changes in the complement of immune cells present in the colonic wall. Inflammation is followed by the accumulation of connective tissue and in-growing neovasculature commonly found in the granulation tissue associated with wound healing. Symptoms and detection are similar to those in Crohn's disease, but because it is limited to the colon, ulcerative colitis is more reliably detected by colonoscopy. Treatment, too, is similar to that for Crohn's disease, and includes corticosteroids, anti-inflammatories, diet, antibiotics, immune suppressants, and surgery. Surgery can be curative if the colon is removed but this drastic measure significantly reduces quality of life.

The ability to detect earlier stages of ulcerative colitis and to treat affected areas locally earlier has the potential to significantly improve patient outcome and wellbeing. CE generally is considered as having a future role to play in the colon and UmTDD may have significant value. However, at least initially, this may come through implementation of a tethered theranostic capsule inserted through the anus and manipulated robotically to reduce the discomfort associated with conventional colonoscopy.

### 8.6.3 Colon Cancer

Colon cancer is the second most common cause of cancer deaths in the United Kingdom.[101] Inflammation can be a predisposing condition and inflammatory changes are common to early and late stage tumours of the GI tract. Colon cancer is commonly detected by colonoscopic screening prompted by a positive faecal occult blood test. Symptoms include iron deficiency, rectal bleeding and abdominal pain, and changes in bowel habits.[102] Information about the extent and staging of the disease can be obtained using barium studies, CT, MRI, positron emission tomography, and US used with contrast agents. These approaches not only provide detailed information about the dimensions and localization of the primary tumour in the colon but, where relevant, may also reveal metastases, which are most common in the liver, lung, and bone.[101]

The most common treatment for colon cancer is surgery.[103] If lesions are small and their position allows, they can be removed during colonoscopy. If patients are not suitable for surgery due to co-morbidities, radiofrequency ablation or cryotherapy may be employed.[103] Chemotherapeutic agents to treat colon cancer include 5-fluorouracil (5-FU), capecitabine, tegafur, oxaliplatin, and irinotecan.[103] In addition, a number of biological agents are now available to target specific signalling pathways that are commonly deregulated in colon cancer.[104]

Surgery alone or in combination with chemotherapy can be curative, but relapse is not uncommon. At present, there are no reliable biomarkers to predict relapse or therapeutic response. Taking into account the common side effects of chemotherapy such as severe nausea, extreme fatigue, immune deficiencies, and diarrhoea, and the reduced quality of life associated with a stoma, which is frequently required to compensate for surgical removal of colonic sections, improved detection and treatment modalities are needed, including the ability to monitor response to therapy. The ability to deliver therapy locally and monitor tissue response again makes a theranostic combination of USCE and UmTDD attractive.

## 8.7 Future Theranostic and Image-guided Opportunities for Ultrasound Capsule Endoscopy

Present methods to detect inflammatory and neoplastic diseases of the GI tract only reveal disease already associated with significant tissue damage, making effective treatment difficult. Furthermore, many existing pharmaceutical treatments are associated with significant side effects due to systemic administration. To be successful, new approaches should deliver earlier detection, targeted delivery to reduce systemic effects, complete remission, and reduced cost, and should ultimately aid in prevention. Specifically, this means for Barrett's oesophagus, fewer conversions to neoplasia and cancer; in Crohn's disease, improved quality of life through more sustained remission; in ulcerative colitis, long term cure without the need for surgery; and in colon and other forms of GI cancer, earlier detection of primary and recurring tumours and local treatment with reduced side effects. Theranostic CE, particularly with US-based diagnosis and treatment offers one pathway forward.

As discussed in this chapter, VCE, μUS and pH are the most promising modalities for localisation of CE through different regions of the GI tract. Changes in pH provide information about the capsule's passage through the GI tract, with VCE and μUS allowing accurate localisation of the treatment site. μUS can also be used to monitor treatment effects with QUS to measure changes consequent upon treatment. IR imaging may also be useful if the necessary devices can be miniaturised for CE deployment. The advantage of using μUS, pH, VCE and IR imaging is that they can be housed within the CE device itself. This means patients can continue their daily lives except during the brief period of ingestion. It also matches the contemporary trends towards cost reduction since no expensive scanner nor staff are required and it alleviates stress for the patient since CE is less invasive even than conventional endoscopy or colonoscopy. Therefore, future work should aim to combine different imaging and other monitoring modalities with UmTDD to produce a theranostic capsule that can identify, localise, treat, and monitor diseases in the GI tract.

Whilst the goals that have been outlined can be based on adaptations of existing diagnostic techniques such as QUS and therapeutic approaches such as pharmaceuticals activated thermally or by cavitation, a longer-term goal potentially accessible with US-enabled TCE relates to nanoparticles and small molecules. Delivering such reagents into living cells in tissues could allow monitoring of specific molecular processes. Small molecule probes are now available that can measure specific enzymatic activities, *i.e.* ubiquitylation,[105,106] by detecting the abundance of biological molecules modified by specific enzymes to allow reporting on the activity of specific signalling pathways in cells. Being able to deliver such reagents into cells in tissue could revolutionize the ability to screen drugs in physiologically relevant settings such as the GI tract and could allow measurement of the effect of therapies on specific signalling.

## Acknowledgements

The financial support from the UK Engineering and Physical Sciences Research Council under Grant EP/K034537 (Sonopill Programme) is gratefully acknowledged.

## References

1. E. V. Loftus, *Gastroenterology*, 2004, **126**, 1504–1517.
2. J. Ferlay, E. Steliarova-Foucher, J. Lortet-Tieulent, S. Rosso, J. Coebergh, H. Comber, D. Forman and F. Bray, *Eur. J. Cancer*, 2013, **49**, 1374–1403.
3. A. Fatehullah, S. Sharma, I. Newton, A. Langlands, H. Lay, S. Nelson, R. McMahon, N. McIlvenny, P. Appleton and S. Cochran, *Sci. Rep.*, 2016, **6**, 29570.
4. F. Stewart, B. Cox, J. Vorstius, A. Verbeni, Y. Qiu and S. Cochran, *Ultrasonics Symposium (IUS)*, 2015, pp. 1–4.
5. D. A. Warrell, T. M. Cox, J. D. Firth and E. J. Benz, *Oxford Textbook of Medicine: Vol. 1*, Oxford University Press, 2003.
6. N. A. Wright and M. Alison, *The Biology of Epithelial Cell Populations*, Oxford University Press, USA, 1984.
7. C. Tomasetti and B. Vogelstein, *Science*, 2015, **347**, 78–81.
8. W. H. Organization, *GLOBOCAN 2012: Estimated Cancer Incidence, Mortality and Prevalence Worldwide in 2012*, 2013, accessed 27th February 2017.
9. G. Iddan, G. Meron, A. Glukhovsky and P. Swain, *Nature*, 2000, **405**, 417.

10. C. Van de Bruaene, D. De Looze and P. Hindryckx, *World J. Gastrointest. Endosc.*, 2015, **7**, 13.
11. G. Imaging, *PillCam Capsule Endoscopy*, 2015, accessed 18th July 2016.
12. A. Koulaouzidis, E. Rondonotti and A. Karargyris, *World J. Gastroenterol.*, 2013, **19**, 3726–3746.
13. G. S. Raju, L. Gerson, A. Das and B. Lewis, *Gastroenterology*, 2007, **133**, 1697–1717.
14. S. L. Triester, J. A. Leighton, G. I. Leontiadis, S. R. Gurudu, D. E. Fleischer, A. K. Hara, R. I. Heigh, A. D. Shiff and V. K. Sharma, *Am. J. Gastroenterol.*, 2006, **101**, 954–964.
15. P. Sharma, S. Wani, A. Rastogi, A. Bansal, A. Higbee, S. Mathur, R. Esquivel, L. Camargo and R. E. Sampliner, *Am. J. Gastroenterol.*, 2008, **103**, 525–532.
16. A. Wang, S. Banerjee, B. A. Barth, Y. M. Bhat, S. Chauhan, K. T. Gottlieb, V. Konda, J. T. Maple, F. Murad and P. R. Pfau, *Gastrointest. Endosc.*, 2013, **78**, 805–815.
17. A. Ross, S. Mehdizadeh, J. Tokar, J. A. Leighton, A. Kamal, A. Chen, D. Schembre, G. Chen, K. Binmoeller and R. Kozarek, *Dig. Dis. Sci.*, 2008, **53**, 2140–2143.
18. T. Rokkas and Y. Niv, *Eur. J. Gastroenterol. Hepatol.*, 2012, **24**, 303–308.
19. E. Rondonotti, C. Spada, D. Cave, M. Pennazio, M. E. Riccioni, I. De Vitis, D. Schneider, T. Sprujevnik, F. Villa and J. Langelier, *Am. J. Gastroenterol.*, 2007, **102**, 1624–1631.
20. B. W. Chwirot, S. Chwirot, W. Jedrzejczyk, M. Jackowski, A. M. Raczyńska, J. Winczakiewicz and J. Dobber, *Lasers Surg. Med.*, 1997, **21**, 149–158.
21. N. Uedo, K. Higashino, R. Ishihara, Y. Takeuchi and H. Iishi, *Dig. Endosc.*, 2007, **19**, S134–S138.
22. B. Mayinger, M. Jordan, T. Horbach, P. Horner, C. Gerlach, S. Mueller, W. Hohenberger and E. G. Hahn, *Gastrointest. Endosc.*, 2004, **59**, 191–198.
23. M. Kfouri, O. Marinov, P. Quevedo, N. Faramarzpour, S. Shirani, L. W.-C. Liu, Q. Fang and M. J. Deen, *IEEE J. Sel. Top. Quantum Electron.*, 2008, **14**, 226–234.
24. M. A. Al-Rawhani, J. Beeley and D. R. Cumming, *Sci. Rep.*, 2015, **5**, 18591.
25. S. Roman, F. Mion, F. Zerbib, R. Benamouzig, J. Letard and S. B. des Varannes, *Endoscopy*, 2012, **44**, 270–276.
26. E. A. Johannessen, L. Wang, L. Cui, T. B. Tang, M. Ahmadian, A. Astaras, S. W. Reid, P. S. Yam, A. F. Murray and B. W. Flynn, *IEEE Trans. Biomed. Eng.*, 2004, **51**, 525–535.
27. D. Evans, G. Pye, R. Bramley, A. Clark, T. Dyson and J. Hardcastle, *Gut*, 1988, **29**, 1035–1041.
28. M. Sivak, *Gut*, 2006, **55**, 1061–1064.
29. F. Memon, G. Touma, J. Wang, S. Baltsavias, A. Moini, C. Chang, M. F. Rasmussen, A. Nikoozadeh, J. W. Choe and E. Olcott, *Ultrasonics Symposium (IUS)*, 2016, pp. 1–4.
30. J. H. Lee, G. Traverso, C. M. Schoellhammer, D. Blankschtein, R. Langer, K. E. Thomenius, D. S. Boning and B. W. Anthony, *Ultrasonics Symposium (IUS)*, 2014, pp. 734–737.

31. C. Liu, Y. Yan, L. Sun, Y. Chen, J. Dai and W. Qiu, *Ultrasonics Symposium (IUS)*, 2013, pp. 2045–2048.
32. G. Cummins, H. Lay, B. Cox, V. Seetohul, Y. Qiu, F. Stewart, J. Faerber, V. Mitrakos, M. Al Rawhani and J. Beeley, *Proceedings of the 6th Joint Workshop on New Technologies for Computer/Robot Assisted Surgery*, 2016, pp. 12–14.
33. W. H. Crosby, U. Army and H. W. Kugler, *Am. J. Dig. Dis.*, 1957, **2**, 236–241.
34. J. Smith, J. Mayberry, I. Ansell and R. Long, *Clin. Chim. Acta*, 1989, **183**, 317–321.
35. M. Simi, G. Gerboni, A. Menciassi and P. Valdastri, *Proceedings of ASME Design of Medical Devices Conference*, 2012, pp. 10–12.
36. B. O. Limited, *BioMe – A Transformation in Gut Health*, 2016, accessed 27th February 2017.
37. I. Wilding, P. Hirst and A. Connor, *Pharm. Sci. Technol. Today*, 2000, **3**, 385–392.
38. A. F. Parr, E. P. Sandefer, P. Wissel, M. McCartney, C. McClain, U. Ryo and G. A. Digenis, *Pharm. Res.*, 1999, **16**, 266–271.
39. S. P. Woods and T. G. Constandinou, *IEEE Trans. Biomed. Eng.*, 2013, **60**, 945–953.
40. P. Valdastri, C. Quaglia, E. Susilo, A. Menciassi, P. Dario, C. Ho, G. Anhoeck and M. Schurr, *Endoscopy*, 2008, **40**, 979–982.
41. J.-M. Escoffre and A. Bouakaz, *Therapeutic Ultrasound: From Biophysics Concepts to Clinical Applications*, Springer International Publishing, 2016.
42. S. Hernot and A. L. Klibanov, *Adv. Drug Delivery Rev.*, 2008, **60**, 1153–1166.
43. K. Ferrara, R. Pollard and M. Borden, *Annu. Rev. Biomed. Eng.*, 2007, **9**, 415–447.
44. J. Alter, C. A. Sennoga, D. M. Lopes, R. J. Eckersley and D. J. Wells, *Ultrasound Med. Biol.*, 2009, **35**, 976–984.
45. H. Yu and L. Xu, *J. Controlled Release*, 2014, **174**, 151–160.
46. Y. Z. Zhao, Y. K. Luo, C. T. Lu, J. F. Xu, J. Tang, M. Zhang, Y. Zhang and H. D. Liang, *J. Drug Targeting*, 2008, **16**, 18–25.
47. I. Lentacker, I. De Cock, R. Deckers, S. C. De Smedt and C. T. W. Moonen, *Adv. Drug Delivery Rev.*, 2014, **72**, 49–64.
48. V. Lariccia, M. Fine, S. Magi, M.-J. Lin, A. Yaradanakul, M. C. Llaguno and D. W. Hilgemann, *J. Gen. Physiol.*, 2011, **137**, 111–132.
49. B. D. M. Meijering, L. J. M. Juffermans, A. van Wamel, R. H. Henning, I. S. Zuhorn, M. Emmer, A. M. G. Versteilen, W. J. Paulus, W. H. van Gilst, K. Kooiman, N. de Jong, R. J. P. Musters, L. E. Deelman and O. Kamp, *Circ. Res.*, 2009, **104**, 679–687.
50. V. Lionetti, A. Fittipaldi, S. Agostini, M. Giacca, F. A. Recchia and E. Picano, *Ultrasound Med. Biol.*, 2009, **35**, 136–143.
51. S. R. Sirsi and M. A. Borden, *Adv. Drug Delivery Rev.*, 2014, **72**, 3–14.
52. R. Karshafian, S. Samac, P. D. Bevan and P. N. Burns, *Ultrasonics*, 2010, **50**, 691–697.

53. M. X. M. Tang, H. Mulvana, T. Gauthier, A. K. P. Lim, D. O. Cosgrove, R. J. Eckersley and E. Stride, *Interface Focus*, 2011, **1**, 520–539.
54. C. M. H. Newman and T. Bettinger, *Gene Ther.*, 2007, **14**, 465–475.
55. H. Chen, A. A. Brayman, W. Kreider, M. R. Bailey and T. J. Matula, *Ultrasound Med. Biol.*, 2011, **37**, 2139–2148.
56. H. Chen, W. Kreider, A. A. Brayman, M. R. Bailey and T. J. Matula, *Phys. Rev. Lett.*, 2011, **106**, 034301.
57. R. S. Leow, J. M. F. Wan and A. C. H. Yu, *Membrane Blebbing as a Recovery Manoeuvre in Site-specific Sonoporation Mediated by Targeted Microbubbles*, 2015.
58. Y. Zhou, R. E. Kumon, J. Cui and C. X. Deng, *Ultrasound Med. Biol.*, 2009, **35**, 1756–1760.
59. A. Bouakaz, M. Versluis and N. de Jong, *Ultrasound Med. Biol.*, 2005, **31**, 391–399.
60. N. de Jong, P. J. A. Frinking, A. Bouakaz, M. Goorden, T. Schourmans, X. Jingping and F. Mastik, *Ultrasound Med. Biol.*, 2000, **26**, 487–492.
61. X. Chen, J. Wang, M. Versluis, N. de Jong and F. S. Villanueva, *Rev. Sci. Instrum.*, 2013, **84**, 063701.
62. A. A. Doinikov and A. Bouakaz, *IEEE Trans. Ultrason. Ferroelectr. Freq. Control*, 2011, **58**, 981–993.
63. P. Prentice, A. Cuschieri, K. Dholakia, M. Prausnitz and P. Campbell, *Nat. Phys.*, 2005, **1**, 107–110.
64. J. Collis, R. Manasseh, P. Liovic, P. Tho, A. Ooi, K. Petkovic-Duran and Y. Zhu, *Ultrasonics*, 2010, **50**, 273–279.
65. A. Novell, J. Collis, A. A. Doinikov, A. Ooi, R. Manasseh and A. Bouakazaz, *Ultrasonics Symposium (IUS)*, 2011, pp. 1482–1485.
66. M. F. Hamilton and D. T. Blackstock, *Nonlinear Acoustics*, Academic Press, 1998.
67. C. E. M. Démoré, P. M. Dahl, Z. Yang, P. Glynne-Jones, A. Melzer, S. Cochran, M. P. MacDonald and G. C. Spalding, *Phys. Rev. Lett.*, 2014, **112**, 174302.
68. H. J. Vos, F. Guidi, E. Boni and P. Tortoli, *IEEE Trans. Ultrason. Ferroelectr. Freq. Control*, 2007, **54**, 1333–1345.
69. H. Mulvana, R. J. Browning, Y. Luan, N. de Jong, M.-X. Tang, R. J. Eckersley and E. Stride, *IEEE Trans. Ultrason. Ferroelectr. Freq. Control*, 2016, **64**, 232–251.
70. K. Tachibana, T. Uchida, K. Ogawa, N. Yamashita and K. Tamura, *Lancet*, 1999, **353**, 1409.
71. A. Rahim, S. L. Taylor, N. L. Bush, G. R. ter Haar, J. C. Bamber and C. D. Porter, *Ultrasound Med. Biol.*, 2006, **32**, 1269–1279.
72. I. De Cock, E. Zagato, K. Braeckmans, Y. Luan, N. de Jong, S. C. De Smedt and I. Lentacker, *J. Controlled Release*, 2015, **197**, 20–28.
73. A. Delalande, C. Leduc, P. Midoux, M. Postema and C. Pichon, *Ultrasound Med. Biol.*, 2015, **41**, 1913–1926.
74. J. P. Kilroy, A. L. Klibanov, B. R. Wamhoff, D. K. Bowles and J. A. Hossack, *Ultrasound Med. Biol.*, 2014, **40**, 2458–2467.

75. F. Yan, L. Li, Z. Deng, Q. Jin, J. Chen, W. Yang, C.-K. Yeh, J. Wu, R. Shandas and X. Liu, *J. Controlled Release*, 2013, **166**, 246–255.
76. B. Methachan and K. Thanapprapasr, *Ultrasound Med. Biol.*, 2017, **43**, 69–82.
77. M. C. Cochran, J. Eisenbrey, R. O. Ouma, M. Soulen and M. A. Wheatley, *Int. J. Pharm.*, 2011, **414**, 161–170.
78. Y. Lei, N. Nosoudi and N. Vyavahare, *J. Controlled Release*, 2014, **196**, 79–86.
79. S. R. Sirsi, C. Fung, S. Garg, M. Y. Tianning, P. A. Mountford and M. A. Borden, *Theranostics*, 2013, **3**, 409–419.
80. S. Tinkov, G. Winter, C. Coester and R. Bekeredjian, *J. Controlled Release*, 2010, **143**, 143–150.
81. S. Sirsi and M. Borden, *Bubble Sci., Eng., Technol.*, 2009, **1**, 3–17.
82. J. S.-M. Yeh, C. A. Sennoga, E. McConnell, R. Eckersley, M.-X. Tang, S. Nourshargh, J. M. Seddon, D. O. Haskard and P. Nihoyannopoulos, *PLoS One*, 2015, **10**, e0129681.
83. E. Fahrländer, S. Schelhaas, A. Jacobs and K. Langer, *Nanotechnology*, 2015, **26**, 145103.
84. M. W. Miller, D. L. Miller and A. A. Brayman, *Ultrasound Med. Biol.*, 1996, **22**, 1131–1154.
85. S. Ibsen, M. Benchimol, D. Simberg and S. Esener, *Nano-biotechnology for Biomedical and Diagnostic Research*, Springer, 2012, pp. 145–153.
86. C. M. Tempany, N. J. McDannold, K. Hynynen and F. A. Jolesz, *Radiology*, 2011, **259**, 39–56.
87. O. Al-Bataineh, J. Jenne and P. Huber, *Cancer Treat. Rev.*, 2012, **38**, 346–353.
88. J. R. McDaniel, M. W. Dewhirst and A. Chilkoti, *Int. J. Hyperthermia*, 2013, **29**, 501–510.
89. H. Chen and J. H. Hwang, *J. Ther. Ultrasound*, 2013, **1**, 10.
90. R. M. Staruch, Hyperthermia mediated drug delivery using thermosensitive liposomes and MRI-controlled focused ultrasound, Doctoral Dissertation, University of Toronto, 2013.
91. D. Xu, Targeted drug delivery with cyclodextrin-based nanocarriers and focused ultrasound triggering, Doctoral Dissertation, University of Dundee, 2014.
92. M. Thanou and W. Gedroyc, *J. Drug Delivery*, 2013, **2013**, 616197.
93. S. Vaezy and V. Zderic, *Image-guided Therapy Systems*, Artech House, 2009.
94. R. Clarke, N. Bush and G. Ter Haar, *Ultrasound Med. Biol.*, 2003, **29**, 127–135.
95. A. Molckovsky, L.-M. W. K. Song, M. G. Shim, N. E. Marcon and B. C. Wilson, *Gastrointest. Endosc.*, 2003, **57**, 396–402.
96. M. Banić, D. Kolarić, N. Borojević, Ž. Ferenčić, S. Pleško, L. Petričušić, T. Božin and S. Antonini, *Period. Biol.*, 2011, **113**, 439–444.
97. M. H. Johnston, *Barrett Esophagus Diagnosis*, 2016, accessed 17th February 2017.

98. J. H. Rubenstein and N. J. Shaheen, *Gastroenterology*, 2015, **149**, 302–317.e301.
99. C. R. UK, *Oesophageal Cancer - Survival*, 2016, accessed 3rd February 2017.
100. M. J. Gora, J. S. Sauk, R. W. Carruth, K. A. Gallagher, M. J. Suter, N. S. Nishioka, L. E. Kava, M. Rosenberg, B. E. Bouma and G. J. Tearney, *Nat. Med.*, 2013, **19**, 238–240.
101. C. R. UK, *Bowel Cancer Mortality*, 2015, accessed 16th February, 2017.
102. B. C. UK, *Symptoms of Bowel Cancer*, 2016, accessed 16th February, 2017.
103. T. Dragovich, *Colon Cancer Treatment and Management*, 2016, accessed 16th February 2017.
104. M. Yu and W. M. Grady, *Ther. Adv. Gastroenterol.*, 2012, **5**, 319–337.
105. K.-C. Pao, M. Stanley, C. Han, Y.-C. Lai, P. Murphy, K. Balk, N. T. Wood, O. Corti, J.-C. Corvol and M. M. Muqit, *Nat. Chem. Biol.*, 2016, **12**, 324–331.
106. M. S. Ritorto, R. Ewan, A. B. Perez-Oliva, A. Knebel, S. J. Buhrlage, M. Wightman, S. M. Kelly, N. T. Wood, S. Virdee and N. S. Gray, *Nat. Commun.*, 2014, **5**, 4763.

CHAPTER 9

# Microwave Imaging and the Potential of Contrast Enhancing Agents for Theranostics Use

R. LAHRI[a], M. RAHMAN[a], T. KALLOS[b], P. KOSMAS[c] AND M. THANOU*[a]

[a]King's College London, Institute of Pharmaceutical Science, 150 Stamford Street, London, SE19NH, UK; [b]Mediwise, 42 New Rd, London E1 2AX, UK; [c]King's College London, Department of Informatics, Strand, London, WC2R 2LS, UK
*E-mail: maya.thanou@kcl.ac.uk

## 9.1 Introduction

Microwave imaging (MWI) has emerged recently as a new imaging modality, as it has the potential to image certain parts of the body.[1] The use of non-ionizing electromagnetic (EM) waves at microwave frequencies for medical diagnosis shows a number of advantages, such as low cost and low health risk compared to ionizing radiation based imaging.[2,3] Microwave imaging offers a non-invasive way for the diagnosis of biological tissues as well as pathological tissue conditions.[4] Its general principle is depicted in Figure 9.1, microwave signals that are transmitted from an antenna are received by

Drug Discovery Series No. 63
Theranostics and Image Guided Drug Delivery
Edited by Maya Thanou

Published by the Royal Society of Chemistry, www.rsc.org

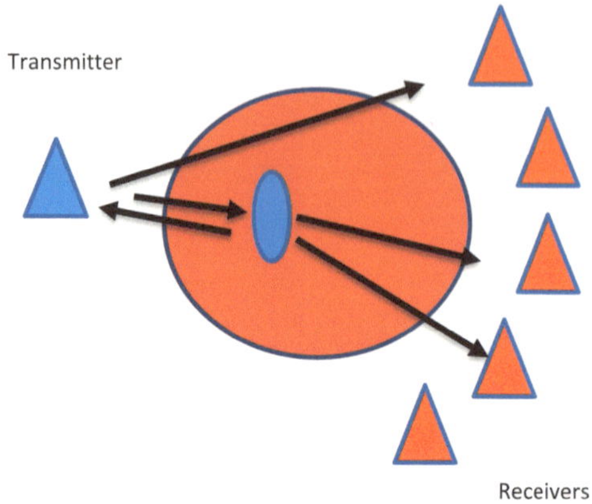


**Figure 9.1** Scheme representing the microwave signal pathway in an organ having a lesion. Transmitted signal from the antenna (blue triangle) is scattered and received by the same or other receiver antennas (red triangles).

the same or separate antenna after signal reflection or scattering by tissues that are different in their morphology.[3]

Using this sensing technique, it is possible to identify changes in the microwave signal due to the difference of the accumulation of water in pathological tissue compared to healthy tissue. This is the case in sensing applications applied to cerebral oedema or tumours.[5,6] When the microwave (MW) signals are used for imaging applications, a two-dimensional (2-D) or three-dimensional (3-D) image is generated.[7,8] These images can be quantitative, by estimating the different tissue electrical properties (microwave tomography), or qualitative, for which the location and shape of an object that is a strong "scatterer", usually a tumour lesion, are identified (radar-based MWI).[9] The hardware that MW systems require to process these signals are usually built in a simple and cost effective way compared to other medical imaging scanners.[10] These systems can be of low cost, and hence, can be applicable globally and offer rapid initial diagnosis of various life threatening conditions when time is critical in the development of the disease. The hardware of MW systems may consist of a microwave transmitter/detector (an antenna), and a system that generates and records EM signals, such as a vector network analyser (VNA).[10] Depending on the sensing approach, the MW system configuration may use an antenna array, connected to a radio-frequency switch matrix connected to VNA (console) to acquire multiple channel data.[11] With various advances in electronics, the cost of manufacturing of such systems is becoming lower every year, leading to MW scanning systems potentially being available at various points-of-care, for the monitoring of patients or assessing critical cases and emergencies (such as in an ambulance).[12]

The development of MWI systems has largely focused on breast imaging. Due to the structurally varying dielectric properties of the breast, breast cancer detection provides an excellent paradigm for the development of microwave imaging techniques.[6] The breast has different tissue layers, and each layer is composed of varying levels of water. Higher concentrations of water affect the electrical properties of the respective tissue. The electrical (or dielectric) properties of interest are the relative permittivity (or dielectric constant) $\varepsilon_r$, and the effective conductivity $\sigma$.[13] Water is a key factor in determining these parameters,[14] as there are significant differences in these properties between low and high water containing tissues. In particular, low water content tissues, such as fat, have lower dielectric properties compared to high water content tissues such as muscle and skin.[13] These properties are frequency dependent,[15] due to the structure of water molecules ultimately being a version of a permanent dipole.

As a result of the different dielectric properties, different types of breast tissue will cause different amounts of scattering from the incident EM waves. In a typical radar-based MW system[16] for the breast, an antenna array sends a pulsed ultra-wideband (UWB) signal to a focused point in the tissue. The scattered signal is recorded by the same or another antenna array. Then an algorithm operates on the received signals, exploiting the time delay for a round-trip from the transmitter antenna to the focused point. Using this method, detection of tumour sizes has been recorded with tumours sizes of 9 mm.[17] High resolution is achievable with radar systems if the transmitted signal is a short (UWB) pulse (low energy for short range sensing). However, the use of high frequencies in UWB radar methods results in increased signal attenuation inside the tissue, which in turn requires receivers with a high dynamic range.[18]

The tomography imaging approach employs an array of antennas, where a transmitter antenna illuminates the body region being imaged, and the remaining antennas receive the transmitted waves.[19] The process is repeated sequentially for each of the antennas in the array.[20] The received signals are used to produce a quantitative image using an "inverse scattering problem" which estimates the spatial distribution of the tissue dielectric properties.[21] Simply put, the "inverse scattering problem" is a method to obtain the characteristics of a real object based on data of how it scatters incoming radiation. The nature of this inverse scattering problem makes image recovery a difficult task, because it requires an ever-increasing set of computing tools for the efficient solution of these algorithms.[22] The proposed methodology of developing algorithms for MW imaging has varied according to desired imaging properties such as robustness of image reconstruction or accuracy.[20,23,24]

## 9.2 Dielectric Theory and Tissue Electrical Properties

When a permanent dipole, has an electric field applied to it, the molecule becomes polarised with polarisation vector $\overline{\mathbf{P}}$ as presented in Figure 9.2.

This creates an induced internal field that goes against the external field. The electrical field will exercise a torque on the molecule that will cause the

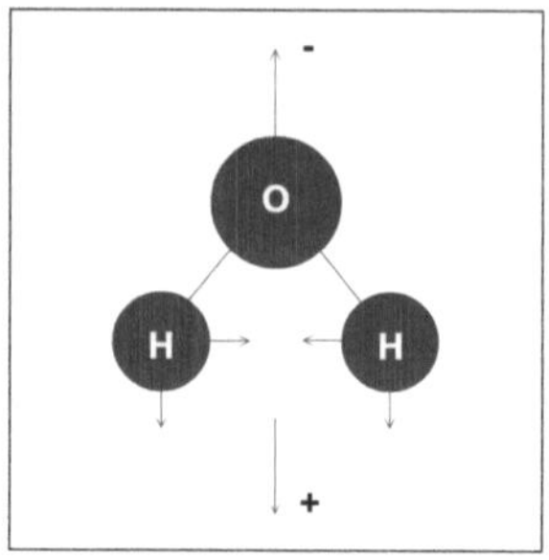


**Figure 9.2** Lewis structure of a water molecule. It follows a bent structure due to the 2 pairs of electrons on the oxygen atom forcing repulsion to the more positive hydrogen atoms. Oxygen is more electronegative than hydrogen. The hydrogen atoms have a net charge that consists of a vertical and horizontal component. The horizontal component cancels each other out, as they are the same magnitude but have opposite direction. While the vertical components of each atom are in the same direction, and are thus constructive.

dipole to rotate to align with the direction of the electric field. For lossy, anisotropic materials (materials that cause electromagnetic radiation attenuation) like breast tissues, when an external electric field is applied, the total field is the sum of the applied and induced fields. In eqn (9.1), the electric flux displacement $\bar{\mathbf{D}}$ expresses the electric field. Polarization may occur with a time delay. The permittivity of free space is represented by $\varepsilon_0$ F m$^{-1}$.

All materials are lossy to a degree depending on the frequency dependent imaginary part of the complex dielectric constant.[13] When the applied electric field has a low frequency the dipoles in a material follow the direction of the applied field, however at higher frequencies, the dipoles cannot align to the rapid change of the electric field, which causes rotational loss.

$$\bar{\mathbf{D}} = \varepsilon_0 \bar{\mathbf{E}} + \bar{\mathbf{P}} \ldots \text{Cm}^{-2} \tag{9.1}$$

All materials have variant dielectric properties, which can be represented as a function of frequency. Eqn (9.2) refers to the frequency dependent complex dielectric constant. The real part $\varepsilon'$, represents how much energy the material stores, the imaginary part $\varepsilon''$, refers to how much energy is lost, and is also related to the conductivity property of the material.

$$\varepsilon_r^*(\omega) = \varepsilon_r' - j\varepsilon_r'' = \varepsilon_r' - j\frac{\sigma}{\omega\varepsilon_0} \tag{9.2}$$

In vector notation, the real and imaginary part of the complex dielectric constant are out of phase by 90°. The magnitude vector of the complex value forms an angle $\delta$, with the real axis. The $\tan\delta$ is defined as the ratio of the energy lost with respect to the energy stored and is characteristic of a material. In eqn 9.3, $Q$ represents the quality factor, which is the reciprocal of the loss tangent. For low loss materials, the angle $\delta$ is very small, so the magnitude is close to the real component, however at microwave frequencies, biological tissues with high water content such as the breast tissue have a noticeable imaginary component leading to a higher loss tangent.

$$\tan\delta = \frac{\varepsilon_r''}{\varepsilon_r'} = \frac{1}{Q} \tag{9.3}$$

Various healthy tissues were characterised and their properties were summarised by Gabriel *et al.*[15] The measured dielectrical properties of various healthy and malignant tissues in the range of 50–900 MHz, were also assessed.[25] This study suggested that the difference between the normal and the malignant tissue appeared rather large in the case of breast tissue.[25] Studies by Lazebnik *et al.* showed that the difference in the dielectric properties between malignant and normal adipose-dominated tissues in the breast is considerable, as large as 10 : 1, while the contrast in the microwave-frequency dielectric properties between malignant and normal glandular/fibroconnective tissues in the breast is no more than about 10%.[26–28]

A certain challenge that is related to the tissue to be imaged and assessed is the "tissue loss" due to the increased conductivity in high water content tissues. There are tissues that show significant conductivity, which means significant signal attenuation that results in limited penetration depth. For example when the microwave power penetrates beyond the skin the thickness of tissue along the propagation path is sufficient to attenuate a 10 GHz signal by more than 80 dB.[13]

Microwaves have so far been investigated for applications where the lesion is at a depth of down to 10 cm, for instance in breast cancer. Imaging of tumours lying deeper in the body may not possible using microwaves.

### 9.2.1 Microwave Imaging Methods

Several different methods of detection have been studied to translate microwave imaging, such as quantitative and qualitative imaging. Quantitative imaging is the process that extracts numerical data and expresses it as a physical property, which can relate it to the physiological condition of the tissue to be imaged, and by default two or three dimensional images would reveal structural information as physical properties are tissue specific. In this scenario, the complex dielectric constant is the property that indicates tissue health.

Microwave tomography generates an image that maps the electrical properties and their distribution in tissues, based on signals acquired from the antennas that surround the organ to be imaged. These dielectrical properties are translated by the received parameters from the multiport imaging system that surrounds the organ to be imaged. Experimental EM measurements are susceptible to noise and errors from various sources, such us errors caused from the source (VNA), mechanical error, model discrepancies, background noise and lossy materials in the measurement region. Some errors may be averaged, while other errors need structural modifications; dielectric inclusions are also necessary to decrease the amount of signal loss. To control the amount of loss, various liquids (coupling mediums) can be used in between the antennas and the organ, to improve impedance matching between the boundaries and thus decrease loss in a manner similar to ultrasound gel for ultrasound imaging techniques. The coupling medium and the extent of loss in tissues has been described in the review by Nikolova.[29] This "coupling medium" is selected to be a homogenous fluid with chosen dielectric values that reduces significantly the reflection due to the air–body interface. The electrical properties of various living tissues are very different from that of air; for example, the skin has a relative permittivity around 30 and a conductivity of about 3 S $m^{-1}$ when measured in the low GHz range usually used for MW imaging. Therefore, the skin acts as the first barrier of the body against the penetration of EM waves in the MW range.[30] The fat in the subcutaneous adipose and muscle tissues add to the diminishing of the microwave power *via* reflection and dissipation, similar effects might appear for other tissues and this has been considered when breast tissues are modelled.[21,31]

To evaluate MW methods prior to clinical prototypes engineering, tailor-made phantoms are usually prepared with liquids and materials that mimic the relative permittivity of the various components.[32–34] The electrical properties of the various tissues exhibit strong dependence on frequency (named as frequency dispersion) in the radio frequency (RF) and MW bands due to the significant amount of water in biological tissues. This is due to its relative dependence of dominant oriental polarisation effect on water molecules. In particular, the relative permittivity of distilled (pure) water at 1 GHz is about 80, while at 10 GHz it is reduced to as low as 50.[29] While the tissue permittivity depends on water content, its conductivity depends mostly on the concentration of ions (ionic strength) due to a variety of dissolved substances, *e.g.*, salts or ionised molecules in tissues. Effective coupling of the microwave signal into the tissue can be achieved with the immersion of the antennas and the organ to be imaged in what is referred to as a "coupling liquid" this in some cases being water.[35] The antennas are fitted into a tank containing the liquid, and the imaged object is immersed in the tank.

Microwave imaging resolution is still a challenge. Resolution is defined as the ability to distinguish two objects by imaging and the distance that is between these two objects, in other words the ability to distinguish the two objects and not depict them as one. The resolution depends strongly on the wavelength in the microwave radiation through the diffraction limit (1/2 or 1/4 of a wavelength). This is the size of the smallest object that can be captured with sufficient detail *e.g.* regarding the detail in an objects shape. In MW imaging the resolution limit is dependent on the frequency range of the microwave radiation used. However, in microwave imaging, a short wavelength also means small penetration into the tissue. Therefore, in MW imaging the problem lies in identifying the optimal frequency that allows for the required resolution and at the same time is offering a signal sufficiently higher than the noise. In breast cancer diagnosis, microwave systems aim to achieve resolution at the centimetre or subcentimetre level. In the case of 3D reconstruction, the diffraction limit is an issue and wavelengths as short as 1 cm should be used.[13,29]

### 9.2.2 Microwave Tomography Equipment and Image Processing

Tomography aims to allow the complete reconstruction of tissue using the permittivity and conductivity values of the imaged object. Tomography methods allow successful reconstruction of images of 3-D phantoms.[8] Three-dimensional imaging through optimization tools is also being investigated.[36] Inverse scattering algorithms used for this optimisation can be categorised into two major methods (a) linearization and (b) iterative algorithms. 3-D Tomography uses all available data simultaneously to produce 3-D permittivity and conductivity distribution of the imaged object.

The first microwave imaging system, which was tested in the clinic, was the tomographic system developed by Dartmouth College[37–39] (Figure 9.3).

Another imaging system had been long developed by Klemm *et al.*[40] The authors used mechanical array rotation (angular rotation around the array centre) to extract the tumour response from raw measured data. Using a physical 3D curved breast phantom, the authors demonstrated the ability to screen small (4 mm and 6 mm) tumours. The team had concluded that such a system could be used for a breast cancer detection clinical trial. A study published recently presented the clinical system MARIA M4 from the same team performing successful MW imaging of the breast. This microwave imaging system was operating in the 3- to 8-GHz region using multistatic data collection.[41] There were 86 patients recruited and who were scanned with a prototype design. The resultant three-dimensional images were compared with available ultrasound and mammogram images to determine the efficiency of this prototype. Images showed the location of the strongest signal, and this corresponded in both older and younger women, with sensitivity of >74%, found in dense breasts. The results of this pivotal clinical trial indicate that these MW imaging systems can find rapid application and give promise for the detection of breast tissue abnormalities.

An important parameter on the development of MW tomography is the use of numerical models. Shea *et al.* introduced a 3D MW imaging realistic numerical breast phantom.[21] The authors suggested that a low-cost, non-ionizing, 3-D tomographic imaging modality can exploit the strong property of the contrast between dense fibroglandular tissue and less dense adipose tissue could be a method for routine breast screening. They used low-power microwaves to reconstruct the spatial distribution of breast tissue dielectric properties and to evaluate the modality for application to breast density characterization. Test phantoms included one sample from

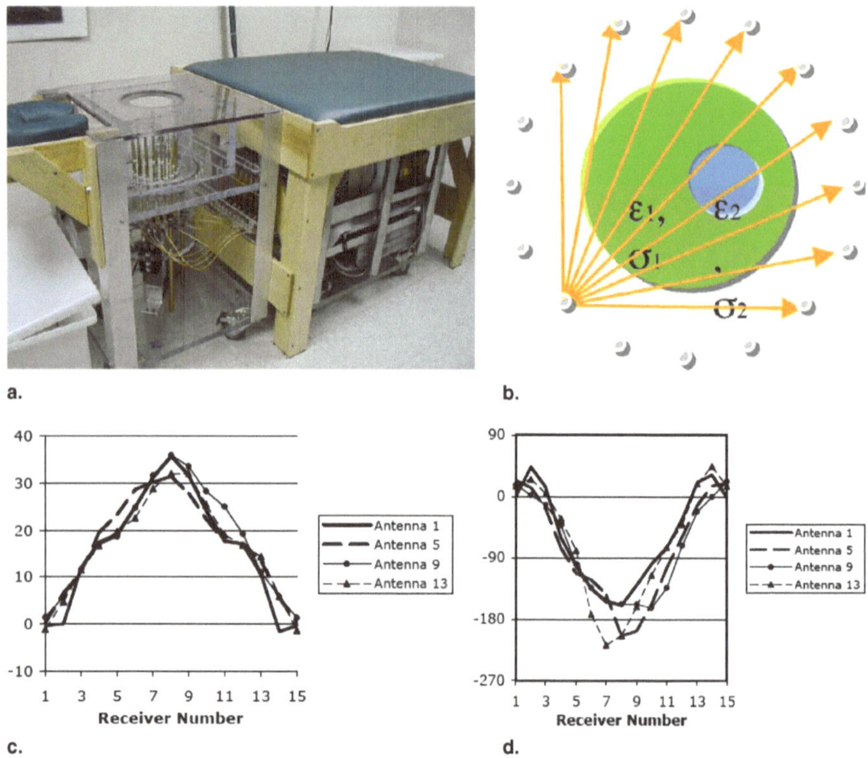


**Figure 9.3** (a) Photograph of the clinical microwave breast imaging prototype showing the illumination tank, examination platform and electronics cart (underneath bed); (b) 2D schematic diagram of the illumination and reception configuration; and representative (c) amplitude and (d) phase projections for a set of measured clinical data for a single imaging plane at 1300 MHz (data for only 4 of the 16 illuminations are shown). Reprinted from *Academic Radiology*, **14**, P. M. Meaney, M. W. Fanning, T. Raynolds *et al.*, Initial Clinical Experience with Microwave Breast Imaging in Women with Normal Mammography, 207–218,[38] Copyright (2007), with permission from Elsevier.

each of four classes of mammographic breast density. The distorted Born iterative imaging method was applied to simulated array measurements of the numerical phantoms and used forward solver in the imaging algorithm that employed the finite-difference time-domain (FDTD) method of solving the time domain Maxwell's equations. The imaging results for each numerical phantom showed robustness of the method relative to tissue density. It was shown that the distribution of fibroglandular tissues was well represented in the resulting images of these numerical phantoms.[21]

Further work by the group of Kosmas *et al.* focused on developing a novel distorted Born iterative method (DBIM) algorithm based on the two-step iterative shrinkage/thresholding method. The authors tested several strategies to increase the algorithm's robustness. Examples of reconstructions based on numerical breast phantoms taken from MRI images illustrated the benefits of these optimization strategies, which outperformed previous implementations for microwave breast imaging in terms of resolution and robustness (Figure 9.4).[24]

## 9.3 The Potential of Microwaves to Induce Thermal Therapy

Thermal therapy is a procedure to fatally heat malignant tumour cells above 37 °C, to 40–44 °C.[42] Also known as hyperthermia, it has the ability to shrink tumours. Other cancer treatments are being combined with induced hyperthermia to increase the therapeutic effect.[43] Thermal therapy can be applied to the whole body or to a localised region. In the microwave frequency spectrum heating is achievable through the phenomenon of dielectric heating because of dispersive losses. When an alternating electric field is applied to the dielectric material that is made up of these molecules, these particles become polarized and align with the dynamic direction of the field. However, the intermolecular forces between each molecule act as a resistive force to polarization and cause losses, which dissipate as heat.[13] These forces also introduce a time delay for dipoles to return to equilibrium, called relaxation time, $\tau_0$. It is the time for $\bar{\mathbf{P}}$ to decrease 1/e of the steady state value. The difference in phase of the electric and polarization vector can be expressed using the complex dielectric constant.

$$W_0 = \omega\varepsilon_0\varepsilon'' E^2 = \varepsilon_0 \frac{(\varepsilon_s - \varepsilon_\infty)\omega^2{\tau_0}^2}{1 + \omega^2{\tau_0}^2} E^2 \, \mathrm{Wm}^{-3} \tag{9.4}$$

Power absorbed by the dielectric can be defined using eqn (9.4), which is the electrical energy absorbed in to a material. The magnitude of the electric field strength is represented by $E$, the imaginary part of the complex permittivity is $\varepsilon''$, and $\omega$, is the angular frequency.[13] Designs of thermal applicators have varied, with invasive and non-invasive techniques.

Antenna needles, operating at 2.45 GHz have been assessed for heating by inserting the applicator inside the region of interest in the body. Huang *et al.* showed increase in temperature of pigs liver being illuminated from one plane by 5–6 °C, and similar specific absorption rate values in simulation and in experiment.[44] Non-invasive waveguide applicators have been researched for thermal therapy in the neck. This type of applicator works by restricting the propagation of the electromagnetic waves to one dimension to reduce power loss.[45] Non-invasive v-slot patch antennas have been designed in simulation for hyperthermia with a single applicator and in an array. The antenna system is setup using saline solution to increase impedance matching between the antenna and the phantom, and has shown

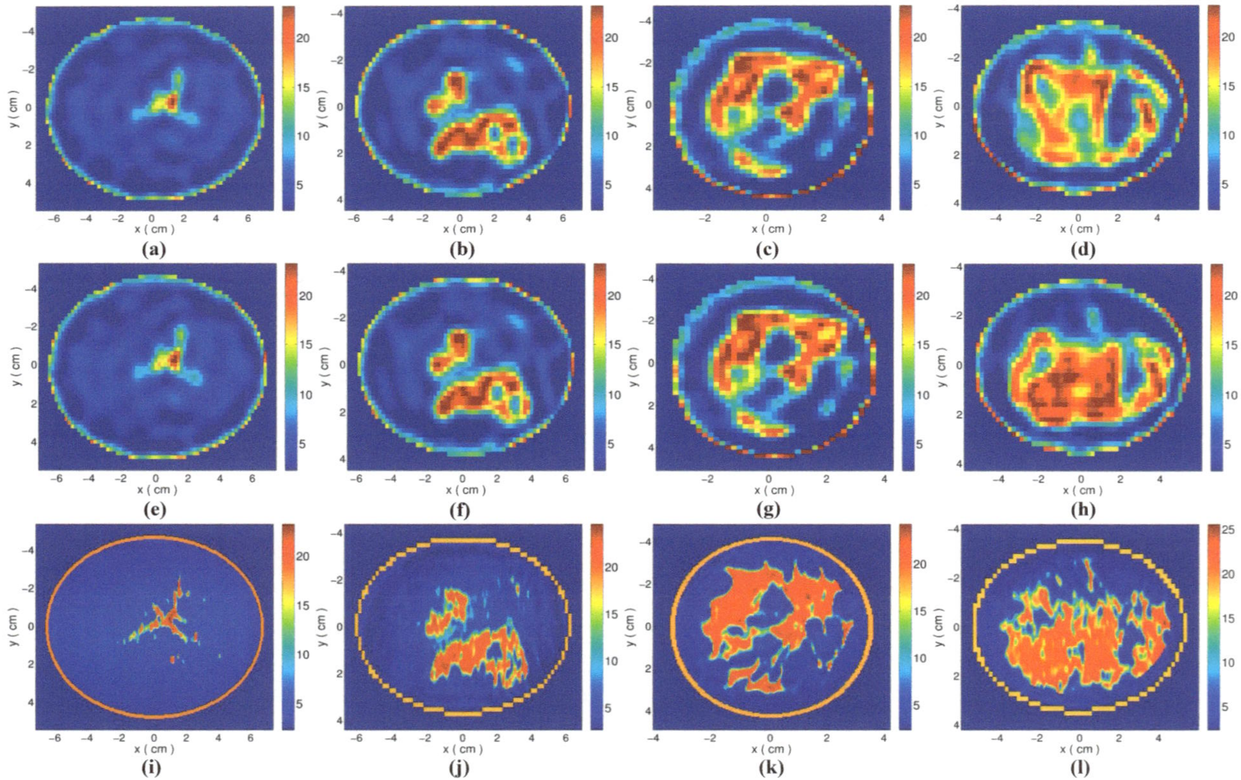


**Figure 9.4** Reconstructed images of ε∞ without and with optimization of initial guess for 4 phantoms in 2 mm. (a)–(d) Reconstructions with a fixed initial guess representative of normal breast tissue average properties ($\varepsilon_\infty = 11.27$, $\Delta\varepsilon = 5.51$, $\sigma_s = 0.0802$); (e)–(h) reconstructions with an optimized initial guess; (i)–(l) original 2-D images of 4 breast phantoms (mostly fatty, scattered fibroglandular, heterogeneously dense and very dense, from left to right). © 2017 IEEE. Reprinted, with permission, from Z. Z. Miao and P. Kosmas, *Ieee T Antenn Propag*, 2017, **65**, 2507–2516.[24]

good coupling of −20 Db.[46] A modified version of this antenna has been researched for the purpose of microwave imaging, with similar reflection coefficients.[47]

## 9.4 Use of Nanoparticles as Contrast Agents and Their Potential as Microwave Theranostics for Microwave Cancer Treatment

All medical imaging modalities have limitations to resolution and imaging depth to certain stages. The notion of contrast was introduced to increase the detail in images and make clear and decisive diagnosis. Contrast enhancing agents that are based on nanomaterials and/or nanoparticles find application in tumour diagnosis due to their preferred accumulation in tumours.[48] Medical imaging modalities exploit certain tissue natural properties, which alter the transmitted signal from one part of the body differently from other parts. For MW imaging this can be the dielectric properties of the tissues. Permittivity values, in the breast vary from $5 \leq \varepsilon_r' \leq 60$ where high water content tissues represent high dielectric constant and low water content tissues represent low dielectric constant.[26] The region of interest is the area that needs extra contrast, which is the tumour because the dielectric difference for low water content tissues like adipose and tumour tissue is 10 : 1, while the dielectric difference between fibro-glandular tissue and tumour tissue is only 10%.[27] Therefore contrast agents are required to improve the difference as we have previously stated that the tumours are more frequent in this region.

Nanoparticles are appropriate at acting as contrast enhancement as well as being nanocarriers of therapeutic drugs, due to the enhanced retention permeability effect (EPR) of small molecules.[49,50] This phenomenon describes how cancer cells grow abnormally, and begin to produce blood vessels to survive, these blood vessels become leaky allowing macromolecules and nanoparticles to diffuse within the tumour. The dependency to oxygen supply forces aggregation of nanoparticles, hence it is ideal for imaging and therapeutic applications.[51]

The concept of contrast enhancement on MW imaging was considered almost a decade ago. The use of microbubbles as a dielectric contrast agent for enhancing the sensitivity of breast cancer detection by MW was suggested.[52] Mashal *et al.*, used microbubbles that were air-filled glass spheres (iM30K, 3M) with an average diameter of 18 μm and density of 0.6 g mL$^{-1}$. Ethylene glycol was used as a dispersant as it has dielectric properties that are in the range of the particular biological tissues at microwave frequency (3 GHz). Mixtures of ethylene glycol with 0%, 5%, 10%, 20% and 30% by weight concentration of microbubbles were prepared. Properties were measured from a frequency range of 0.5–6 GHz by open ended coaxial probe technique. Results suggested that microbubbles significantly lowered the dielectric properties. Also the stability of the dispersion had large effects on the dielectric characterisation. Since microbubbles are less dense that the background medium, they float to the top of the surface, therefore measurements were taken with respect to time.[52]

Measurements at $t_0$ & $t_{2.5\ \mathrm{min}}$ were compared. Table 9.1 shows that a large weight percentage of microbubbles produces the same/similar electrical properties at $t_0$ at a lower weight percentage at $t_{2.5\ \mathrm{min}}$. Therefore dielectric measurements with the coaxial probe are highly dependent on the colloidal stability of the material under test in question, as the measurement depth is very small.[53] Hence measurements with the dielectric probe must be dispersed completely to produce comparable results.

Later studies by the same authors were performed on single walled carbon nanotubes (SWCNTs) and they have shown to increase the dielectric contrast of a background medium, and hence have the ability to act as a contrast agent for microwave imaging. They have also shown to increase the temperature while being illuminated by a high intensity microwave signal. This is a new field that may show promise. Mashal *et al.*, reported that SWCNTs manufactured by Cheap Tubes Inc. using a chemical vapour deposition technique and purified using an acid treatment, have the potential to enhance dielectric contrast between malignant and normal tissue for microwave detection of breast cancer.[54] Pristine carbon nanotubes may have high conductivity and therefore could be used as nanomaterials that could affect the dielectric properties of tumours.[55] Mashal *et al.* used tissue mimicking materials in their study to replicate the dielectric properties of tumour tissue over a broadband frequency range, were mixed with varying concentrations of SWCNTs.[54] Their dielectric properties were characterized over a frequency range of 0.6–20 GHz (Figure 9.5a and b). They were measured by a well-established coaxial-ended probe method. At SWCNT concentrations of less than 0.5% by weight, significant increase in the relative dielectric constant and effective conductivity was observed.

Table 9.2 shows the dielectric constant and the effective conductivity of a basic oil in gel phantom and same consistency phantoms with varying concentrations of SWCNTs. Probe measurements showed that the SWCNTs

**Table 9.1** Summary of data for microbubbles assessed at 3 GHz. Adapted from ref. 52.

| Solution | Initial microbubble concentration (% by weight) at 0 min | Estimated actual concentration (% by weight) at 2.5 min | Average $\varepsilon_r$ (3 GHz) | Average $\sigma$ (3 GHz) (S m$^{-1}$) |
|---|---|---|---|---|
| 1 | 0 | 0 | 14.03 | 2.32 |
| 2 | 5 | 20 | 9.66 | 1.33 |
| 3 | 10 | 30 | 8.34 | 1.11 |
| 4 | 20 | 35 | 7.38 | 0.95 |
| 5 | 30 | 40 | 6.86 | 0.84 |
| 6 | 100 | 100 | NA[a] | NA[a] |

[a]NA: not available.

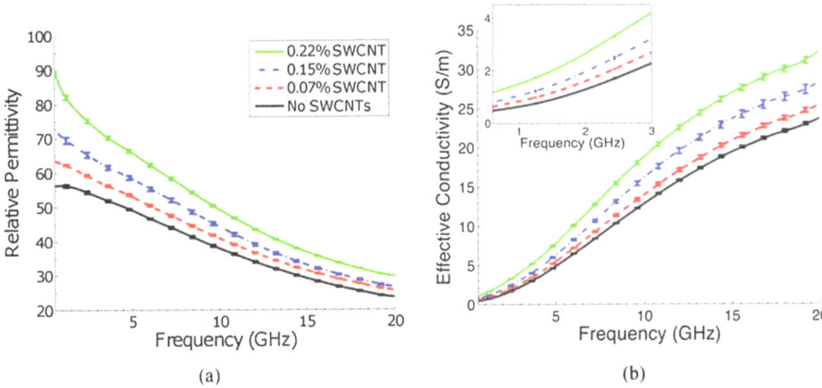


**Figure 9.5** (a) Relative permittivity and (b) effective conductivity of tissue mimicking materials with varying concentrations of single walled carbon nanotubes, measured over the frequency range from 0.6 to 20 GHz. Each curve represents the average properties of three different samples of the same concentration. © 2010 IEEE. Reprinted, with permission, from *IEEE Trans Biomed Eng*, 2010, **57**, 1831–1834.[54]

**Table 9.2** Electrical properties of SWCNT dispersions at 3 GHz, adapted from ref. 54.

| % Weight concentration of SWCNTs | Average $\varepsilon_r$ (3 GHz) | Average $\sigma_C$ (3 GHz) |
|---|---|---|
| None | 53.3 | 2.3 S m$^{-1}$ |
| 0.07% | 57.9 | 2.7 S m$^{-1}$ |
| 0.15% | 63.7 | 3.2 S m$^{-1}$ |
| 0.22% | 73.0 | 4.2 S m$^{-1}$ |

increase the dielectric constant of the medium that they were dissolved in. The average dielectric constant of a pure phantom with no SWCNTs, 1 mg mL$^{-1}$ (0.07%), 2 mg mL$^{-1}$ (0.15%) and 3 mg mL$^{-1}$ (0.22%) mg mL$^{-1}$ of SWNTs was 53.3, 57.9, 63.7 and 77.3 respectively at 3 GHz (Table 9.2). Therefore, not only showing increase in dielectric contrast from the background medium but also increase dependency with increase in NP concentration. Results suggest that even at very low concentrations, SWCNTs have the potential to significantly increase the dielectric properties. Therefore, SWCNTs could be utilized to be used as a contrast agent to enhance the contrast for microwave imaging for breast cancer detection.

Heating response was also measured by Mashal *et al.*, a tumour mimicking gel mixed with and without the dispersion of SWCNTs at the same concentration from dielectric characterisation was set into a tube connected to a fibre optic thermometer. This tube was placed inside an S-band (WR-284) rectangular waveguide with cross-sectional dimensions of 72 mm × 34 mm that was connected to a Agilent 83623B signal generator, and an amplifier (Mini-Circuits, ZHL-42W) produced 1 W of continuous microwave power at 3 GHz to the waveguide to heat the gel for 3 min with a 5 min cool down time. Results show that the temperature of the gel is increased from 2–8 °C depending on the SWNT concentration to the basic gel phantom. It was also shown that the temperature increase in each phantom scaled relatively linear with the effective conductivity and those from previous experiments with SWCNTs in colloidal dispersion showed similar heating trends, as well as similar proportionality with temperature increase and the effective conductivity.[54]

Xie *et al.*,[56] were able to achieve exceptionally high microwave frequency dielectric properties with stable high concentration (~20 mg mL$^{-1}$) of 50:50 mixture of single- and double-walled carbon nanotubes (CNTs).[56] The method involved functionalising CNTs (Cheaptubes, Inc.) *via* sonication in nitric and sulphuric acid ($H_2SO_4$:$HNO_3$ 3:1). Samples for dielectric measurements were prepared by dispersing treated CNTs in deionised water with mannitol (50 mg mL$^{-1}$) and the dielectric properties were characterised by dielectric probe at frequency range of 0.5–6 GHz. Properties of CNTs were also measured before and after passing them through a 25 gauge needle.[56] The study indicated the effect of functionalisation on either the dielectric properties or the dispersion, and subsequently on their assessed permittivity.

The dielectric properties of CNT dispersions were shown to be significantly higher than those of the control. The dielectric constant of CNT dispersion before passing through the needle was increased by 14% at 10 mg mL$^{-1}$ and 68% at 20 mg mL$^{-1}$ and effective conductivity was increased by 66% and 404% for 10 and 20 mg mL$^{-1}$ samples. The authors suggested that such high increase in dielectric properties was due to the successful functionalization of CNTs with carboxylic and sulfonic groups after the acid treatment on the outer CNT walls. This was responsible for their stabilisation and their increased dispersion and colloidal stability was attributed to the stronger length-shortening effect.[56]

Xie *et al.*[57] also used acid purification methods (hydrochloric acid/hydrogen peroxide method and mitric acid reflux method) for synthesis of single walled carbon nanotubes, Co-CNTs (CNTs synthesized by catalytic chemical vapor deposition) and Fe-CNTs (CNTs synthesized by high pressure CO method) and their dielectric properties were studied using a slim form probe between 0.5–6 GHz.[57] For microwave frequency dielectric measurements, 3 mg mL$^{-1}$ of SWCNTs were dispersed in 0.1% aqueous PL-PEG$_{5400}$-NH2 (Phospholipid-Polyethyleneglycol-Amine) or DSPE-PEG (1,2-Distearoyl-*sn*-glycero-3-phosphoethanolamine-Polyethyleneglycol) and sonicated for 12 hours. These lipids and amphiphiles are widely used to improve the colloidal stability and biocompatibility of nanomaterials. The authors observed a large increase in the dielectric properties for pristine (non-purified) Co-SWCNT compared to control (SWCNT free solution). For example, at 3 GHz, the average $\varepsilon_r$ (dielectric constant) increases by 10% and the average $\sigma_C$ (effective conductivity) increases by 45%. The dielectric properties of the (HCl + $H_2O_2$)-purified Co-SWCNT dispersion were slightly lower than those of the pristine dispersion, indicating an effect of possible oxidation affecting the

SWCNT dielectric properties. At 3 GHz, the average $\varepsilon_r$ decreases by 1.6% and the average $\sigma_e$ decreases by 8.7%. The dielectric properties of the $HNO_3$-purified Co-SWCNT dispersion were closer to those of the control sample than the pristine SWCNT dispersion. These results show that the $HNO_3$ purification method more severely reduces the microwave-frequency dielectric properties enhancement effect of Co-SWCNTs in dispersion than the HCl + $H_2O_2$ purification method.

Their study demonstrated the following: (1) Aqueous dispersions of pristine Co-SWCNTs and Fe-SWCNTs synthesized by two distinct methods, with significant differences in morphology, show similarly elevated microwave dielectric properties compared to a control: (2) acid-purification-based changes in SWCNT structure and composition decreases the length and fraction of metallic nanotubes, as well as increases in amorphous carbon and covalent functionalization of SWCNTs depending on the purification method and dramatically reduces the microwave dielectric properties of SWCNT aqueous dispersions.[57] These findings have important implications in the development of SWCNT-based formulations of microwave-frequency theranostic agents for biomedical applications.

Li *et al.*,[58] measured the effective dielectric properties of PL-PEG functionalised SWCNTs at a concentration of 2 mg $mL^{-1}$. Dielectric properties were measured from 0.5–10 GHz by using a precision open-ended coaxial probe technique. They were shown to have a 90% increase in effective conductivity and 10% in dielectric constant compared to the properties of PL-PEG in water at 1 GHz (Figure 9.5). SWCNT (0.5 μm length) colloidal dispersions (0.5 and 1 mg $mL^{-1}$) produced by Stony Brook University were also stated to increase the dielectric constant compared to aqueous background and that other forms of carbon such as buckyballs show no effective contrast. The large difference of aspect ratio between the SWCNTs and the buckyballs was stated as the reason why SWCNTs increase the dielectric constant.[58]

In our group we also aim to investigate the potential of nanomaterials for MW contrast enhancing effect. We have measured the dielectric properties of SWCNT-OH (Cheap Tubes Inc, Brattleboro, VT). According to the manufacturer's specifications the nanotubes were 1–4 nm in diameter and 5–30 μm in length and were mainly composed of SWCNT (>90% by weight with traces of double walled carbon nanotubes). Measurements were taken from a frequency range of 1–20 GHz. The dielectric properties show a dependency on the concentration of NPs (Figure 9.6); *i.e.* the higher the concentration of nanoparticles, the higher are the dielectric properties of the suspension. Our results are comparable with previously reported data.[58] In previously reported data, SWCNT were functionalised with a biocompatible PEG.

The -OH functionalized SWCNTs showed an average change of 9.72% increase in the real component at 2 mg $mL^{-1}$ at 1 GHz, which is comparable with the previously reported 10% increase in the dielectric constant at similar conditions. There is variation observed within the conductivity, which

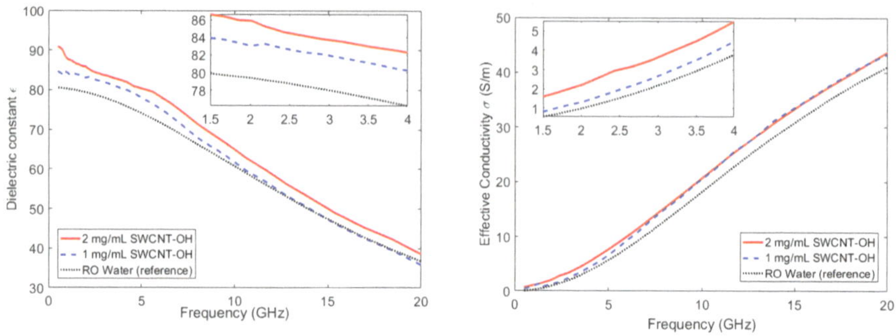


**Figure 9.6** Dielectric constant (left) and effective conductivity (right) of colloidal dispersion of CNT-OH in RO water with 1% Pluronic characterised between 1–20 GHz. Both permittivity and conductivity increase with an increase in the SWCNT-OH concentration.

could be because of the different types of functionalisation methods used for SWCNTs. These results indicate the importance of dispersion of the nanoparticles when studies focus on their dielectric properties.

Metallic nanoparticles were also characterised; dielectric properties of colloidal dispersions consisting of 30% volume fraction of copper nanoparticles (384 nm) in ethylene glycol, by using both wideband open-ended coaxial probe technique and narrowband cavity perturbation technique were performed. There was no significant change in the effective conductivity, however, a 20% increase in the dielectric constant compared to background medium at 2 GHz was observed. This percentage increase is only related to the background medium $\left(\text{dielectric change \%} = \frac{\varepsilon_{\text{Background}} - \varepsilon_{\text{Colloid}}}{\varepsilon_{\text{Background}}} \times 100\right)$, and that a large percentage of NP was dispersed in the solution, compared to the characterisation of CNT colloids.[58] This characterisation experiment shows that different measurement techniques have good agreement with each other, with respect to a stable and fully dispersed medium that is isotropic.

Other contrast agents, such as gadolinium based compounds and iron oxide have also been explored. Ogunlade *et al.*[59] used a resonant cavity perturbation technique to characterise the electrical properties of superparamagnetic carboxydextran coated iron oxide (EM1301 which contains 0.5 M of iron as iron oxide NPs) and paramagnetic gadolinium based contrast agents; Magnevist®, Dotarem® and Prohance® at concentration of 0.5 M at 3 GHz (Table 9.3).[59]

The authors stated that the MW contrast originating from gadolinium compounds was not because of gadolinium itself but it was due to the dissociation of gadolinium salt which increased the ionic conductivity.[59] Iron oxide particles were also found to produce low but measurable dielectric

**Table 9.3** Dielectric properties of Magnetic resonance contrast agents at 0.5 M at 3 GHz adapted from ref. 59.

| Contrast agent | $\varepsilon_r$ | $\sigma_c$ |
|---|---|---|
| Magnevist® | 49.75 | 0.568 |
| Dotarem® | 56.65 | 0.435 |
| Prohance® | 60.54 | 0.0504 |
| EM1301 | 74.70 | 0.0097 |
| Water | 77.76 | 0.0002 |

**Table 9.4** Dielectric properties of ferroelectric nanoparticles at 1 and 2.45 GHz.[60]

| | 1 GHz | | 2.45 GHz | |
|---|---|---|---|---|
| | $\varepsilon'$ | $\varepsilon''$ | $\varepsilon'$ | $\varepsilon''$ |
| $CaTiO_3$ | 135 | 52 | 97 | 45 |
| $BaTiO_3$ | 128 | 122 | 67 | 72 |

contrast due to dielectric polarisation loss. This was due to the carboxylated dextran coating on the NPs and not because of iron oxide itself. They concluded that gadolinium based contrast agents and iron oxide have little merit as contrast agents.[59]

Semenov *et al.*,[60] measured the dielectric properties of certain ferroelectrics as potential contrast agents for microwave tomography. Barium titanate ($BaTiO_3$) and calcium titanate ($CaTiO_3$) with a particle size of 50 nm were suspended in glycerol. Measurements were conducted using a well-established coaxial probe technique. Results suggested that with the presence of ferroelectric NPs (Table 9.4), the dielectric contrast observed is much higher compared to biological tissue such as muscle which has a dielectric constant of 56 at 1 GHz. This contrast with ferroelectrics is even more evident if compared with lower water content tissues such as fat.[60]

It is demonstrated that the dielectric constant of the nanocomposite can be significantly improved by increasing the aspect ratio of $BaTiO_3$ NWs, while the loss tangent shows little variation with value around 0.04 at 1 kHz under different of aspect ratio and volume fraction of $BaTiO_3$ NWs. The loss tangent of PVDF matrix is around 0.06 at 1 kHz, which indicates that integration of $BaTiO_3$ NWs decrease the loss tangent of the film. This study was able to determine the relation between the aspect ratio of fillers and dielectric constant of the nanocomposites. It shows that the dielectric constant of the nanocomposites can be improved by higher aspect ratio fillers.

Tang *et al.*,[61] demonstrated that the dielectric constant of the nanocomposite increases with the aspect ratio of barium titanate ($BaTiO_3$) nanowires. The aspect ratio of BaTiO3 nanowires can be successfully controlled by adjusting the temperature of the hydrothermal growth from 150 to 240 °C,

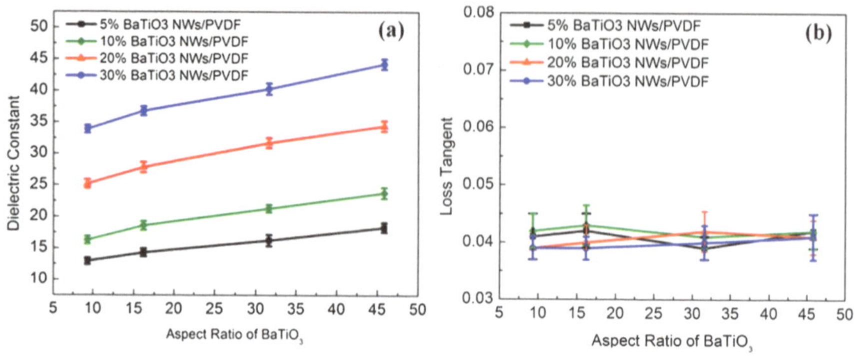


**Figure 9.7** (a) Dielectric constant and (b) loss tangent of the nanocomposites measured at 1 kHz as a function of aspect ratio and volume fraction of BaTiO3 NWs. Reprinted from ref. 61. Copyright (2014) American Chemical Society.

corresponding to aspect ratios from 9.3 to 45.8, respectively.[61] Figure 9.7 clearly shows that the aspect ratio of $BaTiO_3$ nanowires highly depend on the hydrothermal temperature. The dielectric constant of the nanocomposites was measured with an Agilent 4980A LCR meter at a frequency of 1 kHz. The dependency of the nanocomposite dielectric constant on the concentration and aspect ratio of the fillers is shown in Figure 9.7. The dielectric constant of the nanocomposites increases with increasing volume fraction of the fillers due to the higher dielectric permittivity of $BaTiO_3$ as compared with the PVDF polymer (9.8 at 1 kHz).

Magnetic nanoparticles (MNPs) have gained an immense interest in biomedicine because of their biocompatibility, easy functionalisation and magnetic properties. In MWI, MNPs have been used to induce magnetic contrast within the tumour. In principle, MNP-enhanced MWI has the potential of significantly reducing the occurrence of false positives and false negatives due to low dielectric contrast between healthy and fibroglandular tissues.

Bucci *et al.*,[62,63] carried out a study in which MNPs were dispersed in a biocompatible environment, such as water or phosphate buffer saline (PBS), known as water-based ferrofluids (WFF). The concentration of MNPs used are achievable in practice (few tens of mg $mL^{-1}$ of magnetic material). WFF is a colloidal suspension of magnetite nanoparticles dispersed in water and coated with anionic surfactant. The colloidal suspension was diluted in PBS to a MNP concentration of about 25 mg $mL^{-1}$.

The investigated frequency range was 0.1–8 GHz and $S_{11}$ and $S_{21}$ were measured at 601 equispaced points with a vector network analyser (Planar 804/1 from copper mountain technologies) connected to the measurement coaxial cell. This consists of N-male and N-female connectors together to obtain a sealed cell of fixed length. The cell is made of non-magnetic metal. The WFF is injected into the cell through two small holes on the lateral

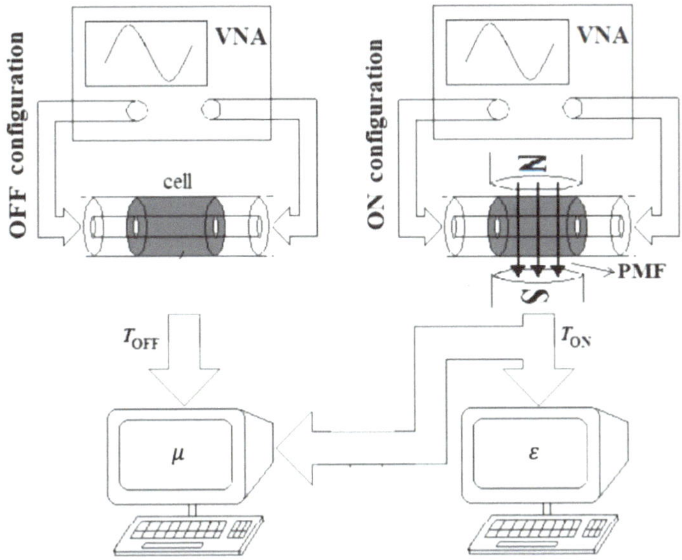


(a)

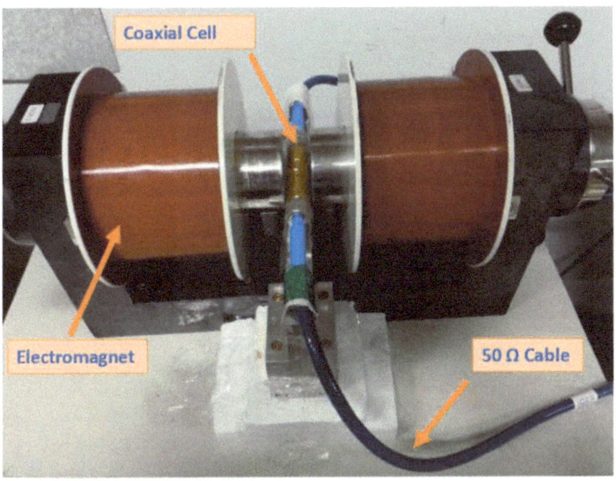


(b)

**Figure 9.8** (a) Conceptual scheme of the adopted measurement approach. (b) Picture of the measurement setup. © 2017 IEEE. Reprinted with permission from *IEEE Transactions on Magnetics*, 2017, **53**, 1–8.[62]

wall of the cell. The cell is then put between the poles of an electromagnet (EMU 75 SES Instrument PVt. Ltd.) generating the polarising magnetic field (Figure 9.8).

In this paper,[62] the results of an experimental broadband characterization, in the presence of the polarising magnetic field, of a sample of water based ferrofluids with a low concentration of magnetite MNPs in PBS have been reported and discussed. Characterization of a diluted WFF sample has never been presented in the literature before, and represents an essential prerequisite for the implementation and the subsequent optimization of all the applications exploiting the response of MNPs to microwave and electromagnetic fields.

To this end, an *ad hoc* measurement strategy and setup have been exploited. In order to investigate in detail the nature of the measured response, as well as to obtain an analytical description of the magnetic properties of the investigated sample in the considered ranges of frequency and polarising magnetic field intensity, the measurement results have been fitted with the simplest dispersion model reflecting the basic physical requirements, namely, a superposition of resonant and relaxation responses.

The above mentioned studies indicate that nanomaterials can be used as contrast enhancing agents for MW imaging. It is likely that these nanomaterials can have therapeutic properties. For instance some of the nanoparticles could provide ablative hyperthermia in tumours when combined with EM radiation. Alternatively these MW contrast enhancing nanomaterials could be combined with therapeutics and promote their targeted drug delivery effect in lesions. The development of such systems is yet to be seen. Nevertheless MW imaging and hyperthermia systems are a promising new tool that when combined with nanomaterials could offer a safe and accessible method of diagnosis and treatment.

## 9.5 Conclusions

Microwave systems can provide sensing, imaging and hyperthermic therapies. Research efforts on the effects of EM radiation at MW frequencies and the development of MW sensing equipment indicate an emerging research area that aims to provide systems for diagnosis and therapy that are safe (non-ionising radiation) and accessible (low cost). Nanomaterials such as carbon nanotubes and iron oxide nanoparticles could function as contrast enhancing agents for MW imaging due to their advantage of affecting the dielectric properties of the medium that they are dispersed within. Their effect on the dielectric constant of living tissue indicates that they can function as contrast enhancing agents for MW imaging. These nanomaterials show a preference in localising in cancer lesions due to their extravasation from the leaky angiogenic blood vessels into the tumour. Nevertheless their properties in *in vivo* MW imaging experiments is yet to be seen. More screening work is required for the identification of nanomaterials that can affect the dielectric properties of tissues.

## Acknowledgement

BBSRC is acknowledged for the support of the King's College London team.

## References

1. S. Kwon and S. Lee, *Int. J. Biomed. Imaging*, 2016, **2016**, 5054912.
2. H. E. Davies, C. G. Wathen and F. V. Gleeson, *BMJ*, 2011, **342**, d947.
3. R. Chandra, H. Zhou, I. Balasingham and R. M. Narayanan, *IEEE Trans. Biomed. Eng.*, 2015, **62**, 1667–1682.
4. S. Semenov, *Philos. Trans. R. Soc., A*, 2009, **367**, 3021–3042.
5. M. J. Clarke and J. C. Lin, *Invest. Radiol.*, 1983, **18**, 245–248.
6. E. C. Fear, *Technol. Cancer Res. Treat.*, 2005, **4**, 69–82.
7. S. Y. Semenov, R. H. Svenson, A. E. Boulyshev, A. E. Souvorov, V. Y. Borisov, Y. Sizov, A. N. Starostin, K. R. Dezern, G. P. Tatsis and V. Y. Baranov, *IEEE Trans. Biomed. Eng.*, 1996, **43**, 869–877.
8. S. Y. Semenov, R. H. Svenson, A. E. Bulyshev, A. E. Souvorov, A. G. Nazarov, Y. E. Sizov, A. V. Pavlovsky, V. Y. Borisov, B. A. Voinov, G. I. Simonova, A. N. Starostin, V. G. Posukh, G. P. Tatsis and V. Y. Baranov, *IEEE Trans. Biomed. Eng.*, 1999, **46**, 937–946.
9. M. Klemm, J. A. Leendertz, D. Gibbins, I. J. Craddock, A. Preece and R. Benjamin, *IEEE Antennas Wireless Propag.*, 2009, **8**, 1349–1352.
10. J. J. Mohr and T. Rubæk, *An Introduction to Microwave Imaging for Breast Cancer Detection*, Springer International Publishing, 1st edn, 2016.
11. S. Semenov, J. Kellam, P. Althausen, T. Williams, A. Abubakar, A. Bulyshev and Y. Sizov, *Phys. Med. Biol.*, 2007, **52**, 5705–5719.
12. A. T. Mobashsher, K. S. Bialkowski, A. M. Abbosh and S. Crozier, *PLoS One*, 2016, **11**, e0152351.
13. A. Vander Vorst, A. Rosen and Y. Kotsuka, *RF/Microwave Interaction with Biological Tissues*, John Wiley & Sons, Inc., 2006, pp. 7–62.
14. K. R. Foster and H. P. Schwan, *CRC Crit. Rev. Bioeng.*, 1989, **17**, 25–104.
15. S. Gabriel, R. W. Lau and C. Gabriel, *Phys. Med. Biol.*, 1996, **41**, 2271–2293.
16. E. C. Fear, X. Li, S. C. Hagness and M. A. Stuchly, *IEEE Trans. Biomed. Eng.*, 2002, **49**, 812–822.
17. M. Klemm, I. Craddock, J. Leendertz, A. Preece and R. Benjamin, *2008 IEEE Antennas and Propagation Society International Symposium*, 2008, vol. 1–9, pp. 1–4.
18. M. Bassi, M. Caruso, M. S. Khan, A. Bevilacqua, A. D. Capobianco and A. Neviani, *IEEE Trans. Microwave Theory*, 2013, **61**, 2108–2118.
19. A. H. Golnabi, P. M. Meaney, N. R. Epstein and K. D. Paulsen, *Conf. Proc. IEEE Eng. Med. Biol. Soc.*, 2011, **2011**, 5730–5733.
20. K. Purkait, S. Mondal, S. N. Bose, R. Roy, S. Halder and K. Adhikary, *2015 Third International Conference on Computer, Communication, Control and Information Technology (C3it)*, 2015.
21. J. D. Shea, P. Kosmas, S. C. Hagness and B. D. Van Veen, *Med. Phys.*, 2010, **37**, 4210–4226.
22. P. Kosmas and Z. Z. Miao, *2014 Ieee Mtt-s International Microwave Workshop Series on Rf and Wireless Technologies for Biomedical and Healthcare Applications (Imws-bio)*, 2014.

23. M. T. Bevacqua, L. Crocco, L. Di Donato and T. Isernia, *IEEE Trans. Antennas Propag.*, 2015, **63**, 601–610.
24. Z. Z. Miao and P. Kosmas, *IEEE Antennas Wireless Propag.*, 2017, **65**, 2507–2516.
25. W. T. Joines, Y. Zhang, C. X. Li and R. L. Jirtle, *Med. Phys.*, 1994, **21**, 547–550.
26. M. Lazebnik, L. McCartney, D. Popovic, C. B. Watkins, M. J. Lindstrom, J. Harter, S. Sewall, A. Magliocco, J. H. Booske, M. Okoniewski and S. C. Hagness, *Phys. Med. Biol.*, 2007, **52**, 2637–2656.
27. M. Lazebnik, D. Popovic, L. McCartney, C. B. Watkins, M. J. Lindstrom, J. Harter, S. Sewall, T. Ogilvie, A. Magliocco, T. M. Breslin, W. Temple, D. Mew, J. H. Booske, M. Okoniewski and S. C. Hagness, *Phys. Med. Biol.*, 2007, **52**, 6093–6115.
28. M. O'Halloran, D. Byrne, R. C. Conceição, E. Jones and M. Glavin, in *An Introduction to Microwave Imaging for Breast Cancer Detection*, ed. R. C. Conceição, J. J. Mohr and M. O'Halloran, Springer International Publishing, Cham, 2016, pp. 5–16.
29. N. K. Nikolova, *IEEE Microw. Mag.*, 2011, **12**, 78–94.
30. B. Maklad and E. C. Fear, *Conf. Proc. IEEE Eng. Med. Biol. Soc.*, 2008, **2008**, 21–24.
31. E. Porter, A. Santorelli, M. Coates and M. Popovic, *Technol. Cancer Res. Treat.*, 2013, **12**, 131–143.
32. G. Ruvio, A. Cuccaro, R. Solimene, A. Brancaccio, B. Basile and M. J. Ammann, *Healthc. Technol. Lett.*, 2016, **3**, 218–221.
33. A. Kletsov, A. Chernokalov, A. Khripkov, J. Cho and S. Druchinin, *Healthc. Technol. Lett.*, 2015, **2**, 108–111.
34. M. Lazebnik, E. L. Madsen, G. R. Frank and S. C. Hagness, *Phys. Med. Biol.*, 2005, **50**, 4245–4258.
35. L. E. Larsen and J. H. Jacobi, *Med. Phys.*, 1978, **5**, 500–508.
36. S. Y. Semenov, R. H. Svenson, A. E. Bulyshev, A. E. Souvorov, A. G. Nazarov, Y. E. Sizov, V. G. Posukh, A. Pavlovsky, P. N. Repin, A. N. Starostin, B. A. Voinov, M. Taran, G. P. Tatsis and V. Y. Baranov, *IEEE Trans. Biomed. Eng.*, 2002, **49**, 55–63.
37. D. Li, P. M. Meaney, T. Raynolds, S. A. Pendergrass, M. W. Fanning and K. D. Paulsen, *Rev. Sci. Instrum.*, 2004, **75**, 2305–2313.
38. P. M. Meaney, M. W. Fanning, T. Raynolds, C. J. Fox, Q. Fang, C. A. Kogel, S. P. Poplack and K. D. Paulsen, *Acad. Radiol.*, 2007, **14**, 207–218.
39. S. P. Poplack, T. D. Tosteson, W. A. Wells, B. W. Pogue, P. M. Meaney, A. Hartov, C. A. Kogel, S. K. Soho, J. J. Gibson and K. D. Paulsen, *Radiology*, 2007, **243**, 350–359.
40. M. Klemm, I. J. Craddock, J. A. Leendertz, A. Preece and R. Benjamin, *IEEE Antennas Wireless Propag.*, 2009, **57**, 1692–1704.
41. A. W. Preece, I. Craddock, M. Shere, L. Jones and H. L. Winton, *J. Med. Imaging (Bellingham)*, 2016, **3**, 033502.
42. J. van der Zee, *Ann. Oncol.*, 2002, **13**, 1173–1184.

43. P. Wust, B. Hildebrandt, G. Sreenivasa, B. Rau, J. Gellermann, H. Riess, R. Felix and P. M. Schlag, *Lancet Oncol.*, 2002, **3**, 487–497.
44. C. F. Huang, C. F. Li and H. Y. Chao, *2015 Asia-Pacific Symposium on Electromagnetic Compatibility (APEMC)*, 2015.
45. F. Ondrej, M. Ilja, V. Lucie and V. Jan, *2014 44th European Microwave Conference*, 2014.
46. H. Dobšíček Trefná, J. Vrba and M. Persson, *Int. J. Hyperthermia*, 2010, **26**, 185–197.
47. S. Ahsan, Z. Guo, I. Gouzouasis, E. Kallos and P. Kosmas, *2016 10th European Conference on Antennas and Propagation (EuCAP)*, 2016.
48. C. Rancoule, N. Magne, A. Vallard, J. B. Guy, C. Rodriguez-Lafrasse, E. Deutsch and C. Chargari, *Cancer Lett.*, 2016, **375**, 256–262.
49. H. Maeda, *J. Controlled Release*, 2012, **164**, 138–144.
50. M. Thanou and W. Gedroyc, *J. Drug Delivery*, 2013, **2013**, 616197.
51. K. Greish, *Methods Mol. Biol.*, 2010, **624**, 25–37.
52. A. Mashal, J. H. Booske and S. C. Hagness, *Phys. Med. Biol.*, 2009, **54**, 641–650.
53. P. M. Meaney, A. P. Gregory, N. R. Epstein and K. D. Paulsen, *BMC Med. Phys.*, 2014, **14**, 3.
54. A. Mashal, B. Sitharaman, X. Li, P. K. Avti, A. V. Sahakian, J. H. Booske and S. C. Hagness, *IEEE Trans. Biomed. Eng.*, 2010, **57**, 1831–1834.
55. L. Liu, Z. Niu and J. Chen, *Chem. Soc. Rev.*, 2016, **45**, 4340–4363.
56. S. X. Xie, F. Gao, S. C. Patel, J. H. Booske, S. C. Hagness and B. Sitharaman, *IEEE Trans. Biomed. Eng.*, 2014, **61**, 2718–2723.
57. S. X. Xie, F. Gao, S. C. Patel, J. H. Booske, S. C. Hagness and B. Sitharaman, *Appl. Phys. Lett.*, 2013, **103**, 133114.
58. X. Li, A. V. Sahakian, P. Messersmith, S. C. Hagness, J. H. Booske and I. Knezevic, *Nanoparticle Contrast Agents for Enhanced Microwave Imaging and Thermal Treatment of Breast Cancer*, Northwestern Univ, Evanston IL, U.S., 2010.
59. O. Ogunlade and P. Beard, *Med. Phys.*, 2015, **42**, 170–181.
60. S. Semenov, N. Pham and S. Egot-Lemaire, in *World Congress on Medical Physics and Biomedical Engineering, September 7–12, 2009, Munich, Germany: Vol. 25/8 Micro- and Nanosystems in Medicine, Active Implants, Biosensors*, ed. O. Dössel and W. C. Schlegel, Springer Berlin Heidelberg, Berlin, Heidelberg, 2010, pp. 311–313.
61. H. Tang, Z. Zhou and H. A. Sodano, *ACS Appl. Mater. Interfaces*, 2014, **6**, 5450–5455.
62. O. M. Bucci, G. Bellizzi and G. G. Bellizzi, *IEEE Trans. Magn.*, 2017, **53**, 1–8.
63. O. M. Bucci, L. Crocco and R. Scapaticci, *IEEE Trans. Biomed. Eng.*, 2015, **62**, 407–414.

# Subject Index

www.ingramcontent.com/pod-product-compliance
Ingram Content Group UK Ltd.
Pitfield, Milton Keynes, MK11 3LW, UK
UKHW062338170926
8624IPUK00029B/414
* 9 7 8 1 7 8 2 6 2 4 6 6 0 *